T0327555

Advances in Semiconductor Technologies

Advances in Semiconductor Technologies

Selected Topics Beyond Conventional CMOS

Edited by

An Chen
IBM Research – Almaden, CA, USA

IEEE PRESS
WILEY

Published by John Wiley & Sons, Inc., Hoboken, New Jersey.
Published simultaneously in Canada.

For general information on our other products and services or for technical support, please contact our Customer Care Department within the United States at (800) 762-2974, outside the United States at (317) 572-3993 or fax (317) 572-4002.

Wiley also publishes its books in a variety of electronic formats. Some content that appears in print may not be available in electronic formats. For more information about Wiley products, visit our web site at www.wiley.com.

Library of Congress Cataloging-in-Publication Data:

Names: Chen, An (Electronics engineer), editor.
Title: Advances in semiconductor technologies : selected topics beyond
    conventional CMOS / An Chen.
Description: Hoboken, New Jersey : Wiley-IEEE Press, [2023] | Includes
    bibliographical references and index.
Identifiers: LCCN 2022018684 (print) | LCCN 2022018685 (ebook) | ISBN
    9781119869580 (cloth) | ISBN 9781119869597 (adobe pdf) | ISBN
    9781119869603 (epub)
Subjects: LCSH: Semiconductors.
Classification: LCC TK7871.85 .A3565 2023 (print) | LCC TK7871.85 (ebook)
    | DDC 621.3815/2 – dc23/eng/20220627
LC record available at https://lccn.loc.gov/2022018684
LC ebook record available at https://lccn.loc.gov/2022018685

Cover image: © Blue Andy/Shutterstock
Cover design: Wiley

Set in 9.5/12.5pt STIXTwoText by Straive, Chennai, India

Contents

Preface

Since the invention of the solid-state transistors, the semiconductor technologies have advanced at an exponential pace and become the foundation for numerous industries, e.g. computing, communication, consumer electronics, autonomous systems, and defense. Guided by Moore's law, the scaling of transistors has provided new generations of chips every one to two years, with ever-increasing density and better performance. Today, silicon transistors are approaching some fundamental limits of dimensional scaling. The semiconductor industry has also transformed through several phases and foundational technologies. The emergence of Internet of Things (IoT), big data, artificial intelligence (AI), and quantum computing has created new opportunities for advanced semiconductor technologies. The complementary metal-oxide-semiconductor (CMOS) technology dominates the semiconductor industry today, but there are numerous technologies and active research beyond conventional CMOS. Although semiconductors are often associated with high-performance computing chips such as central processing unit (CPU) and graphics processing unit (GPU), there is a wide range of applications beyond computing for semiconductor products, e.g. sensors, displays, and power electronics. Silicon (Si) is the most important semiconductor, but the semiconductor research also covers a variety of materials, e.g. germanium (Ge), III–V compounds, organic materials, carbon nanotube, 2D materials, magnetic materials, and topological materials.

This book is a collection of articles reviewing advanced semiconductor technologies beyond conventional Si CMOS for various applications. These articles written by the experts in the fields can be read independent of each other. The variety of topics reflects the breadth of the semiconductor R&D and applications today, but these articles only cover a very small fraction of semiconductor technologies.

With the transistor scaling approaching the fundamental limits, heterogeneous integration is a promising direction to sustain the improvement of performance and functionalities without relying on reducing transistor sizes. Chapter 1, "Heterogeneous Integration at Scale," provides a comprehensive review of technologies, design/architecture considerations, reliability issues, applications, and future directions of large-scale heterogeneous integration.

While technology innovation has been a primary driver for the semiconductor industry, the future of semiconductor systems will increasingly resort to novel computing paradigms. Chapter 2, "Hyperdimensional Computing: An Algebra for Computing with Vectors," presents an example of entirely new ways of computing inspired by the information processing in the brain. Instead of traditional model of computing with numbers, hyperdimensional (HD) computing encodes information in a holographic representation with wide vectors and unique operations. HD computing is extremely robust against noise, matches well with 3D circuits, and is uniquely suitable to process a variety of sensory signals without interference with each other.

The majority of semiconductor chips are digital circuits; however, analog and mixed-signal circuits are crucially important. The physical world is analog; therefore, analog circuits are always needed to connect digital chips with real world, e.g. sensory data, power management, and communication. Although digital circuit design is highly automated, analog circuit design still relies on manual effort. Chapter 3, "CAD for Analog/Mixed-Signal Integrated Circuits," reviews the progress toward automated computer-aided design (CAD) of analog and mixed-signal circuits.

Modern computers are built based on the von Neumann architecture with separate logic/computing units and memory/storage units. Emerging memory devices not only provide new technologies to improve memory systems but also enable novel computing architectures, e.g. in-memory computing. One of the most promising emerging memories is based on magnetic materials and properties. Chapters 4 and 5 focus on a so-called magnetoelectric field effect transistor (MEFET) based on the programming of the polarization in a 2D semiconductor channel with large spin-orbit coupling, via the proximity effect of a magnetoelectric gate. Chapter 4, "Magnetoelectric Transistor Devices and Circuits with Steering Logic," presents various logic gate designs based on a one-source two-drain MEFET configured with a steering function. Chapter 5, "Nonvolatile Memory Based Architectures Using Magnetoelectric FETs," describes MEFET memory designs with the performance and size suitable to fulfill the application space between static random-access-memory (SRAM) and dynamic random-access-memory (DRAM).

Novel materials beyond Si, Ge, and III–V compounds may enable new semiconductor products and applications. Among them, organic semiconductors are promising materials for low-cost, flexible, and bio-compatible electronics. Chapter 6, "Organic Electronics," discusses the opportunities of organic semiconductors for large-area flexible electronics, including organic light-emitting diode (OLED), organic displays, organic solar cells, and thin-film transistors. Chapter 7, "Active-Matrix Electroluminescent Displays," delves into the details of flat panel electroluminescent displays based on light-emitting diodes (LEDs) that have been utilized in a wide range of applications including smart phones, tablets, laptops, and TVs. Various underlying LED technologies, associated circuits, and design considerations are reviewed. Another interesting application of organic materials is memory. Chapter 8, "Organic and Macromolecular Memory – Nanocomposite Bistable Memory Devices," discusses the mechanisms, characteristics, and current status of organic memories. One of the advantages of organic materials is their low-cost processing and the potential to stack up multiple layers. Chapter 9, "Next Generation of High-Performance Printed Flexible Electronics," summarizes different printing technologies for flexible electronics, showcases the state-of-the-art printed flexible electronic circuits, and discusses the challenges and future directions of large-scale cost-effective printed electronics. The vision of integrating electronic components onto polymer foils leads to the flexible electronics version of systems-on-chip (SoC), known as systems-in-foil (SiF). A wide range of applications can benefit from SiF, e.g. smart labels, intelligent electronic skin, and implanted devices. Chapter 10 "Hybrid Systems-in-Foil" reviews the opportunities of SiF and challenges in materials, integration, and testing.

The electronic systems need an interface with the physical world. Semiconductor chips rely on sensors to "see," "hear," and "smell." Optical sensing is utilized in a wide range of applications, e.g. camera, fiber optics and communication, light source and laser, data storage, medical monitoring and diagnostics, and manufacturing. Chapter 11, "Optical Detectors," reviews the photodiodes based on Si, III–V, and emerging materials as the essential components for highly sensitive detectors for a broad spectrum of wavelengths. Chapter 12, "Environmental Sensing," covers comprehensively different air pollution sources, air quality metrics, and various sensing approaches for particulate matters and volatile organic compounds. The advancement of semiconductor technologies contributes to the miniaturization of the sensing equipment and the improvement of their performance.

Unlike computer chips operating with very low voltage and current, power electronics handle very high voltage (e.g. thousands of volt or higher) and current required to operate machinery, vehicles, appliances, etc. Special device designs and unique material properties are required to sustain such high voltage and current in semiconductor chips. Chapter 13, "Insulated Gate Bipolar Transistors (IGBTs)," reviews an important high-power device known as Si insulated gate bipolar transistors (IGBTs). IGBT not only dominates power electronics today but also continues to be innovated for further gains in power density and efficiency. At the same time, significant progress has been made on wide bandgap semiconductors. Chapter 14, "III–V and Wide Bandgap," reviews promising materials (e.g. diamond, GaN) and their applications in high-frequency power conversion and high-temperature electronics. While wide bandgap power modules may be combined with Si-based control circuits in near-term solutions, considerable effort is made to advance integrated circuits based on wide bandgap semiconductors. Chapter 15, "SiC MOSFETs," reviews SiC-based power semiconductor devices including diodes and transistors. SiC is well positioned to fulfill the requirements of power electronics, e.g. energy efficiency, scaling, system integration, and reliability. The unique ability of SiC to form a native SiO_2 as the gate dielectric makes it particularly attractive for power metal-oxide-semiconductor field-effect-transistors (MOSFETs). At the end, Chapter 16, "Multiphase VRM and Power Stage Evolution," provides a detailed overview of the evolution of CPU power delivery technologies and explains the reasons driving the technology shifts.

This book could be considered as a small-scale reference of advanced semiconductor technologies, which may potentially be expanded into a large-scale reference with more comprehensive coverage. It is our wish that this collection of chapters will provide useful tutorials on selected topics of advanced semiconductor technologies.

An Chen
IBM Research – Almaden, CA, USA
September 2022

List of Contributors

Mohamed B. Alawieh
Department of Electrical and Computer
Engineering
The University of Texas at Austin
Austin, TX
USA

Mohammed Alomari
Institut für Mikroelektronik Stuttgart
Stuttgart
Germany

Shaahin Angizi
Department of Electrical and Computer
Engineering
New Jersey Institute of Technology
Newark, NJ
USA

Ahmet F. Budak
Department of Electrical and Computer
Engineering
The University of Texas at Austin
Austin, TX
USA

Hao Chen
Department of Electrical and Computer
Engineering
The University of Texas at Austin
Austin, TX
USA

Danny Clavette
Infineon Technologies
Americas Corporation
El Segundo, CA
USA

Abhishek S. Dahiya
University of Glasgow
James Watt School of Engineering, Bendable
Electronics and Sensing Technologies (BEST)
Group
Glasgow
UK

Ravinder Dahiya
University of Glasgow
James Watt School of Engineering, Bendable
Electronics and Sensing Technologies (BEST)
Group
Glasgow
UK

Li'ang Deng
Department of Electronic Engineering
Shanghai Jiao Tong University
China

Peter A. Dowben
Department of Physics and Astronomy
Jorgensen Hall, University of Nebraska
Lincoln, NE
USA

Mourad Elsobky
Institut für Mikroelektronik Stuttgart, Sensor
Systems
Stuttgart
Germany

Peter Friedrichs
IFAG IPC T, Infineon Technologies AG
Neubiberg
Germany

Xiaojun Guo
Department of Electronic Engineering
Shanghai Jiao Tong University
China

Subramanian S. Iyer
Electrical Engineering Department
University of California
Los Angeles, CA
USA

Pentti Kanerva
University of California at Berkeley
Redwood Center for Theoretical Neuroscience
Berkeley, CA
USA

Hagen Klauk
Max Planck Institute for Solid State Research
Stuttgart
Germany

Tihomir Knežević
Faculty of Electrical Engineering
Mathematics & Computer Science
MESA+ Institute of Technology
University of Twente
Enschede
The Netherlands

Yogeenth Kumaresan
University of Glasgow
James Watt School of Engineering, Bendable
Electronics and Sensing Technologies (BEST)
Group
Glasgow, UK

Thomas Laska
Infineon Technologies AG
Germany

Mingjie Liu
Department of Electrical and Computer
Engineering
The University of Texas at Austin
Austin, TX
USA

Andrew Marshall
Department of Electrical Engineering, The
Erik Johnson School of Engineering and
Computer Science
University of Texas at Dallas
Richardson, TX
USA

Lis Nanver
Faculty of Electrical Engineering
Mathematics & Computer Science
MESA+ Institute of Technology
University of Twente
Enschede
The Netherlands

Arokia Nathan
Darwin College
University of Cambridge
Cambridge
UK

David Z. Pan
Department of Electrical and Computer
Engineering
The University of Texas at Austin
Austin, TX
USA

Shashi Paul
Emerging Technologies Research Centre
De Montfort University
Leicester
UK

Wei Shi
Department of Electrical and Computer
Engineering
The University of Texas at Austin
Austin, TX
USA

Xiyuan Tang
Department of Electrical and Computer
Engineering
The University of Texas at Austin
Austin, TX
USA

Boris Vaisband
McGill University
Electrical and Computer Engineering
Department
Montreal, QC
Canada

Tarek Zaki
Munich, Germany

Shuhan Zhang
Department of Electrical and Computer
Engineering
The University of Texas at Austin
Austin, TX
USA

Keren Zhu
Department of Electrical and Computer
Engineering
The University of Texas at Austin
Austin, TX
USA

Deliang Fan
Department of Electrical, Computer and
Energy Engineering
Arizona State University
Tempe, AZ
USA

1

Heterogeneous Integration at Scale

Subramanian S. Iyer[1] and Boris Vaisband[2]

[1]*Electrical Engineering Department, University of California, Los Angeles, CA, USA*
[2]*McGill University, Electrical and Computer Engineering Department, Montreal, QC, Canada*

1.1 Introduction

Microelectronics has made tremendous progress over the last several decades adhering to what is popularly called Moore's Law. One measure of Moore's law is the scaling factor of minimum features on a silicon integrated circuit (IC). This trend is shown in the dark gray (left-hand y-axis) curve in Figure 1.1 [1] exhibiting over a 1,000-fold decrease in minimum feature size, corresponding to a million-fold transistor density improvement. This improvement corresponds to reduction in power per function as well as reduction of cost and price per function. Nonetheless, until recently, packaging did not scale as seen in the light gray (right-hand y-axis) curve in Figure 1.1. For example, in 1967, when flip-chip bonding was first introduced, the bump pitch was 400 μm. Even today, the pitch of the bump (die to laminate) has scaled to about 130 μm, while ball grid array (BGA) pitch and trace pitch on laminates and printed circuit boards (PCBs) have not fared better. However, in the last few years, we have seen an acceleration of these metrics as shown in the inset in Figure 1.1. Note that the silicon (Si) scale is in nanometers, while the packaging scale is in micrometers. There are two key factors that have mediated this acceleration: (i) the adoption of silicon-like processing materials and methods to achieve scaling, including silicon interposers, and, importantly, (ii) fan-out wafer-level packaging (FOWLP).

This trend has manifested itself in two ways: (i) The extensive use of interposers, which is an additional level in the packaging hierarchy as shown in Figure 1.2. At a basic level, interposers provide a first-level platform for the integration of several (eight) heterogeneous dielets on a thinned silicon substrate that is then further packaged on a laminate and attached to a PCB. This allows the dielets on the interposer to communicate intimately within the interposer, though communication outside the interposer is more conventional. (ii) Three-dimensional (3D) integration, where dies are stacked one on top of the other, typically face to back with through silicon vias (TSVs) or alternatively face to face through surface connections. These face-to-face connections can be at high bandwidth and low latency. Both of these techniques have transformed packaging, especially when it comes to the memory subsystem. A roadmap using the memory subsystem as a paradigm for advanced packaging is depicted in Figure 1.3. Another area in which interposers and 3D integration can play a big role is the integration of analog and mixed signal functions. Moore's law scaling does a good job in scaling digital logic, but is at best marginal when it comes to analog and mixed signal functions. This is shown in Figure 1.4, where the analog/mixed signal components can occupy an increasing percentage of real estate at finer geometries. In these cases, retaining the

Advances in Semiconductor Technologies: Selected Topics Beyond Conventional CMOS, First Edition. Edited by An Chen.
© 2023 The Institute of Electrical and Electronics Engineers, Inc. Published 2023 by John Wiley & Sons, Inc.

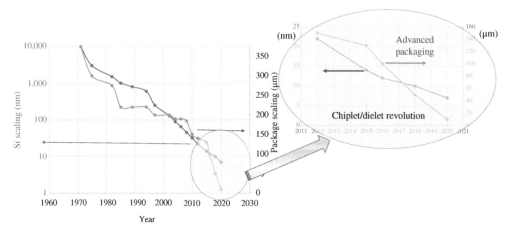

Figure 1.1 Scaling trends for CMOS features as well as package features. Package scaling has lagged significantly as compared to Si scaling. Adoption of silicon-like technology for packaging has somewhat accelerated scaling.

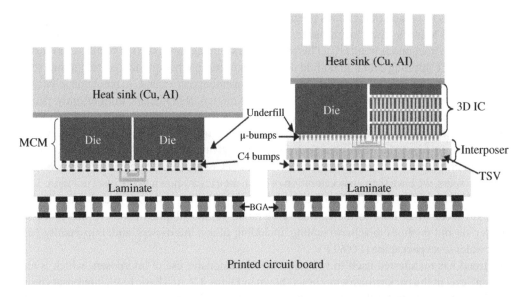

Figure 1.2 Current packaging hierarchy. Left: die-laminate-PCB. Right: die-interposer-laminate-PCB.

analog/mixed signal functions in an older node makes sense as long as one can provide compatible voltage domains and ensure low latency as well as low analog signal distortion. These are not very difficult to do on interposers.

Why is it important to scale packaging? Packaging dimensions determine the size of the system especially since the scale is 10–100 times larger than chip dimensions. Power too is a major consideration. Communication power between chips accounts for 30–40% of total system power. So for size, weight, and power (SWaP), as well as cost, scaling the package has advantages. The key parameters that affect SWaP are dominated by packaging metrics, and scaling the package has greater impact on SWaP than additional Si scaling. For flexible hybrid electronics (FHE), form factor and power play a critical role. Most FHE devices are mobile and dependent on battery power. As such, FHE packages will benefit immensely from scaling.

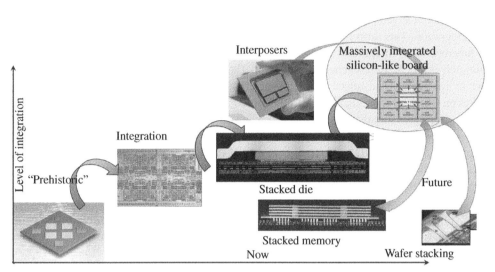

Figure 1.3 Packaging evolution – the memory paradigm.

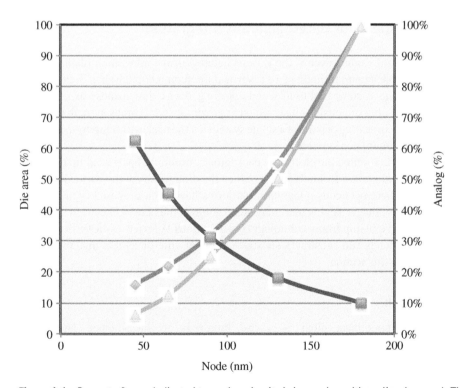

Figure 1.4 Percent of area dedicated to analog circuits is increasing with scaling (squares). Thus, practical die area (normalized to 180 nm technology) is increasing with scaling (rhombuses) as compared to ideal die area (triangles).

Another aspect of advanced packaging that has the focus of attention in recent years is heterogeneous integration. This term requires some clarification. Most packaging constructs do in fact achieve heterogeneous integration via the integration of diverse packaged chips on an extended substrate such as a PCB. Heterogeneous integration, therefore, in general and in itself, is not new. However, in the context of advanced packaging, heterogeneous integration refers to the integration of bare dies on a first-level packaging substrate. This could be an organic, ceramic, or silicon interposer. The key features that distinguish heterogeneous integration from classical or conventional packaging are the pitch of the connections between the bare die and the substrate, the number of connections between the interconnected bare dies, the size of the dies, and hence a significant simplification in the communication protocols of interdie signaling. It is generally accepted that for bump pitches <50 μm, interdie spacing of <2 mm, and trace pitches (wiring between the dies) of <5 μm, the integration is considered in the regime of advanced packaging.

Finally, "chiplets" and "dielets" are another feature of advanced packaging. A complex system or large chip design is fragmented into smaller entities called chiplets and then instantiated in Si as dielets. These dielets are then intimately reintegrated at fine pitch (bump and trace) as well as short interdie spacing, as previously described, to synthesize a subsystem or a module. This construct can be further assembled on a PCB or, in the case of wafer scale systems, and can represent the entire system [2, 3].

1.2 Technology Aspects of Heterogeneous Integration

Technology innovation is the main driver of the various heterogeneous integration platforms. In the past several decades, packaging technology (e.g. vertical interconnect pitch) has been scaling at a significantly slower rate as compared to IC technology (e.g. device dimensions), as shown in Figure 1.1. Specifically, on-chip dimensions have scaled approximately 200 times more than package features. This disparity in scaling of parts of a single system led the package hierarchy to become the bottleneck of modern integrated systems.

Nonetheless, in the last five years, the electronic packaging community has picked up the pace proposing various novel integration technologies to reduce the dimensions of the packaging hierarchy. Specifically, novel heterogeneous integration platforms have been proposed, significantly driving down the features of the packaging hierarchy. Realization of the heterogeneous integration concept is predicated on several important technology considerations. A review of the vertical interconnect pitch (between dielets and substrate), substrate material, interdielet spacing, and dielet termination is provided in this section.

1.2.1 Interconnect Pitch

Typical package-level interconnect pitch is several hundreds of micrometers. These are solder-based BGA or land grid array (LGA) connections between the package and the PCB. Connections between dies and package laminate, i.e. solder-based C4/pillar, exhibit a smaller pitch (~50 to 100 μm). Nonetheless, comparing to the last metal levels on the die (pitch of 2–10 μm), the package-level interconnects are approximately between 10 and 500 times larger. In fact, the main purpose of the packaging hierarchy is to fan-out the interconnect pitch from the small die-level pitch, to the large PCB-level pitch, and then vice versa when connecting to the neighboring packaged die. Solder has the advantage of deforming at low temperature and pressure, to accommodate warpage of the PCB, laminate, and die. A typical Cu pillar capped with

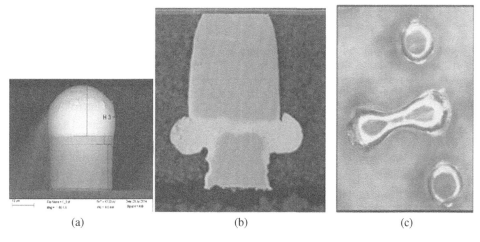

| (a) | (b) | (c) |

Figure 1.5 (a) A 50 mm diameter Cu pillar capped with solder. (b) After mass reflow with compression showing the solder extrusions. (c) Micrograph showing extrusions that cause adjacent pillars to become electrically shorted. Source: Photo courtesy: Eric Perfecto.

solder is shown in Figure 1.5a, a cross section after mass reflow is shown in Figure 1.5b, and a micrograph of two shorted pillars is shown in Figure 1.5c. Shorting of neighboring solder balls is the main challenge to the continuous scaling of solder-based interconnects below about 50 μm.

Another possible integration approach is to attach dielets directly to a substrate without solder (or other intermetallics), using direct metal-to-metal thermal compression bonding (TCB). In this integration approach, two metals are bonded together by applying pressure and temperature for a certain amount of time. After this process, a very strong low-contact resistance bond is formed. To ensure high-quality connections using TCB, the surfaces of the metals must be pristine and atomically smooth. Additional conditions must be met when bonding Cu for example, such as surface preparation to avoid oxidation of the Cu prior to bonding (by using plasma or formic acid).

Alignment is another challenge in small pitch connections for heterogeneous integration. Solder-based interconnects are easier to align due to the larger dimensions and since they exhibit a form of self-alignment property due to surface tension at solder melting temperature. Unlike solder, direct metal-to-metal integration requires a high level of alignment accuracy which is difficult to achieve by optical means since the die and substrate are typically not transparent. A second-order optical alignment is therefore required, where the die and the substrate are aligned to a virtual reference. For example, in the silicon interconnect fabric (Si-IF) technology, an alignment accuracy of ±1 μm has been achieved [4]. Although this alignment accuracy is good, it is still about an order of magnitude worse than die-level conventional optical lithography.

1.2.2 Substrate Material

Different substrate materials are currently used in heterogeneous integration. The most commonly used materials are compared in Table 1.1.

Hybrid substrate materials are also used in industry. For example, the embedded multidie interconnect bridge (EMIB) [5] approach supports the integration of small Si bridges within an organic substrate. The EMIB enables different interconnect pitches, coarse pitch on the organic FR4 and fine pitch on the Si bridge. This technology, however, exhibits increased complexity and therefore higher cost.

Table 1.1 Key structural and thermal properties of Si and other relevant materials.

Material	Young's modulus (GPa)	Tensile strength (MPa)	CTE (ppm)	Thermal conductivity (W/m K)	Warpage	Cost	Notes
Organic (FR4)	0.1–20	2,000–3,000	14–70	0.3–1	High	Low	Large horizontal and vertical interconnect pitch
Glass	50–90	33–3,500	4–9	1–2	Low	High	Low electrical losses. Metallization is difficult
Silicon	130–185	5,000–9,000	3–5	148	Low	Low	High electrical losses
Steel	190–200	400–500	11–13	16–25			
Copper	128	200–350	17	400			

The Si-IF [2] is a silicon wafer-scale platform that supports integration of small dies at fine vertical pitch (2–10 μm). Si is a highly mature substrate that benefits from decades of technology optimization. Furthermore, passive Si with micrometer size interconnects is a relatively inexpensive construct.

1.2.3 Inter-Die Spacing

In conventional packaging, dies are packaged and placed on PCBs. Interposers, an additional hierarchical layer, is often utilized in modern systems as a stepping stone to close integration of heterogeneous components. However, interposers are expensive and limited to only several components. In any of the abovementioned integration approaches, the inter-die spacing is large. In interposers the spacing is a few hundreds of micrometers, whereas the spacing between two packaged dies on a PCB reaches tens of millimeters. The large inter-die spacing significantly increases the latency and power of communication. In standard von Neumann architectures, where processor-memory communication is a key bottleneck, the excessive performance degradation of communication is especially limiting.

In the Si-IF technology, dies are integrated at high proximity (∼50 μm). This is enabled since the dies are not packaged. The only limitations on interdielet spacing in the Si-IF platform are the roughness of the edge of the die (due to wafer dicing), and edge seals and crack stops that are typically present at the edge of the die, effectively increasing the interdie spacing. Novel technologies, such as plasma dicing [6], will enable further reduction of the interdie spacing on the Si-IF.

1.2.4 Die Size Considerations

Modern systems-on-chips (SoCs) are typically very large in area (excluding mobile SoCs). For example, AMD's Rome server SoC includes up to 1,000 mm^2 of cumulative silicon area [7]. The reason for such a large area is the attempt to integrate as many components as possible at the IC level to leverage the small interconnect dimensions and intercomponent spacing. Large SoCs integrated on interposers, are, however, prone to yield degradation and are typically expensive to fabricate. The concept of heterogeneous integration aims to solve this problem by enabling platforms that support IC-level interconnect dimensions at the package level and eliminate the need for complex and expensive integration hierarchy, (e.g. the use of interposers).

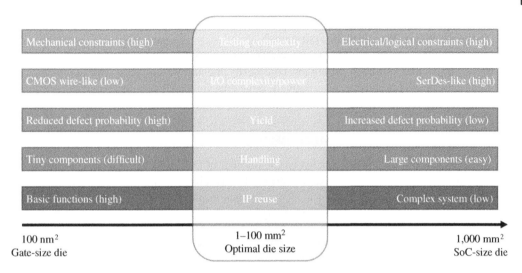

Mechanical constraints (high)	Testing complexity	Electrical/logical constraints (high)
CMOS wire-like (low)	I/O complexity/power	SerDes-like (high)
Reduced defect probability (high)	Yield	Increased defect probability (low)
Tiny components (difficult)	Handling	Large components (easy)
Basic functions (high)	IP reuse	Complex system (low)

100 nm²   1–100 mm²   1,000 mm²
Gate-size die   Optimal die size   SoC-size die

Figure 1.6 Parametric space to determine optimal die size for heterogeneous integration. The key plotted parameters are: intellectual property (IP) reuse, die handling, yield, I/O complexity/power, and testing complexity. Those parameters are plotted as a function of the die size. An optimal parametric space in the center of the figure drives the optimal die size to be 1–100 mm².

Since integration at fine pitch and small inter-component spacing is enabled at the package level by various heterogeneous integration platforms, the components (or dies) need not be very large. Several parameters are key to determine the optimal die size for integration, as shown in Figure 1.6. The optimal die size is represented by the light gray block in the center of Figure 1.6, driven by the optimal parametric space. Following is a discussion of the parameters that determine the die size for heterogeneous integration.

- **IP reuse**: Enables reduced nonrecurring engineering (NRE) cost and faster time to market. Ability to reuse IP significantly increases with smaller die size. The smaller the die, the simpler and potentially more fundamental the function, the higher probability to reuse the die. For example, at the leftmost side of the bar in Figure 1.6, a simple logic block (e.g. multiplexer), has a high probability to be reused many times in a other projects or systems. Alternatively, a very large system (e.g. an entire SoC), will not likely to be reused in other projects or systems. Therefore, according to the IP reuse parameter, the smaller the die, the better.
- **Handling**: This parameter prefers large dies since handling very small dielets (smaller than 1 mm²) is difficult. Special tools, alignment techniques, and handling procedures are required to handle such small dies. On the other hand, handling large dies is easy and established tooling can be used.
- **Yield**: A key parameter and potentially the main driver of any fabrication facility. Statistically, large dies are prone to high probability of defects leading to low yield and therefore high cost. Alternatively, small dies typically exhibit a very high yield in an established process, driving down the cost of the dies and the entire system. Small dies are therefore preferred to optimize the yield parameter.
- **I/O complexity/power**: Rent's rule drives this parameter, i.e. the number of I/Os is related to the complexity of the component (further described in Section 1.2.5). Small components (or dies) require fewer I/Os and therefore less I/O-related power. Integration of small dies supports local highly parallel communication, as compared to large dies that required high-speed

serializer/deserializer (SerDes) communication, which is both power and area-hungry. Smaller are therefore preferred to satisfy this parameter.

- **Testing complexity**: This is a unique parameter that is prohibitive for both very small and very large dies. Small dies, although exhibit low complexity and require simpler testing approach, are difficult to probe and expensive. Alternatively, large dies are highly complex requiring sophisticated testing approaches and significantly limit testing at speed.

1.2.5 Dielet to Substrate Pitch Considerations

Rent's rule [8] determines a relation between the number of I/Os of a chip and the complexity of that chip. Specifically, the number of signal terminals T is related to the number of internal chip components, g (e.g. gates, blocks)

$$T = t \cdot g^p \tag{1.1}$$

where t and p are constants that represent the technology and circuit complexity. Typical values of p are $0 < p < 1$ and approach unity for low-complexity dies. In microprocessors, for example the values of t and p are, respectively, 0.8 and 0.45 [9]. Current system integration technologies do not support the number of I/O terminals that are required according to Rent's rule. Power-hungry SerDes circuitry is used to "bypass" Rent's rule. In [2], the following expression is derived for the pitch P of the I/Os between the die and substrate

$$P = \frac{4\left(\frac{A}{A_T}\right)^p}{t \cdot A^{p-\frac{1}{2}}} \tag{1.2}$$

where A is the area of the die, and A_T is the area per transistor (for the specific technology).

From (1.2) and assuming SerDes circuitry is to be eliminated, an I/O pitch of ~3 to 7 μm will be required for SoCs (in technologies of 45 nm and smaller). Note that this pitch is similar to the fat wire pitches of ICs. Advanced heterogeneous integration platforms will, therefore, have to support a similar range of vertical interconnect pitch between the dies and substrates. The Si-IF platform, for example borrows heavily from standard Si fabrication techniques and supports a fine integration pitch of 2–10 μm. Heterogeneous integration platforms that utilize solder-based vertical interconnects will not be able to reduce the pitch to the required range, as previously discussed. The notion of the fine pitch and small interdie spacing, positions the Si-IF as a natural platform to realize an SoC-like system-on-wafer (SoW).

In addition to the power savings due to the elimination of SerDes, the fine I/O pitch on the Si-IF supports a significant increase in the number of I/Os. The data rate of each I/O can, therefore, be lower as compared to the data rate per I/O in current packaging technologies. The aggregate bandwidth of the IBM POWER9 chip, for example is 1,206 GB/s [10] with 2,359 C4 pads dedicated to differential signaling [11], resulting in a data rate per pad of 8.2 Gb/s. Whereas, if integrated using a fine pitch of 3.5 μm [2], the number of pads dedicated to signaling could be increased by a factor of approximately 35. The increase in the number of pads for signaling will result in reduced data rate requirements per pad of 0.23 Gb/s to support the same aggregate bandwidth. Similar to the signal I/Os, the number of I/Os dedicated to power delivery will significantly increase in the Si-IF platform.

1.2.6 Backward Compatibility

Heterogeneous integration strongly promotes the concept of IP reuse leveraging components-off-the-shelf (COTS). These components may be fabricated in various technologies with a wide range of integration pad pitches and materials. Heterogeneous integration platforms must support integration of these disparate technologies within the same platform. For example, the Si-IF technology supports integration of dies of different height, area, pad pitches, and attachment material, i.e. solder-based and direct metal). This feature supports utilization of hardened legacy IP and eliminates the need to redesign circuits in newer technologies unnecessarily, significantly reducing NRE.

1.3 Design and Architecture of Heterogeneous Integration Platforms

Several heterogeneous integration platforms are at somewhat mature stages of fabrication technology [5, 12]–[15]. Nonetheless, many system-level design and architectural challenges must still be addressed. In this section, we will discuss power delivery and thermal management, floorplanning, and communication for heterogeneous integration systems. These design aspects pose some of the key challenges on the path to functional and efficient heterogeneous systems.

1.3.1 Power Delivery and Thermal Management

While power delivery design has recently shifted toward distributed power delivery schemes with point-of-load (POL) voltage regulation, in modern heterogeneous ICs power is primarily managed in centralized manner, at the expense of limited real-time control and significant power dissipation. In large-scale heterogeneous integration platforms, where many dielets can be potentially integrated, the physical horizontal distances become very large, rendering centralized power delivery and management approaches impractical.

To enable efficient power supply within these large heterogeneous systems, a fundamental change in power delivery and management approach is necessary. (i) Global high-voltage power should be distributed across the wafer and locally converted and regulated within a wide range of dynamically scaled supply voltages and high nonlinear load currents; (ii) Power should be managed locally, with fine spatial, temporal, and voltage granularity; (iii) Local power management decisions should be communicated across the system, optimizing system-wide power efficiency [16]. The distributed power delivery and management approach is, however, area-hungry, requiring space for local power regulators and limiting scaling of interdielet spacing and therefore communication. Three potential power delivery approaches are, therefore, to be considered:

(1) **Vertically integrated miniature voltage regulators**: In this approach, novel high-efficiency and high-current density miniature voltage regulators [17] are integrated on top of functional dielets in a 3D manner. These regulators will receive power from peripheral connectors at high voltage and convert and regulate locally. The Si-IF platform supports such integration as part of the network on interconnect fabric (NoIF) methodology [18], where utility dies (UDs) are floorplanned alongside with the functional dies on the wafer. The NoIF, with UDs as nodes within the network, is expected to provide all of the services required by the heterogeneous system integrated on the Si-IF, including power delivery. The voltage regulator dies can be integrated on top of the UDs with power management circuitry located at the UDs. This approach supports high quality of delivered power and high efficiency (i.e. low resistive losses), since relatively low current is distributed in the long cross-wafer interconnect. For the same reason, heat dissipation is also favorable in this approach. Locally, heat can be managed using heat sinks and active cooling techniques. The Si-IF platform is effectively a Si wafer with significantly better thermal conductivity as compared to classical organic packaging materials. The Si-IF serves,

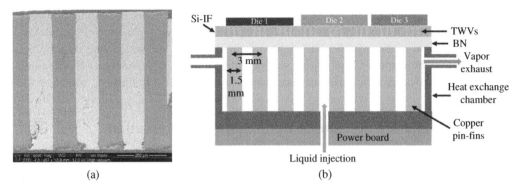

(a)   (b)

Figure 1.7 (a) Cross section of SEM images of TWVs using a two-step DRIE with low-current (50 mA) electroplating [19] M.-H. Liu et al., 2019, IEEE. (b) Schematic of the PowerTherm concept. Source: Based on Ambhore et al. [20].

therefore, as efficient heat spreaders, further alleviating the thermal challenge. Nonetheless, for extremely high-power systems, as potentially expected to be integrated on the Si-IF, a more elaborate cooling system is required. Such system is described in the following power delivery approach.

(2) **Backside distributed power delivery**: Modern systems integrated on interposers utilize TSVs to communicate power and data signals from the IC to the package/board and vice versa. Integrated voltage regulators (IVRs) support higher-power efficiency and robustness of the power delivery system. Nonetheless, for high-power applications, the interposer-based packaging hierarchy is thermally blocked in the direction of the package/board, and cooling is limited in the other vertical direction, significantly restricting scaled integration. In the Si-IF, an elaborate cooling system is possible due to the simple packaging hierarchy. PowerTherm, is a power delivery structure that doubles as a heat dissipation system. In this approach, power is delivered to the Si-IF at low-to-medium voltage using through wafer vias (TWVs) [19], see Figure 1.7a. On the backside of the Si-IF, the current flows through large (1 mm in diameter) low-resistance Cu pins that connect the Si-IF to an external power board. The concept is to disintegrated the large-power converters from the Si-IF to maintain high-area utilization on the front side of the Si-IF. The Cu pins serve as cooling fins enclosed within a cooling chamber that utilizes two-phase cooling [20], see Figure 1.7b.

(3) **Hybrid power delivery**: In this approach, the previous two approaches are combined to provide efficient power delivery and heat dissipation to the heterogeneous system on the Si-IF. In the hybrid scheme, power is delivered from the backside at high voltage, significantly reducing resistive losses and Joule heating, and loosening the constraints on the cooling system. The power is delivered to the front side of the platform using pins and TWVs as before. The power will then reach the vertically IVRs to be locally converted and regulated. This approach is favorable in terms of both power and heat. Furthermore, it supports heterogeneous power requirements, such as various voltage domains. As an additional step, deep trench decoupling capacitors (DTCAPs) are fabricated within the Si-IF [21], under the functional and UDs to provide local charge reservoirs and support the high-frequency power demands of the integrated dielets.

Increasing packaging density exacerbates the thermal challenge, as the heat density increases and the area for thermal mitigation is reduced. It has been a significant barrier to the proliferation of 3D integration beyond low-power memory stacks. In conventional packages, heat is extracted from the top of the die through a heat-spreader and a heat sink that may be air or water cooled.

Very little heat (10%) is extracted from the bottom of the die, due to the poor thermal conductivity of the organic laminate. In contrast, the Si-IF presents a richer thermal scenario where heat may be extracted from either side of the platform.

We consider a high-power die (0.5 W/mm^2) and a low-power die (0.1 W/mm^2) assembled in three configurations: all high-power dies, all low-power dies, and a checkerboard configuration. The equivalent thermal model used in this analysis is shown in Figure 1.8a. The Si-IF behaves as an integrated heat spreader because of the high thermal conductivity of Si, and a fin structure is micro-machined on the backside of the Si-IF. Note that heat can be transferred laterally from a hot die to a cold die through the Si-IF. No heat sink is used on the top of the dies. A convective coefficient of 1,000 W/m^{2}K, typical value for forced air cooling was used. The heat is assumed to enter the Si-IF only through the interconnects. No thermal interface material used [22].

The superior thermal properties of the Si-IF are apparent from Figure 1.8b. The worst case is when all of the dies are operating at high power. In steady state, all dies exhibit a temperature of around 120 °C and a few degrees cooler near the edge. In the checker board case, the temperature is significantly lower, but once again, all dies (including low-power dies) reach the same temperature of about 80 °C. This is due to the isothermal nature of the Si-IF heat-sink. The temperature of the hot and cold dies as a function of the convective heat transfer coefficient is plotted in Figure 1.8c, showing that forced air cooling is an adequate approach for thermal management even for a large SoW.

This preliminary study shows that the use of an all silicon package has significant thermal advantages – leverage the Si-IF as an efficient heat spreader, integrated micromachined heat sinks, elimination of thermal interface material as the heat is transferred through Cu interconnects from the die to the Si-IF (and not through solder intermetallics and organic under fill). Nonetheless, if power density of the heterogeneous system is significantly higher than the values used in this experiment, die side heat sinks and the use of more elaborate cooling systems, such as the previously described PowerTherm, is required.

1.3.2 Floorplanning

The future of heterogeneous systems is in tightly integrated dielets rather than large SoCs. This notion leads to a new system-level design aspect that was not previously considered – system-level floorplanning. Placement of dielets across the integration platform must be performed in an intelligent manner considering parameters that are borrowed from IC floorplanning, such as area, wire length, and heat. Furthermore, additional parameters that were not applicable to on-chip floorplanning, also must be considered in system-level floorplanning, such as placement of UDs, communication, especially for connections that are not simple wires, e.g. optical communication and SerDes, testability, and others.

UD placement is a key design challenge for ultralarge-scale heterogeneous integration. A tradeoff between area dedicated to UDs and the benefits that they provide for the system in terms of power management, communication, testability, etc., must be considered. Similarly to networks on chip (NoCs), the NoIF is a necessary overhead for the system integrated on the Si-IF. The area overhead that the NoIF imposes is evaluated based on the parameters listed in Table 1.2 and plotted as a function of the number of UDs and the area of each UD in Figure 1.9. A typical minimal overhead of NoCs for SoCs at 32 nm is 12% [24], marked for reference in Figure 1.9. Note that the columns for 4,096 UDs at 16 and 25 mm^2 are missing since the values are above 100%. The NoIF exhibits a wide range of area overhead depending on the number and size of UDs. Nonetheless, majority of the UD number-area combinations exhibit an area overhead that is smaller than typical NoCs [23].

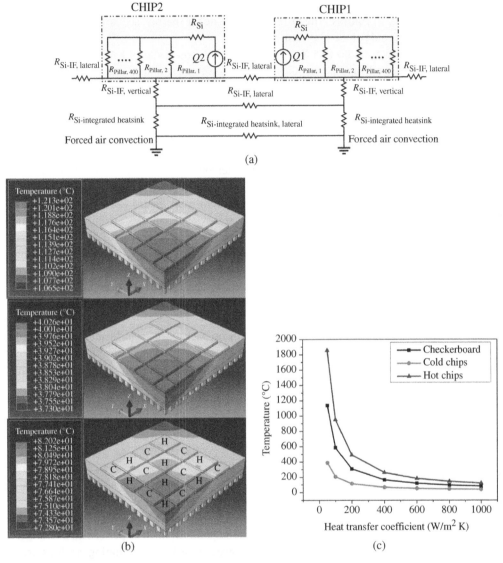

Figure 1.8 (a) Equivalent thermal resistance model of two dielets assembled on the Si-IF with inter-dielet spacing of 100 μm and an integrated Si micromachined heat sink. (b) Temperature maps of the dielets-on-Si-IF assembles in the three studied cases, Case 1: all dielets assembled on Si-IF are operating at a power density of 0.5 W/mm^2 (hot dies), Case 2: all dielets assembled on Si-IF are operating at a power density of 0.1 W/mm^2 (cold dies), and Case 3: dielets are assembled on Si-IF in a checkerboard configuration with alternating hot (H: 0.5 W/mm^2) and cold (C: 0.1 W/mm^2) dielets. (c) Maximum temperature profile of dies assembled on Si-IF for three operational cases: all hot (triangles), all cold (circles), and checkerboard (black/squares) configurations. Source: [22]/IEEE.

1.3.3 Communication

Communication is one of the most important design aspects of any system, and even more so in heterogeneous systems. The different communication specifications, applications, voltage levels, IC technologies, etc., drive the need for significant research into the communication system.

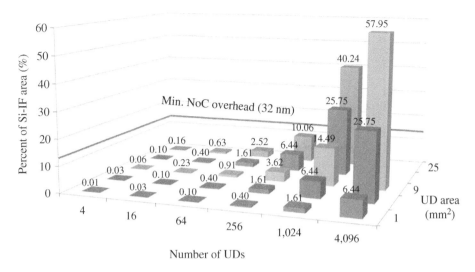

Figure 1.9 Area overhead of NoIF as a function of the number of UDs and the area of each UD. Note, large UDs refer to hierarchical UDs, in other words, UDs that consist of multiple dielets (e.g. control, μ-SerDes, antenna). Source: Vaisband and Iyer [23], 2019 IEE.

Table 1.2 Si-IF and UD parameters used to determine the area overhead of the NoIF.

Parameter	Value
Si-IF diameter (mm)	300
Si-IF area (mm^2)	~70,686
Effective Si-IF area percentage (%)	90
Effective Si-IF area (mm^2)	~63,617
Number of UDs	{4, 16, 64, 128, 256, 1,024, 4,096}
Area of each UD (mm^2)	{1, 4, 9, 16, 25}

For example, neural network applications mostly require local (short-haul) communication, whereas highly heterogeneous systems may benefit from global (long-haul) communication. Realization of the communication system is highly dependent on the application of interest.

Local communication has been addressed by various packaging technologies, whereas the common denominator for these efforts is higher interconnect bandwidth per millimeter of dielet edge. Effectively, we have been trying to reduce the I/O pitch for interdielet communication. For example, EMIB utilizes a Si bridge embedded within the organic package that supports fine pitch communication wherever it is needed. The rest of the signals are routed at a larger pitch and therefore lower bandwidth [5]. The Si-IF supports local communication at extremely high density using the hardware protocol named SuperCHIPS [25, 26]. Alternatively, communicating across the entire wafer (few hundreds of millimeters) is still an unsolved challenge.

The NoIF supports global communication on the Si-IF borrowing heavily from NoCs. Each UD serves as a node with the network and includes circuits to enable packetized routing. Various efficient routing algorithms to prevent deadlocks and enable efficient low-power communication

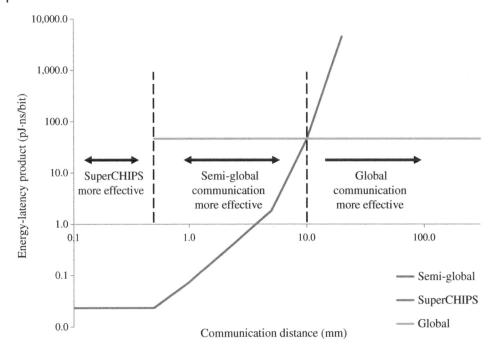

Figure 1.10 Design space of communication on the Si-IF. SuperCHIPS is utilized for local communication, semiglobal communication is achieved by utilizing repeaters and hoping across dies, and global communication is based on SerDes. Source: Vaisband and Iyer [32], 2019 IEEE.

(e.g. XY routing) can also be utilized. Each UD will receive data from local functional dielets, queue the data, and then route the data to the next UD according to the designed algorithm. Communication among UDs is expected to be implemented using μ-SerDes, RF, or optical technology. A preliminary study of the design space for communication, including local, semiglobal, and global communication shows the effective distance of semiglobal communication using repeaters (included in UDs and in functional dielets as an overhead). The energy-latency product was considered in the comparison between SerDes-based [27–29] and repeater-based [30, 31] communication. The design space for the various communication approaches is illustrated in Figure 1.10 [32].

1.4 Reliability of Heterogeneous Integration Systems

Packaging has more than its fair share of reliability failures, as shown in the pie chart in Figure 1.11. Many of these failures stem from the fact that packaging involves the joining of multiple material contacts such as solder joints. Another issue has to do with the fact that today's packaging constructs are pushing the limits of fabrication and as such the limits of reliability. Package failure rates are notoriously difficult to predict and very often difficult to screen out as well. Furthermore, accelerated test methods used in, for example, CMOS reliability estimation, are not easy to implement in package constructs since the physics of failure is not well understood. These principal failure mechanisms have been classified in [33], as shown in Figure 1.12.

As can be seen in Figure 1.12, many reliability issues arise from the intimate contact between materials that either react to form brittle materials such as solder intermetallic compounds, or

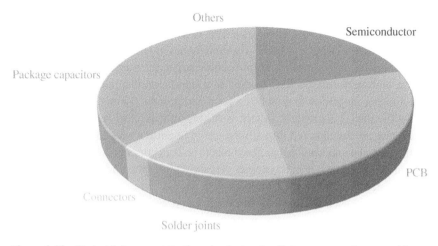

Figure 1.11 Typical failure contributions in electronics. Data are approximate and from multiple sources including anecdotal.

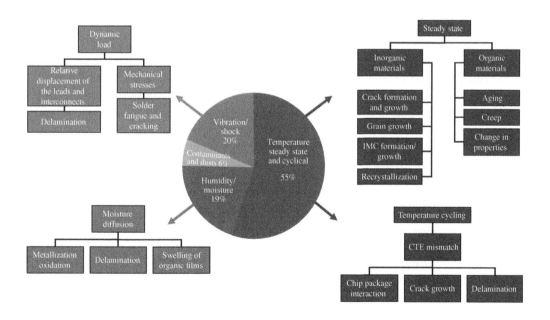

Figure 1.12 Classification of principal failure mechanisms. Source: Modified from Cen and Stewart [33].

exhibit large coefficient of thermal mismatch. Both of these problems form a volatile combination when the assembly is thermally cycled. These issues are exacerbated when the dimension of the parts being assembled become large, and so the assembly issues of joining large chips to laminates is indeed quite significant. In addition, as chips become larger, the number of wiring levels also increases (15–20 in high-performance ICs), leading to a significant increase in the warpage of the chips and thus limiting the ability to establish a good and reliable contact, especially near the edges and corner of the die. This problem, in fact, is the biggest source of reliability failures. To some extent, these issues are ameliorated through the use of underfills which fill up the space between the die and the package and keep things glued together. Unfortunately, when the bumps become smaller or when dies become larger, it becomes very difficult for the underfill to penetrate throughout the space between the die and the laminate.

A limited number of thermal cycling iterations can be done on assembled parts to weed out bad contacts. However, since the assembled systems are quite expensive with multiple integrated chips, diagnosing failure points is necessary, although difficult and expensive. The only recourse is to rework parts by shearing off suspect dies and reattaching new ones. This process, as can be imagined, is very expensive and time-consuming.

Another aspect of the reliability problem is the ingress of corrosive elements into the chip and the connections between the chip. Moisture, for example, can corrode exposed metal and cause electrochemical reactions such as dendritic growth and oxidation. Although, underfill material slows down this process, it does not eliminate it completely as the material is porous and allows, among other effects, moisture ingress.

Fortunately, the advanced packaging techniques we have discussed in this chapter are intrinsically more reliable. By minimizing the use of solder and relying on direct metal-to-metal bonds, the issue of brittle intermetallic formation is completely eliminated. We have, in fact, shown that the shear strength of our direct Cu–Cu TCB die attach is high, enabling the connections to survive multiple thermal cycling and to meet the so-called "mil" standards for die shear after thermal cycling. The extremely short distance between dies precludes the introduction of any kind of underfill. However, misalignment of the die to the Si-IF can potentially expose Cu to the ambient that would be subject to oxidation if there is any moisture ingress. Multiple approaches have been developed to mitigate this problem. These include the use of conformal coatings and, as demonstrated in [34], an atomic layer deposition of a thin layer ($\sim$10 nm) of alumina (Al_2O_3) provides robust protection against moisture ingress.

The attachment of heatsinks, and, in the case of two-phase cooling, the impact of vibrations, will cause additional failure modes. These have to be mitigated by careful design and adequate damping to avoid mechanical resonances. Similarly, connectors to a wafer scale system are a potential source of failure and need to be carefully designed, simulated, and tested. As always, thermal runaway is a constant danger and reliable connection of the heatsink to the assembly is critical. Fortunately, these methods have been developed for classical packaging and can be reused in heterogeneous integration platforms.

To summarize, the intrinsically simplified assembly techniques of advanced packaging and the reduction in utilization of diverse materials, result in a better reliability prognosis for advanced packaging. However, complex assemblies with a large number of dielets that comprise the system, mean that there is statically a greater probability of failure. This, coupled with the fact that rework is not going to be possible, will mandate a system level approach to redundancy and self-repair of these applications. We predict that the judicious use of redundancy and self-repair will indeed decrease the observable failure rates in advanced packaging.

1.5 Application Space of Heterogeneous Integration

Heterogeneous integration not only improves system parameters (power, latency, bandwidth, thermal management) by orders of magnitude but also enables novel applications that were previously impossible. One such class of applications is ultralarge-scale systems, such as neural networks. In this particular class of applications, more is better, in other words, a larger neural network will produce better training, lower generalization error, and higher performance [35]. Emerging heterogeneous integration platforms are highly compatible for such large-scale applications.

A 300 mm diameter Si-IF can support the integration of over 60,000 dielets with an area of 1 mm² (see Table 1.2. Each of these dielets can be a memory or a GPU in the effort to tile an entire wafer as proposed in [3, 36]. The proposed GPU-mem integration on the Si-IF is illustrated in Figure 1.13. To scale this approach further, a 3D wafer-scale-integration (WSI) technology was proposed in [37]. In this approach, multiple Si-IF wafers are stacked on top of each other in a 3D manner. Alternating processing and memory wafers are used. The end result is an ultralarge-scale system with memories located at high proximity to processors. Communication across the Si-IF levels is implemented using TSVs. An illustration of the 3D WSI ultralarge-scale system utilizing the Si-IF as is shown in Figure 1.14 [37].

Alternatively, to avoid classical von Neumann architectures, novel compute-in-memory technologies can be used. One such technology, the charge trap transistor (CTT) is based on the high-k dielectric utilized in recent CMOS technologies (starting 45 nm) [38]. This technology enables integration of analog memory alongside computation in a single process. Utilizing CTTs will improve system performance by orders of magnitude, eliminating most compute-memory external communication, thus providing tremendous reduction in power and latency.

A different class of applications is highly heterogeneous small edge systems, such as edge devices in the Internet of Things (IoT) domain, or wearable and implantable devices. Recent innovation in flexible and biocompatible heterogeneous integration technology supports this class of applications. The FlexTrate is a completely flexible (1 mm bending radius) biocompatible platform recently demonstrated in [39]. Since the introduction of FlexTrate, multiple applications

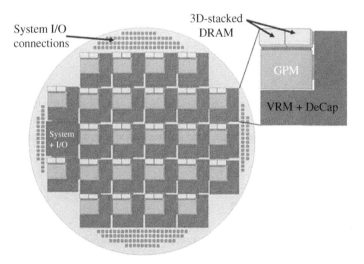

Figure 1.13 Waferscale GPU with 25 GPM units comprising of two 3D-stacked DRAMs per unit, VRM unit, and decoupling capacitors. Source: Pal et al. [3], 2019 IEE.

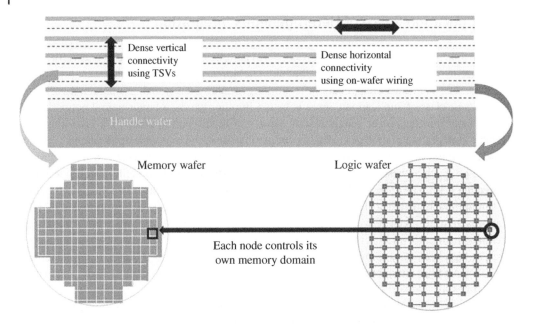

Figure 1.14 One possible embodiment of a neuronal system using 3D-WSI, consisting of separate logic and memory wafers bonded together and connected using high-density TSVs. Source: Kumar et al. [37]/Association for Computing Machinery.

Figure 1.15 An image of fabricated sEMG electrodes on FlexTrate™. Source: [41] A. Alam et al., 2020, IEEE.

have been demonstrated [40–42]. A fabricated FlexTrate used in a surface electromyography system as is shown in Figure 1.15.

1.6 Future of Heterogeneous Integration

We end this chapter with some thoughts on the future of heterogeneous integration. At the very outset, packaging has always been about heterogeneous integration. Before the advent of a viable backend technology and monolithic integration, IBM introduced a very basic integration of small number of devices in what was called Solid Logic Technology (SLT). Since then packaging has evolved and became significantly more complex. This has resulted in the ability to build ultralarge systems that encompass a multiplicity of components, board, and interconnect systems. These legacy systems are indeed heterogeneous. The main difference between today's heterogeneous systems and the legacy approaches is the scale. This includes the integration of a multiplicity of components far more physically and electrically intimately than before. This journey has only just begun. The true advantages of heterogeneous integration will not accrue until we are able to make a packaged system look more like a monolithic system. This goal is at best aspirational today, but certainly achievable in the next several years. We have highlighted the use of wafer scale systems in this chapter. Already, we are seeing a trend of large ICs in industry. Back in 1993, Intel Corporation's most advanced processor was built in 0.8 μm technology and was 300 mm^2 in area. It had a mind boggling (at the time) 3.1 million transistors! Fast forward to 2020. The latest Ampere A100 offering from NVIDIA (Figure 1.16) sports a silicon interposer that is more than two standard reticle fields and a processor die that is 836 mm^2 in area – a full reticle field in 7 nm technology – a 17,000 times increase in the number of transistors. Furthermore, impressively, the A100 includes 1 TB of DRAM that is assembled on the same interposer. The DRAM is 3D stacked using High Bandwidth Memory (HBM). This system represents the most advanced chip and memory technology today – in silicon capability and, more importantly, packaging technology.

Nonetheless, this is just the beginning. We see this large interposer trend continuing and culminating in multiple full wafer systems. Such wafer scale systems built on a 300 mm wafer may have 10–100 times the processing power than the system described above, and potentially 10–50 times the amount of memory. More importantly, the performance of this system will strongly depend on the ability to integrate more memory and the bisection bandwidth of the system (the ability of any processor to quickly access any memory with near uniform accessibility). The ability to deliver 50–70 kW of power to such a system is challenging. At the same time, extracting the heat is an even greater challenge and will require two-phase cooling or other advanced cooling techniques. Typically, we expect an high performance computing (HPC) data center to utilize several such wafers that will be interconnected with extremely high bandwidth using wired, millimeter wave, and photonic interconnects. We have currently started to address these problems. There are two other areas where heterogeneous integration will have a very high impact: (i) the area of flexible biocompatible

Figure 1.16 NVIDIA A100 Tensor Core GPU.

medical electronics, hinted at in Section 1.5. (ii) Another area where heterogeneous integration plays a big role is in the development of millimeter wave and THz systems for 6G and diagnostic systems. As antennas become smaller, they radiate less power and antenna arrays are required. These arrays will need III–V technology for the power amplifiers, SiGe BiCMOS technology for the intermediate frequency, and conventional CMOS elsewhere in an extremely tight footprint. We believe that these systems will also be wafer scale.

There is another aspect of heterogeneous integration that warrants mention. This has to do with the huge NRE cost for designing and building a chip and a system. In some of the more advanced nodes, this can be upward of several hundred million dollars to build a large server grade chip, and few tens of millions of dollars for even a small one. This has priced small players out of the market and led to a very significant consolidation of the players who can afford to break into the market. Furthermore, building a chip from scratch is a long duration endeavor and may take up to three years seriously affecting the time to market and potentially the competitive viability of this chip in the marketplace. This is not conducive to innovation. The approach we have outlined in this chapter can address this problem. We have proposed a model where even a small player can conceive of and build a system quickly at low NRE. This conceptual design is first abstracted and then submitted to a heterogeneous system integrator. This entity has access to a database of hardened IP, i.e. IP that exists as fully verified dielets and available to purchase from distributors, and it intelligently fragments the system into individual readily available dielets. Some very specialized IP may not be readily available and might serve as the "secret sauce" that represents the innovative contribution of the system designer. That would have to be designed as a chiplet and taped out to a foundry and built. This, however, is a significantly smaller effort than taping out the entire chip. The system integrator would then assemble all the dielets. For this to happen, we need systems that can be fragmented and reintegrated at the design tool level, and this will only make sense and be viable if we can integrate quickly (short time to market) and without loss of performance. This last point requires the fine pitch interconnect we have talked about in this chapter. As the reader can see, we do have all the ingredients, but we still do not have the ecosystem. Developing this ecosystem is clearly the biggest challenge to realize the full advantages and potential of heterogeneous integration at scale. This ecosystem includes reusable hardened IP in the form of readily available cost-effective known good dielets, fragmentation and reintegration tools, and open standards such as SuperCHIPS (or equivalent) that support communication among the dielets, as well as longer range protocols as discussed in this chapter. Open-source is the key to develop such an ecosystem, but the development of acceptable open-source protocols is a slow and laborious process. Nonetheless, we are already seeing impressive progress in this area through the activities of open domain-specific architecture (ODSA), the DARPA CHIPS program, and the heterogeneous integration roadmap (HIR). The future of a democratized and innovative subsystems based on heterogeneous integration at scale is certainly work in progress and will evolve over the next several years.

1.7   Summary

This chapter describes the future of heterogeneous integration. We discuss state-of-the-art platforms and focus on the Si-IF as a favorable platform that supports the dielet paradigm shift. Dielets (1–100 mm^2) are the future of microelectronics systems and will eventually replace large SoCs. These dielets must be integrated at high proximity (Si-IF supports ~50 μm interdielet spacing) and fine I/O pitch (Si-IF supports 2–10 μm) to ensure SoC-like systems. These dimensions are supported by Rent's rule for packaging and serve as a marker for the future of scaling at the system level. Furthermore, these dimensions are supported by various designs, fabrications, and handling limitations as described in this study.

Some heterogeneous integration platforms are in relatively advanced state of maturity, nonetheless, many system-level design aspects must still be addressed to practically enable systems on these platforms. Specifically, we discuss the power delivery and thermal management challenges associated with heterogeneous integration. We propose distributed back side power delivery approaches supported by TWVs that enable a range of power specifications as expected in heterogeneous systems. Furthermore, thermal management is expected to be a significant challenge, especially in ultralarge-scale systems. Elaborate cooling approaches are also described. Additional design aspects such as floorplanning and communication are discussed. Ultralarge systems will require a utility network to support system-level services (such as power management, communication, testing) and ensure correct functionality and high performance. The NoIF, based on UDs, is proposed as such a service provider conceptually borrowing from NoCs for large SoCs. Further investigation into the NoIF is required to determine the number of UDs, their floorplanning, dimensions, etc.

Reliability and testing are key aspects of enablement of heterogeneous platforms. Some reliability verification of the Si-IF has been experimentally demonstrated. Testing however, remain a significant challenge. The Si-IF is a nonreworkable system since it is a solderless integration technology. Built-in self-test and repair methodologies need to, therefore, be developed to ensure high yield and low cost of the system.

We discuss the application space of interest for heterogeneous integration. Specifically we focus on ultralarge-scale neural networks and the possibility to scale them out to match the neuron and synapse capacity of a human brain. Various technologies should be utilized to accommodate such a vision, including compute-in-memory devices and 3D integration.

Finally, the future of heterogeneous integration is presented. We believe that the main challenge for heterogeneous integration is the development of an ecosystem to support reusable hardened IP, fragmentation, and re-integration tools, and open standards protocols. Nonetheless, this future is bright and not too far out.

References

1 S. S. Iyer, "Three-dimensional integration: an industry perspective," *Materials Research Society Bulletin*, vol. 40, no. 3, pp. 225–232, March 2018.

2 S. S. Iyer, S. Jangam, and B. Vaisband, "Silicon interconnect fabric: a versatile heterogeneous integration platform for AI systems," *IBM Journal of Research and Development*, vol. 63, no. 6, pp. 5:1–5:16, November/December 2019.

3 S. Pal, D. Petrisko, M. Tomei, P. Gupta, S. S. Iyer, and R. Kumar, "Architecting waferscale processors: a GPU case study," in *Proceedings of the IEEE International Symposium on High Performance Computer Architecture*, pp. 250–263, February 2019.

4 A. A. Bajwa, S. Jangam, S. Pal, N. Marathe, T. Bai, T. Fukushima, M. Goorsky, and S. S. Iyer, "Heterogeneous integration at fine pitch ($\leq$ 10 μm) using thermal compression bonding," in *Proceedings of the IEEE International Electronic Components and Technology Conference*, pp. 1276–1284, May 2017.

5 R. Mahajan, R. Sankman, N. Patel, D.-W. Kim, K. Aygun, Z. Qian, Y. Mekonnen, I. Salama, S. Sharan, D. Iyengar, and D. Mallik, "Embedded multi-die interconnect bridge (EMIB) - a high density, high bandwidth packaging interconnect," in *Proceedings of the IEEE International Electronic Components and Technology Conference*, pp. 557–565, May 2016.

6 N. Matsubara, R. Windemuth, H. Mitsuru, and H. Atsushi, "Plasma dicing technology," in *Proceedings of the Electronics System-Integration Technology Conference*, pp. 1–5, September 2012.

7 D. Suggs, M. Subramony, and D. Bouvier, "The AMD "Zen 2" processor," *IEEE Micro*, vol. 40, no. 2, pp. 45–52, March 2020.

8 M. Y. Lanzerotti, G. Fiorenza, and R. A. Rand, "Microminiature packaging and integrated circuitry: the work of E. F. Rent, with an application to on-chip interconnection requirements," *IBM Journal of Research and Development*, vol. 49, no. 4.5, pp. 777–803, July 2005.

9 W. D. Brown, *Introduction and Overview of Microelectronics Packaging*, Wiley-IEEE Press, 1999.

10 S. Chun, W. D. Becker, J. Casey, S. Ostrander, D. Dreps, J. A. Hejase, R. M. Nett, B. Beaman, and J. R. Eagle, "IBM POWER9 package technology and design," *IBM Journal of Research and Development*, vol. 62, no. 4/5, pp. 12:1–12:10, June 2018.

11 C. Gonzalez, E. Fluhr, D. Dreps, D. Hogenmiller, R. Rao, J. Paredes, M. Floyd, M. Sperling, R. Kruse, V. Ramadurai, R. Nett, S. Islam, J. Pille, and D. Plass, "3.1 POWER9™: a processor family optimized for cognitive computing with 25Gb/s accelerator links and 16Gb/s PCIe Gen4," in *Proceedings of the IEEE International Solid-State Circuits Conference*, pp. 50–51, February 2017.

12 S. Jangam, U. Rathore, S. Nagi, D. Markovic, and S. S. Iyer, "Demonstration of a low latency (<20 ps) fine-pitch (≤ 10 μm) assembly on the silicon interconnect fabric," in *Proceedings of the IEEE International Electronic Components and Technology Conference*, pp. 1801–1805, May 2020.

13 A. O. Watanabe, B. K. Tehrani, T. Ogawa, P. M. Raj, M. M. Tentzeris, and R. R. Tummala, "Ultralow-loss substrate-integrated waveguides in glass-based substrates for millimeter-wave applications," *IEEE Transactions on Components, Packaging and Manufacturing Technology*, vol. 10, no. 3, pp. 531–533, March 2020.

14 M. Ali, A. Watanabe, T. Kakutani, P. M. Raj, R. R. Tummala, and M. Swaminathan, "Heterogeneous integration of 5G and millimeter-wave diplexers with 3D glass substrates," in *Proceedings of the IEEE International Electronic Components and Technology Conference*, pp. 1376–1382, May 2020.

15 S. Dwarakanath, T. Kakutani, D. Okamoto, P. M. Raj, M. Swaminathan, and R. R. Tummala, "Reliability of fine-pitch <5-μm-diameter microvias for high-density interconnects," *IEEE Transactions on Components, Packaging and Manufacturing Technology*, vol. 10, no. 9, pp. 1552–1559, September 2020.

16 I. Vaisband and E. G. Friedman, "Dynamic power management with power network-on-chip," in *Proceedings of the IEEE Midwest Symposium on Circuits and Systems*, pp. 225–228, June 2014.

17 K. Tien, N. Sturcken, N. Wang, J.-W. Nah, B. Dang, E. O'Sullivan, P. Andry, M. Petracca, L. P. Carloni, W. Gallagher, and K. Shepard, "An 82%-efficient multiphase voltage-regulator 3D interposer with on-chip magnetic inductors," in *Proceedings of the IEEE International Symposium on VLSI Circuits*, pp. C192–C193, June 2015.

18 B. Vaisband, A. Bajwa, and S. S. Iyer, "Network on interconnect fabric," in *Proceedings of the IEEE International Symposium on Quality Electronic Design*, pp. 138–143, March 2018.

19 M.-H. Liu, B. Vaisband, A. Hanna, Y. Luo, Z. Wan, and S. S. Iyer, "Process development of power delivery through wafer vias for silicon interconnect fabric," in *Proceedings of the IEEE International Electronic Components and Technology Conference*, pp. 579–586, May 2019.

20 P. Ambhore, U. Mogera, B. Vaisband, M. Goorsky, and S. S. Iyer, "PowerTherm attach process for power delivery and heat extraction in the silicon-interconnect fabric using thermocompression bonding," in *Proceedings of the IEEE International Electronic Components and Technology Conference*, pp. 1605–1610, May 2019.

21 K. Kalappurakal Thankappan, and S. S. Iyer, "Deep trench capacitors in silicon intercon-nect fabric," in *Proceedings of the IEEE International Electronic Components and Technology Conference*, pp. 2295–2301, May 2020.

22 A. Bajwa, S. Jangam, S. Pal, B. Vaisband, R. Irwin, M. Goorsky, and S. S. Iyer, "Demonstration of a heterogeneously integrated system-on-wafer (SoW) assembly," in *Proceedings of the IEEE International Electronic Components and Technology Conference*, pp. 1926–1930, May 2018.

23 B. Vaisband and S. S. Iyer, "Global and semi-global communication on silicon intercon-nect fabric," in *Proceedings of the IEEE/ACM International Symposium on Networks-on-Chip*, pp. 15:1–15:5, October 2019.

24 P. P. Pratim, C. Grecu, M. Jones, A. Ivanov, and R. Saleh, "Performance evaluation and design trade-offs for network-on-chip interconnect architectures," *IEEE Transactions on Computers*, vol. 54, no. 8, pp. 1025–1040, August 2005.

25 S. Jangam, S. Pal, A. Bajwa, S. Pamarti, P. Gupta, and S. S. Iyer, "Latency, bandwidth and power benefits of the SuperCHIPS integration scheme," in *Proceedings of the IEEE International Electronic Components and Technology Conference*, pp. 86–94, May 2017.

26 S. Jangam, A. Bajwa, K. K. Thankappan, P. Kittur, and S. S. Iyer, "Electrical characterization of high performance fine pitch interconnects in silicon-interconnect fabric," in *Proceedings of the IEEE International Electronic Components and Technology Conference*, pp. 1823–1288, May 2018.

27 R. Navid, E. Chen, M. Hossain, B. Leibowitz, J. Ren, C. A. Chou, B. Daly, M. Aleksic, B. Su, S. Li, M. Shirasgaonkar, F. Heaton, J. Zerbe, and J. Eble, "A 40 Gb/s serial link transceiver in 28 nm CMOS technology," *IEEE Journal of Solid-State Circuits*, vol. 50, no. 4, pp. 814–827, April 2015.

28 Z. Feng, Y. Yi, Y. Zongren, P. Chiang, and H. Weiwu, "A low latency transceiver macro with robust design technique for processor interface," in *Proceedings of the IEEE Asian Solid-State Circuits Conference*, pp. 185–188, November 2009.

29 G. R. Gangasani, J. F. Bulzacchelli, T. Beukema, C. Hsu, W. Kelly, H. H. Xu, D. Freitas, A. Prati, D. Gardellini, G. Cervelli, J. Hertle, M. Baecher, J. Garlett, R. Reutemann, D. Hanson, D. W. Storaska, and M. Meghelli, "A 32-Gb/s backplane transceiver with on-chip AC-coupling and low latency CDR in 32 nm SOI CMOS technology," in *Proceedings of the IEEE Asian Solid-State Circuits Conference*, pp. 213–216, November 2013.

30 V. Adler and E. G. Friedman, "Repeater design to reduce delay and power in resistive intercon-nect," *IEEE Transactions on Circuits and Systems II: Express Briefs*, vol. 45, no. 5, pp. 607–616, May 1998.

31 S. X. Shian and D. Z. Pan, "Wire sizing with scattering effect for nanoscale interconnection," in *Proceedings of the IEEE/ACM Asia and South Pacific Design Automation Conference*, pp. 503–508, January 2006.

32 B. Vaisband and S. S. Iyer, "Communication considerations for silicon interconnect fabric," in *Proceedings of the Workshop on System Level Interconnect Prediction*, pp. 1–6, June 2019.

33 Z. Cen and P. Stewart, "Condition parameter estimation for photovoltaic buck converters based on adaptive model observers," *IEEE Transactions on Reliability*, vol. 66, no. 1, pp. 148–160, March 2017.

34 N. Shakoorzadeh, S. Jangam, P. Ambhore, H. Chien, A. Hanna, and S. S. Iyer, "Reliability studies of Si interconnect fabric (Si-IF)," in *Proceedings of the IEEE International Electronic Components and Technology Conference*, pp. 800–805, May 2019.

35 Y. Wu, L. Deng, G. Li, J. Zhu, Y. Xie, and L. Shi, "Direct training for spiking neural networks: faster, larger, better," *Proceedings of the AAAI Conference on Artificial Intelligence*, vol. 33, no. 1, pp. 1311–1318, July 2019.

36 S. Pal, D. Petrisko, R. Kumar, and P. Gupta, "Design space exploration for chiplet-assembly-based processors," *IEEE Transactions on Very Large Scale Integration (VLSI) Systems*, vol. 28, no. 4, pp. 1062–1073, April 2020.

37 A. Kumar, Z. Wan, W. W. Wilcke, and S. S. Iyer, "Towards human-scale brain computing using 3D wafer scale integration," *ACM Journal of Emerging Technologies in Computing*, vol. 13, no. 3, pp. 45:1–45:21, April 2017.

38 F. Khan, M. Han, D. Moy, R. Katz, L. Jiang, E. Banghart, N. Robson, T. Kirihata, J. C. S. Woo, and S. S. Iyer, "Design optimization and modeling of charge trap transistors (CTTs) in 14 nm FinFET technologies," *IEEE Electron Device Letters*, vol. 40, no. 7, pp. 1100–1103, July 2019.

39 T. Fukushima, A. Alam, S. Pal, Z. Wan, S. C. Jangam, G. Ezhilarasu, A. Bajwa, and S. S. Iyer, "FlexTrate™ - Scaled heterogeneous integration on flexible biocompatible substrates," in *Proceedings of the IEEE International Electronic Components and Technology Conference*, pp. 649–654, May 2017.

40 G. Ezhilarasu, A. Hanna, R. Irwin, A. Alam, and S. S. Iyer, "A flexible, heterogeneously integrated wireless powered system for bio-implantable applications using fan-out wafer-level packaging," in *Proceedings of the IEEE International Electron Devices Meeting*, pp. 29.7.1–29.7.4, December 2018.

41 A. Alam, M. Molter, B. Gaonkar, A. Hanna, R. Irwin, S. Benedict, G. Ezhilarasu, L. Macyszyn, M. S. Joseph, and S. S. Iyer, "A high spatial resolution surface electromyography (sEMG) system using fan-out wafer-level packaging on FlexTrate," in *Proceedings of the IEEE International Electronic Components and Technology Conference*, pp. 985–990, May 2020.

42 G. Ezhilarasu, A. Paranjpe, J. Lee, F. Wei, and S. S. Iyer, "A heterogeneously integrated, high resolution and flexible inorganic μLED display using fan-out wafer-level packaging," in *Proceedings of the IEEE International Electronic Components and Technology Conference*, pp. 677–684, May 2020.

2

Hyperdimensional Computing: An Algebra for Computing with Vectors

Pentti Kanerva

University of California at Berkeley, Redwood Center for Theoretical Neuroscience, Berkeley, CA, USA

2.1 Introduction

The digital computer, enabled by semiconductor technology, has become an ever-present part of our lives. The relative ease of programming it for well-specified tasks has us expecting that we should be able to program it for about any task. However, experience has shown otherwise. What comes naturally to us and may appear easy, such as understanding a scene or learning a language, has eluded programming into computers [Mitchell, 2019]. Since such things are accomplished by brains, we look for computing architectures that work somewhat like brains. Computing of that kind would have wide-ranging uses.

This chapter describes computing with vectors that is modeled after traditional (von Neumann) computing with numbers. Its origins are in the artificial neural systems of the 1970s and 1980s that have evolved into today's deep-learning nets, but it is fundamentally different from them.

Similarity to traditional computing relates to how computing is organized. Traditionally, there is an arithmetic/logic unit (ALU) with circuits for number arithmetic and Boolean logic, a large random-access memory (RAM) for storing numbers, and flow control for running a program one step at a time. The same logical organization now refers to vectors as basic objects. The mathematics of the vector operations is the main topic of this chapter.

Why compute with vectors; don't we already have vector processors for tasks heavy with vector operations? The reason has to do with the nature of the vectors and operations on them. The new algorithms rely on truly high dimensionality – a thousand or more – but the vector components need not be precise or ultrareliable. In contrast, the algorithms for vector processors assume high precision and reliability, the engineering requirements of which are very different. Being able to compute with less-than-perfect circuits makes it possible to fully benefit from the miniaturization of circuits and the inclusion of analog components in them. We will be able to build ever larger circuits that operate with very little energy, eventually approaching the efficiency of the brain which does amazing things with a mere 20 W.

Attempts to understand brains in computing terms go back at least to the birth of the digital computer. Early artificial intelligence (AI) consisted of rule-based manipulation of symbols, which mirrors the logic of programming and appeals to our facility for language. However, it fails to explain learning, particularly of language. Artificial neural nets have tried to fill the void by focusing on learning from examples, but has insufficient means for representing and manipulating structure such as the grammar of a language. It is becoming clear that we need systems capable of both statistical learning and computing with discrete symbols. Computing with high-dimensional vectors is aimed at developing systems of that kind.

The first system of the kind was described by Plate in a PhD thesis in 1994, later published in the book *Holographic Reduced Representation* [Plate, 1994, 2003]. It combines ideas from Hinton's [1990] reduced representation, Smolensky's [1990] tensor-product variable binding, and Murdock's [1982] convolution-based memory. Systems that compute with high-dimensional random vectors go by various names: Holographic Reduced Representation, Binary Spatter Code [Kanerva, 1996], MAP (for Multiply–Add–Permute) [Gayler, 1998], Context-Dependent Thinning [Rachkovskij & Kussul, 2001], Vector-Symbolic Architectures (VSA) [Gayler, 2003; Levy & Gayler, 2008], Hyperdimensional Computing [Kanerva, 2009], Semantic Pointer Architecture [Eliasmith, 2013], and Matrix Binding of Additive Terms [Gallant & Okaywe, 2013].

2.2 Overview: Three Examples

Computing is based on three operations. Two correspond to *addition* and *multiplication* of numbers and are called by the same names, and the third is *permutation* of coordinates. All three take vectors as input and produce vectors of the *same dimensionality* as output. The operations are programmed into algorithms as in traditional computing.

The representation of information is very different from what we are used to. In traditional computing, variables are represented by locations in memory and values by the bit patterns in those locations. In computing with vectors, both the variables and the values are vectors of a common high-dimensional space, and variable x having value a is also a vector in that space. Moreover, any piece of information encoded into a vector is *distributed uniformly* over the entire vector, in what is called holographic or holistic representation. We will demonstrate these ideas with three examples: (i) variable x having value a, (ii) a data record for three variables, and (iii) sequence-learning. The dimensionality of the vectors will be denoted by H ($H = 10{,}000$ for example) and the vectors are also called *HD vectors* or *hypervectors*. Variables and values are represented by random H-dimensional vectors $\mathbf{x}, \mathbf{y}, \mathbf{z}, \mathbf{a}, \mathbf{b}, \mathbf{c}$ of equally probable 1s and -1s, called *bipolar*.

2.2.1 Binding and Releasing with Multiplication

Variable x having value a is encoded with multiplication, which is done coordinatewise and denoted with $*$: $\mathbf{p} = \mathbf{x} * \mathbf{a}$, where $p_h = x_h a_h$, $h = 1, 2, \ldots, H$. We can also find the vector $\mathbf{a}$ that is encoded in $\mathbf{p}$ by multiplying $\mathbf{p}$ with the *inverse* of $\mathbf{x}$. The inverse of a bipolar vector is the vector itself, and thus we have that

$$\mathbf{x} * \mathbf{p} = \mathbf{x} * (\mathbf{x} * \mathbf{a}) = (\mathbf{x} * \mathbf{x}) * \mathbf{a} = \mathbf{1} * \mathbf{a} = \mathbf{a}$$

where $\mathbf{1}$ is the vector of H 1s. Combining a variable and its value in a single vector is called *binding*, and decoding the value as "unbinding" or *releasing*.

2.2.2 Superposing with Addition

Combining the values of several variables in a single vector begins with binding each variable–value pair as above. The vectors for the bound pairs are *superposed* by adding them into a single vector $\mathbf{r} = (\mathbf{x} * \mathbf{a}) + (\mathbf{y} * \mathbf{b}) + (\mathbf{z} * \mathbf{c})$. However, the sum is a vector of integers $\{-3, -1, 1, 3\}$. To make it bipolar, we take the sign of each component, to get $\mathbf{s} = \mathrm{sign}(\mathbf{r})$. Ties are a problem when the number of vectors in the sum is even; we will discuss that later.

Can we decode the values of the variables in the composed vector **s**? We can, if the vectors for the variables are orthogonal to each other or approximately orthogonal. For example, to find the value of *x* we multiply **s** with (the inverse of) **x** as above, giving

$$\mathbf{x} * \mathbf{s} = \mathbf{x} * (\text{sign}((\mathbf{x} * \mathbf{a}) + (\mathbf{y} * \mathbf{b}) + (\mathbf{z} * \mathbf{c}))$$
$$= \text{sign}(\mathbf{x} * ((\mathbf{x} * \mathbf{a}) + (\mathbf{y} * \mathbf{b}) + (\mathbf{z} * \mathbf{c})))$$
$$= \text{sign}((\mathbf{x} * \mathbf{x} * \mathbf{a}) + (\mathbf{x} * \mathbf{y} * \mathbf{b}) + (\mathbf{x} * \mathbf{z} * \mathbf{c}))$$
$$= \text{sign}(\mathbf{a} + \text{noise} + \text{noise})$$
$$\approx \text{sign}(\mathbf{a})$$
$$= \mathbf{a}$$

The result is approximate but close enough to **a** to be identified as **a**.

The example relies on two properties of high-dimensional representation: approximate orthogonality and noise-tolerance. Pairs of random vectors are approximately orthogonal and so the vectors for the variables can be chosen at random, and the vectors **x** * **y** * **b** and **x** * **z** * **c** are meaningless and act as random noise.

The example demonstrates *distributivity* of multiplication over addition and the need for an *associative memory* that stores all known vectors and outputs vectors that match the input the best. The memory is also called *item memory* and *clean-up memory*. Figure 2.1 shows the encoding and decoding of a data record for three variables step by step.

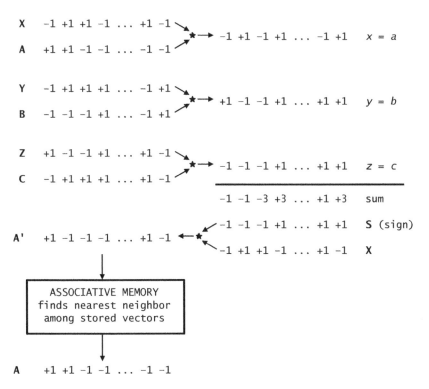

Figure 2.1 Encoding {*x* = *a*, *y* = *b*, *z* = *c*} as **S** and releasing **A** with (the inverse of) **X**. * denotes coordinatewise multiplication.

2.2.3 Sequences with Permutation

The third operation is *permutation*; it reorders vector coordinates. Permutation is shown as $\rho(\mathbf{x})$ and its inverse as $\rho^{-1}(\mathbf{x})$, or simply as $\rho\mathbf{x}$ and $\rho^{-1}\mathbf{x}$. Permutations are useful for encoding sequences as seen in the following example on language identification. We look at text, letter by letter, without resorting to dictionaries – the identification is based on letter-use statistics peculiar to each language [Joshi, Halseth & Kanerva, 2017].

For each language, we compute a high-dimensional *profile vector* or *prototype* from about a million bytes of text. We use the same algorithm to compute a profile vector for a test sentence, which is then compared to the language profiles, and the most similar one is chosen as the system's answer.

The profiles are based on three-letter sequences called *trigrams*, and they are computed as follows. The 26 letters and the space are assigned H-dimensional random bipolar seed vectors like the ones for the variables above. The same *letter vectors* are used with all languages and test sentences. The letter vectors are used to make *trigram vectors* with permutation and multiplication. For example, the vector for the trigram *the* is computed by permuting the *t*-vector twice, permuting the *h*-vector once, taking the *e*-vector as is, and multiplying the three coordinatewise: $\rho(\rho(\mathbf{t})) * \rho(\mathbf{h}) * \mathbf{e}$. This produces an H-dimensional trigram vectors of randomly placed ± 1s. Finally, the trigram vectors are added together into a *profile vector* by stepping through the text one trigram at a time. The result is an H-dimensional vector of integers. The cosine of profile vectors is used to measure their similarity.

Such an experiment with 21 European Union languages gave the following results [Joshi, Halseth & Kanerva, 2017]. All vectors were 10,000-dimensional. The language profiles clustered according to language families: Baltic, Germanic, Romance, Slavic. When test-sentence profiles were compared to the language profiles, the correct language was chosen 97% of the time, and when the language profile for English was queried for the letter most often following *th*, the answer was *e*.

The three examples illustrate computing with vectors, superposing them, and learning from data. Next we go over the operations in detail. Much of what traditional neural networks do – and fail to do – can be analyzed and understood in terms of the three vector operations, and of vector similarity and associative memory.

2.3 Operations on Vectors

We will continue with *bipolar* vectors because of the ease of working with them: components with mean $= 0$ and variance $= 1$ make it easy. However, the basic idea is the same when computing with high-dimensional random binary, real, or complex vectors. Thus, we are dealing with *general* properties of high-dimensional representation. Computing with vectors is just as natural and equally justified as computing with numbers.

- The bipolar **space of representations** rangers over H-dimensional vectors of 1s and -1s, with H a thousand or more: $\mathbf{a}, \mathbf{b}, \mathbf{c}, \ldots \in \{1, -1\}^H$. On occasion, we consider also vectors of integers, $\mathbb{Z}^H$.
- **Associative memory** is the "RAM" for high-dimensional vectors. It stores the vectors known to the system and recognizes or retrieves them from their noisy versions in what is called "clean up." The other use is that of an ordinary RAM: given an address, store, or retrieve the vector associated with that exact address or one most similar to it. We can also think of it as a memory for key–value pairs where the keys can be noisy. The actual making of such a memory will be discussed below.

- The **similarity** ($\sim$) of vectors **a** and **b** implies a distance between them and is computed via their dot product $\mathbf{a} \cdot \mathbf{b}$. For H-dimensional bipolar vectors, it varies from H when the vectors are the same, to $-H$ when they are opposites. Similarity is expressed conveniently with the cosine, given by $\cos(\mathbf{a}, \mathbf{b}) = (\mathbf{a} \cdot \mathbf{b})/H$ for the bipolar. Dot product or cosine $= 0$ means that the vectors are orthogonal, i.e. unrelated, uncorrelated, dissimilar: $\mathbf{a} \nsim \mathbf{b}$. Computing with vectors tries to capture similarity of meaning in the similarity of vectors.

  The distribution of distances between high-dimensional vectors is remarkable. Given any vector, nearly all others are approximately orthogonal to it [Widdows & Cohen, 2015]. This is called *concentration of measure*, and it means that large collections of random vectors – billions when $H = 10{,}000$ – include no similar pairs. The easy availability of approximately orthogonal vectors is paramount to computing with vectors.

  The somewhat imprecise terms "approximately equal" ($\approx$), "similar" ($\sim$), "dissimilar" ($\nsim$), and "approximately orthogonal" need clarification. Approximately equal vectors are like noisy copies of each other and have the same meaning, similar vectors arise from constructions with common constituents (see Addition), dissimilar is used for vectors that are orthogonal or approximately orthogonal. Each bipolar vector has an opposite vector, but the two are not considered representing opposite meanings.

- **Addition** is coordinatewise vector addition, and it *commutes*. The sum is a vector of integers in $\mathbb{Z}^H$ and is *normalized* by the sign function, with 0s mapped to 1s and -1s at random. Normalizing the sum is shown with brackets: $[-2, -6, 0, 0, 2, 4, \ldots] = (-1, -1, 1, -1, 1, 1, \ldots)$ for example.

  The sum is *similar* to its inputs: $[\mathbf{a} + \mathbf{b}] \sim \mathbf{a}, \mathbf{b}$. For example, $\cos(\mathbf{a}, [\mathbf{a} + \mathbf{b}]) = 0.5$ for random **a** and **b**. Addition is *associative* before the sum is normalized but only approximately associative after:

  $$[[\mathbf{a} + \mathbf{b}] + \mathbf{c}] \sim [\mathbf{a} + [\mathbf{b} + \mathbf{c}]]$$

  Likewise, it is *invertible* before normalization, but only approximately invertible after: $[[\mathbf{a} + \mathbf{b}] + (-\mathbf{b})] \sim \mathbf{a}$. Information is lost each time a sum is normalized and so normalizing should be delayed whenever possible. Addition is also called *bundling* and *superposing*.

- **Multiplication** is done coordinatewise, known as Hadamard product, and it *commutes*. The product $\mathbf{a} * \mathbf{b}$ of bipolar vectors is also bipolar and thus ready for use as input in subsequent operations. Multiplication is *invertible* – a bipolar vector is its own inverse – it *distributes* over addition: $\mathbf{x} * [\mathbf{a} + \mathbf{b}] = [(\mathbf{x} * \mathbf{a}) + (\mathbf{x} * \mathbf{b})]$; and it *preserves similarity*: $(\mathbf{x} * \mathbf{a}) \cdot (\mathbf{x} * \mathbf{b}) = \mathbf{a} \cdot \mathbf{b}$, which also means that a vector can be moved across the dot: $\mathbf{a} \cdot (\mathbf{b} * \mathbf{c}) = (\mathbf{a} * \mathbf{b}) \cdot \mathbf{c}$. The product is *dissimilar* to its inputs: $\mathbf{a} * \mathbf{b} \nsim \mathbf{a}, \mathbf{b}$. See Figure 2.1 for examples of addition and multiplication.

- **Permutations** reorder vector coordinates. The number of permutations is enormous, $H!$ overall. Permutations are *invertible*: $\rho^{-1}(\rho(\mathbf{a})) = \mathbf{a}$; they *distribute* over both addition and multiplication: $\rho[\mathbf{a} + \mathbf{b}] = [\rho(\mathbf{a}) + \rho(\mathbf{b})]$ and $\rho(\mathbf{a} * \mathbf{b}) = \rho(\mathbf{a}) * \rho(\mathbf{b})$, in fact, permutations distribute over all coordinatewise operations, such as Booleans; they *preserve similarity*: $\rho(\mathbf{a}) \cdot \rho(\mathbf{b}) = \mathbf{a} \cdot \mathbf{b}$; but most permutations do *not commute*: $\rho(\sigma(\mathbf{a})) \neq \sigma(\rho(\mathbf{a}))$. The output of a random permutation is *dissimilar* to the input: $\rho(\mathbf{a}) \nsim \mathbf{a}$.

  Permutations themselves are not elements of the vector space. In linear algebra, they are represented by matrices and so ρ can be thought of as a permutation matrix; here they are *unary* operations on vectors. They are potentially very useful by incorporating all finite groups up to size H into the vector math. The permutation $\sigma(\rho(\mathbf{a}))$ is commonly abbreviated to $\sigma\rho\mathbf{a}$ and it equals $(\sigma\rho)\mathbf{a}$.

The operations and their properties for bipolar vectors are summarized in Table 2.1, as an example of things to consider when setting up a system of computing with vectors. A system for

Table 2.1 Summary of bipolar vector operations

Property	Dot product $a \cdot b$	Sum $a + b$	Normal'd sum $[a + b]$	Product $a * b$	Permutation $\rho(a)$
Associative	n/a	Yes	Approx[a]	Yes	Yes
Commutative	Yes	Yes	Yes	Yes	No
Invertible	n/a	Yes	Approx[a]	Yes	Yes
Similar to inputs, increases similarity	n/a	Yes	Yes	No	No
Preserves similarity, randomizes	n/a	No	No	Yes	Yes
Distributes over addition	Yes	n/a	n/a	Yes	Yes
Distributes over multiplication	n/a[b]	No	No	n/a	Yes

a) Partly true
b) However, $a \cdot (b * c) = (a * b) \cdot c = (a * b * c) \cdot 1$

binary vectors is equivalent to the bipolar when 1 is replaced by 0, -1 by 1, multiplication by Exclusive-Or (XOR), and the sign function by coordinatewise majority, and when similarity is based on the Hamming distance.

2.4 Data Structures

Computer programming consists of laying and tracing pathways to data, and then doing arithmetic and logic operations on the data. The pathways are called data structures, and they include sets, sequences, lists, queues, stacks, arrays, graphs, heaps, and so forth. The data are the values attached to the structure, but to the computer the structure itself is also data. It is impossible to draw a sharp boundary between data and structure, ever more so when computing in holographic representation.

Since data structures are an essential part of programming and computing, we need to look at how to encode and operate with them in superposed vectors.

- **Seed vectors**: Computing begins with the selection of vectors for basic entities such as variables and values. Bipolar seed vectors are made of random, independent, equally probable 1s and -1s. They are also called *atomic vectors* and *elemental vectors* because they are the stuff from which everything else is built. For example, in working with text, each letter of the alphabet can be represented by a seed vector. A set of seed vectors is called *alphabet* or *vocabulary* or *codebook*.

  Because of high dimensionality, randomly chosen (seed) vectors are *approximately orthogonal* – the superabundance of approximately orthogonal vectors and the relative ease of making them is a primary reasons for high dimensionality. Orthogonality allows multiple vectors to be encoded into a single vector and subsequently decoded, making it possible to analyze and interpret the results of computing in superposition. The selection of a seed vector corresponds to assigning a memory location – an address – to a variable, or choosing a representation for its value.

- **Bound pairs** encode a *variable* and its *value* – or a *role* and a *filler*, or a *key* and a *value* – in a single vector. If binding is done with an invertible operation, the value can be recovered by decoding. An example of binding with multiplication is shown in the Introduction where $x = a$ is encoded with $\mathbf{x} * \mathbf{a}$.
- **Sets** and **multisets** name their members but do not specify their order: $\{a, a, b, c\} = \{a, b, a, c\}$. They can be encoded with addition because it commutes. Since the sum is similar to the vectors in it, it is possible to query whether a specific vector is included in the set or the multiset. That works reliably for small sets, but the adding of vectors makes the sum less similar to any one of them, and normalizing the sum makes it even less similar.

  To decode a sum, we look for vectors similar to it in the associative memory. Once a vector is assumed to be in a sum, it can be subtracted out and the remaining sum queried for further vectors. Peeling off vectors one at a time works with sums that have not been normalized, but poorly with normalized sum vectors because of information that has been lost (hence, invertibility is called "partly true" in Table 2.1).

  Multiplication commutes and therefore also it can encode a set, but not a multiset because bipolar vectors are their own inverses and cancel multiple copies of themselves. The product is *dissimilar* to its inputs and can therefore be used as a *label* for a set. Decoding a product is problematic, however. There is no efficient way to do it in general, but if the inputs come from known sets of dissimilar vectors, and there are not too many of them, a product can be broken down into its inputs with an iterative search, as discussed below under factorization.
- **Sequences** are ordered multisets, e.g. $(a, a, b, c) \neq (a, b, a, c)$. They can be encoded with permutations. If the application needs only one permutation, rotation of coordinates (cyclic shift) is usually most convenient. The sequence (a, b, c) can be encoded as a sum $\mathbf{s}_3 = \rho^2\mathbf{a} + \rho\mathbf{b} + \mathbf{c}$ or as a product $\mathbf{p}_3 = \rho^2\mathbf{a} * \rho\mathbf{b} * \mathbf{c}$, and extended to include $\mathbf{d}$ by permuting $\mathbf{s}_3$ and adding $\mathbf{d}$: $\mathbf{s}_4 = \rho\mathbf{s}_3 + \mathbf{d} = \rho(\rho^2\mathbf{a} + \rho\mathbf{b} + \mathbf{c}) + \mathbf{d} = \rho^3\mathbf{a} + \rho^2\mathbf{b} + \rho\mathbf{c} + \mathbf{d}$; and similarly $\mathbf{p}_4 = \rho\mathbf{p}_3 * \mathbf{d}$. The successive powers of the permutation act like an *index* into the sequence. Either kind of sequence can be extended recursively with only two operations because permutations distribute over both addition and multiplication.

  Decoding the ith vector of the sequence uses the inverse permutation. For example, the first vector in $\mathbf{s}_3$ is found by searching the associative memory for the vector most similar to $\rho^{-2}\mathbf{s}_3$ because that equals $\rho^{-2}(\rho^2\mathbf{a} + \rho\mathbf{b} + \mathbf{c}) = \mathbf{a} + \rho^{-1}\mathbf{b} + \rho^{-2}\mathbf{c} = \mathbf{a} + \text{noise} + \text{noise} = \mathbf{a}' \approx \mathbf{a}$. A sequence encoded with multiplication is harder to decode because products don't resemble their inputs. To decode $\mathbf{a}$ from $\mathbf{p}_3$ requires that $\mathbf{b}$ and $\mathbf{c}$ are already known, in which case $\rho^{-2}(\mathbf{p}_3 * (\rho\mathbf{b} * \mathbf{c})) = \mathbf{a}$ because $\rho\mathbf{b}$ and $\mathbf{c}$ in $\mathbf{p}_3$ cancel out: $\rho^{-2}((\rho^2\mathbf{a} * \rho\mathbf{b} * \mathbf{c}) * (\rho\mathbf{b} * \mathbf{c})) = \rho^{-2}(\rho^2\mathbf{a}) = \mathbf{a}$.
- **Binary trees** can be encoded with two independent random permutations, ρ_1 and ρ_2, that do not commute – most permutations don't. If we encode the pair (a, b) with $\rho_1\mathbf{a} + \rho_2\mathbf{b}$, then the two-deep tree $((a, b), (c, d))$ can be encoded as $\mathbf{t} = \rho_1(\rho_1\mathbf{a} + \rho_2\mathbf{b}) + \rho_2(\rho_1\mathbf{c} + \rho_2\mathbf{d}))$, which equals $\rho_1(\rho_1\mathbf{a}) + \rho_1(\rho_2\mathbf{b}) + \rho_2(\rho_1\mathbf{c}) + \rho_2(\rho_2\mathbf{d})$, and can be written as $\rho_{11}\mathbf{a} + \rho_{12}\mathbf{b} + \rho_{21}\mathbf{c} + \rho_{22}\mathbf{d}$, where ρ_{ij} is the permutation $\rho_i\rho_j$. To decode a tree, we follow the indices and apply the inverse permutations in the reverse order. For example,

$$\rho_1^{-1}(\rho_2^{-1}(\mathbf{t})) = \rho_1^{-1}(\rho_2^{-1}(\rho_1(\rho_1\mathbf{a}) + \rho_1(\rho_2\mathbf{b}) + \rho_2(\rho_1\mathbf{c}) + \rho_2(\rho_2\mathbf{d})))$$
$$= (\rho_1^{-1}\rho_2^{-1}\rho_1\rho_1)\mathbf{a} + (\rho_1^{-1}\rho_2^{-1}\rho_1\rho_2)\mathbf{b} + (\rho_1^{-1}\rho_2^{-1}\rho_2\rho_1)\mathbf{c}$$
$$+ (\rho_1^{-1}\rho_2^{-1}\rho_2\rho_2)\mathbf{d}$$
$$= \text{noise} + \text{noise} + \mathbf{c} + \text{noise}$$
$$\approx \mathbf{c}$$

where $\rho_2^{-1}(\mathbf{t})$ extracts the right half of the tree and ρ_1^{-1} extracts its left branch. An alternative encoding of binary trees uses one permutation and two seed vectors. Successive powers of the permutation encode depth, and they are multiplied by the two vectors that mean left and right [Frady et al., 2020].

- **Graphs** consist of a set of nodes and connecting links. Of the different kind, we consider directed graphs with loops, i.e. where nodes can have links to themselves. Graphs are used to depict relations between entities, for example, "parent of," "hears from," and "communicates with." The nodes can also represent the set of states $s_i \in S$ of a Markov chain or a finite-state automaton, and the links $t_{ij} \in T$ its state transitions, $T \in S \times S$. We will encode the states by random seed vectors $\mathbf{s}_i$ – they name or label the states. The transition $t_{ij} = (s_i, s_j)$ can then be encoded with permutation and multiplication as $\mathbf{t}_{ij} = \rho \mathbf{s}_i * \mathbf{s}_j$, and the graph with the sum of all its transitions:

$$\mathbf{g} = \sum_{t_{ij} \in T} \mathbf{t}_{ij} = \sum_{t_{ij} \in T} \rho \mathbf{s}_i * \mathbf{s}_j.$$

Given the vector $\mathbf{g}$ for the graph, we can ask whether the state s_j can be reached from the state s_i in a single step. The answer is contained in the vector $\mathbf{g}_i = \rho \mathbf{s}_i * \mathbf{g}$ and it is found by comparing $\mathbf{g}_i$ to $\mathbf{s}_j$: it is "yes" if the two are similar and "no" if they are dissimilar. This follows from the distributivity of multiplication over addition, and from the sum being similar to its inputs: the vector $\mathbf{g}$ is a sum of transitions, and multiplying it by $\rho \mathbf{s}_i$ releases $\mathbf{s}_j$ if $\rho \mathbf{s}_i * \mathbf{s}_j$ is included in the sum. Notice that $\mathbf{g}_i$ releases (it brings to the surface) *all* the states (their labels) that can be reached from s_i in a single step; $\mathbf{g}_i$ also includes a noise vector for every transition from states other than s_i and so $\mathbf{g}_i$ is a noisy representation of the *set* of states one step from s_i.

We can go further and look for the (multi)set of states two steps from s_i. The answer is contained in – it is similar to – $\rho(\rho \mathbf{s}_i * \mathbf{g}) * \mathbf{g} = \rho^2 \mathbf{s}_i * \rho \mathbf{g} * \mathbf{g}$; or three steps from s_i: $\rho^3 \mathbf{s}_i * \rho^2 \mathbf{g} * \rho \mathbf{g} * \mathbf{g}$, but it gets noisier at each step, overpowering the signal.

Linear algebra gives us an exact answer in terms of the $|S| \times |S|$ state-transition matrix $\mathbf{T}$, where links are represented by 1s (and nonlinks by 0s). The set of states reached from s_i in a single step is given by the $|S|$-dimensional vector $\mathbf{iT}$, where $\mathbf{i}$ is an $|S|$-dimensional indicator vector whose ith component equals 1. The (multi)set of states reached in exactly three steps is given by $\mathbf{iT}^3$. We can see that $\rho^2 \mathbf{g} * \rho \mathbf{g} * \mathbf{g}$ serves a function similar to $\mathbf{T}^3$, and also that its form is similar to the encoding of a sequence with multiplication, as shown in the discussion of sequences.

Examples of computing with graphs include graph isomorphism [Gayler & Levy, 2009] and a finite automaton [Osipov, Kleyko & Legalov, 2017].

2.5 Vector Sums Encode Probabilities

Probabilities can be included in and inferred from high-dimensional vectors without explicit counting and bookkeeping. This opens the door to statistical learning from data, which is traditionally the domain of artificial neural nets. It is clearly seen in the unnormalized representation of a multiset, i.e. when the vectors of the multiset are simply added into a sum vector $\mathbf{f}$.

The dot product of a bipolar H-dimensional vector $\mathbf{x}$ with itself is H, i.e., $\mathbf{x} \cdot \mathbf{x} = H$. If $\mathbf{f}$ is the sum of k copies of $\mathbf{x}$, the dot product $\mathbf{x} \cdot \mathbf{f} = kH$. If other vectors are added to $\mathbf{f}$ and they all are orthogonal to $\mathbf{x}$, the dot product $\mathbf{x} \cdot \mathbf{f}$ still is kH, and it is approximately kH if the other vectors are approximately orthogonal to $\mathbf{x}$. Thus, the dot product of a bipolar vector $\mathbf{x}$ with a sum vector $\mathbf{f}$, divided by H, is an estimate of the number of times $\mathbf{x}$ has been added into the sum. This explains the identification of languages from their profile vectors in example 3 of the overview. It goes as follows [Joshi, Halseth & Kanerva, 2017].

Each language and each test sentence is represented as a sum of trigrams that have been encoded as sequences of three letter vectors, e.g. $\rho^2 \mathbf{t} * \rho \mathbf{h} * \mathbf{e}$. The letter vectors are dissimilar to each other, and since the outputs of permutation and multiplication are dissimilar to their inputs, the trigram vectors are also dissimilar – approximately orthogonal. A profile vector $\mathbf{f}$, which is their sum, can then be used to estimate the frequencies of the trigrams in the text. The trigram statistics for different languages apparently are different enough to allow test sentences to be identified correctly as to language, but similar enough within language families to produce Baltic, Germanic, Romance, and Slavic clusters. Numerically, 27 letters give rise to $27^3 = 19,683$ possible trigrams, and so the algorithm projects a histogram of 19,683 trigram frequencies randomly to 10,000 dimensions ($H = 10,000$ was used in the example). The experiment was repeated with tetragrams and gave a slightly better result (97.8% vs. 97.3% correct). In that case a histogram of $27^4 = 531,441$ possible frequencies is projected randomly into 10,000 dimensions.

Finally, the letter most often following *th* in English is found by multiplying the profile for English, $\mathbf{f}_{\text{English}}$, with (the inverse of) $\rho^2 \mathbf{t} * \rho \mathbf{h}$. The multiplication distributes over every trigram vector added into $\mathbf{f}_{\text{English}}$ and cancels out the initial *th* wherever it occurs. In particular, it releases $\mathbf{e}$ from $\rho^2 \mathbf{t} * \rho \mathbf{h} * \mathbf{e}$. It also releases every other letter that comes after *th*, but since e is the most frequent, $\rho^2 \mathbf{t} * \rho \mathbf{h} * \mathbf{f}_{\text{English}}$ has a higher dot product with $\mathbf{e}$ than with any other letter vector. The dot product is the same as between $\mathbf{f}_{\text{English}}$ and the vector for the trigram *the*: $(\rho^2 \mathbf{t} * \rho \mathbf{h} * \mathbf{f}_{\text{English}}) \cdot \mathbf{e} = (\rho^2 \mathbf{t} * \rho \mathbf{h} * \mathbf{e}) \cdot \mathbf{f}_{\text{English}}$ – as vectors of a product move across the dot. Its expected value is the number of times e occurs after *th*, multiplied by 10,000.

The language example suggests the possibility of representing a Markov chain as a high-dimensional vector learned from data – in this case a second-order chain. A language profile made of trigrams allows us to estimate letter frequencies following a pair of letters such as *th*. The vector $\mathbf{e}$ was found by searching the 27 letter vectors for the best match to the query $\mathbf{q} = \rho^2 \mathbf{t} * \rho \mathbf{h} * \mathbf{f}_{\text{English}}$. If the letter vectors are treated as a $H \times 27$ matrix $\mathbf{A}$ for the alphabet, then multiplying $\mathbf{q}$ with $\mathbf{A}^{\mathsf{T}}$ approximates each latter's relative weight in candidacy for the next letter. However, since the alphabet vectors and the trigram vectors are only approximately orthogonal, the weights are approximate. Yet the transition frequencies have been captured by the model and can govern the probability of choosing the next letter: choose it in proportion to the 27 elements of $\mathbf{A}^{\mathsf{T}}\mathbf{q}$, adjusted for the randomness due to orthogonality being approximate. How actually to choose the next letter based on the probabilities or their estimates, other than by traditional programming, remains an open question.

The language example demonstrates the power of computing with vectors when it is based on a comprehensive arithmetic of vectors. It uses all three operations: a language profile vector is a *sum of products of permutations*. The algorithm for "training" is the *same* as for making profiles of test sentences. It is simple and easily adapted to classification problems at large, and it works in one pass over the data, meaning that the algorithm is *incremental*. Frequencies and probabilities can be recovered approximately from a profile vector by inverting the operations used to encode a profile: the representation is *explainable*.

2.6 Decoding a Product

Unlike a sum vector which is similar to its inputs, the product vector is dissimilar. Thus, $\mathbf{x} * \mathbf{y}$ gives us no clue as to its originating from $\mathbf{x}$ and $\mathbf{y}$, nor that $\mathbf{w} = \rho^2 \mathbf{t} * \rho \mathbf{h} * \mathbf{e}$ consists of $\mathbf{t}$, $\mathbf{h}$, and $\mathbf{e}$. In fact, any product vector can be factored to possible input vectors in countless ways. However, if we know that the vector $\mathbf{w}$ represents a sequence of three letters, we can examine all 27^3 possible sequences systematically to see which of them yields $\mathbf{w}$. That is up to 19,683 tests. As an alternative to a systematic search, we can search for the answer through successive approximations or educated guesses of $\mathbf{t}$, $\mathbf{h}$, and $\mathbf{e}$ with an algorithm called the *resonator*. We will explain the algorithm by referring to the product $\mathbf{p}$ of three vectors, $\mathbf{p} = \mathbf{x} * \mathbf{y} * \mathbf{z}$, drawn from three different dictionaries or codebooks, $\mathbf{X}, \mathbf{Y}$, and $\mathbf{Z}$ – their columns are the codevectors [Frady et al., 2020].

If the product and all its inputs but one are known, the "unknown" input is simply the product of the known vectors, e.g. $\mathbf{x} = \mathbf{p} * \mathbf{y} * \mathbf{z}$. If $\mathbf{y}$ and $\mathbf{z}$ are noisy, however, the "unknown" $\mathbf{x}' = \mathbf{p} * \mathbf{y}' * \mathbf{z}'$ is even more noisy on the average. However, it can be used to compute a new estimate $\mathbf{x}''$ that has a higher probability of being one of the vectors in $\mathbf{X}$. The vector $\mathbf{x}''$ is computed as the weighted sum of the codevectors in $\mathbf{X}$, normalized to bipolar, where the similarity of $\mathbf{x}'$ to the vectors in $\mathbf{X}$ serve as the weights. This can be expressed as $\mathbf{x}'' = [\mathbf{X}(\mathbf{X}^\mathrm{T}\mathbf{x}')]$, where $\mathbf{X}^\mathrm{T}\mathbf{x}'$ are the weights and $[\cdots]$ makes the result bipolar. If $\mathbf{x}''$ is in the codebook $\mathbf{X}$ we accept it and continue to search for the remaining inputs. If it is not in $\mathbf{X}$, we continue to search for the inputs one codebook at a time by computing $\mathbf{y}'' = [\mathbf{Y}(\mathbf{Y}^\mathrm{T}\mathbf{y}')] = [\mathbf{Y}(\mathbf{Y}^\mathrm{T}(\mathbf{x}'' * \mathbf{p} * \mathbf{z}'))]$, computing $\mathbf{z}'' = [\mathbf{Z}(\mathbf{Z}^\mathrm{T}\mathbf{z}')] = [\mathbf{Z}(\mathbf{Z}^\mathrm{T}(\mathbf{x}'' * \mathbf{y}'' * \mathbf{p}))]$, back to computing $\mathbf{x}'''$ from $\mathbf{x}''$ as before, and so on.

We still need to choose vectors $\mathbf{y}'$ and $\mathbf{z}'$ to get started. Recalling that a sum vector is similar to each of its inputs, the normalized sums of the codevectors in $\mathbf{Y}$ and $\mathbf{Z}$ are used. The probability and rate of convergence to the correct vectors depend on the number of inputs in the product, the sizes of the codebooks, and the dimensionality H [Kent et al., 2020].

2.7 High-Dimensional Vectors at Large

The idea of computing with high-dimensional vectors is simplest to convey with the bipolar, and bipolar vectors are also useful in applications. However, the idea is general and depends more on the abundance of nearly orthogonal vectors, than on the nature of the vector components. The abundance comes from high dimensionality. It also matters greatly to have a useful set of operations on the vectors, akin to add, multiply, and permute for bipolar vectors. In fact, the corresponding addition and multiplication of numbers constitute an algebraic field. The vector math adds to it all finite groups up to size H.

We have already commented on the equivalence of the *binary* with the bipolar when coordinate-wise multiplication is replaced by bitwise XOR. To convert a binary sum vector to the exact bipolar sum vector, and vice versa, we also need to keep count of the vectors in the sum.

The original Holographic Reduced Representation [Plate, 1994] is based on *real* vectors with random independent normally distributed components with mean $= 0$ and variance $= 1/H$. Addition is vector addition followed by normalization (to Euclidean length 1), and multiplication is by circular convolution; its approximate inverse is called "circular correlation." Similarity of vectors is based on the Euclidean distance, dot product, or cosine, all of them being essentially the same when the vectors are normalized to unit length.

Holographic Reduced Representation with *complex* vectors uses random phase angles as vector components. Addition is by vector addition followed by normalization (coordinatewise projection to the unit circle), multiplication is by coordinatewise addition of phase angles (i.e. complex multiplication), and similarity is based on the magnitude of the difference between H-dimensional complex vectors.

All these frameworks are related and their properties are essentially the same. The binary and the bipolar are equivalent, the complex becomes the bipolar when the phase angles are restricted to $0°$ and $180°$, and the real and the complex are related by Fourier transform. The choice of representation can depend on a variety of factors. For example, binary vectors are the simplest to realize in hardware, and complex vectors (phase angles) provide a model for computing with the timing of spikes.

Computing with vectors, as describes in this chapter, assumes *dense* vector: half 1s and half −1s (or half 0s and half 1s for binary). Addition and multiplication automatically tend toward dense vectors, which is mathematically convenient but may not be desirable otherwise and will be commented on below.

Computing can also be based on Boolean operations on bit vectors and on permutations. These operations are common in hashing for distributing data in a high-dimensional space. They are also used in *Context-Dependent Thinning* [Rachkovskij & Kussul, 2001] to encode structure with sparse binary vectors. *Geometric Algebra* offers a further possibility to compute with high-dimensional vectors, called multivectors [Aerts et al., 2009].

The activity of neurons in the brain is *very sparse*, which is partly responsible for the remarkable energy efficiency of brains. Sparse representation is also the most efficient for storing information, so why not compute with sparse vectors? The answer is simply that we have not found operations for sparse vectors that work as well as the combination of add, multiply, and permute for dense vectors. This is a worthy challenge for mathematicians to take on.

2.8 Memory for High-Dimensional Vectors

Computing with vectors is premised on the dimensionality H remaining constant. The choice of H can vary over a wide range, however, and the exact value is not critical (e.g. $1,000 \leq H \leq 10,000$). Mainly, it needs to be large enough to give us a sufficient supply of random, approximately orthogonal vectors. That number grows exponentially with H [Gallant & Okaywe, 2013].

Whatever the dimensionality within a reasonable range, a single vector can reliably encode only a limited amount of information. In psychologists' models of cognition, such a vector is called a working memory or a short-term memory, implying the existence also of a long-term memory. That distinction agrees with the traditional organization of a computer where the ALU and its "active" registers comprise the working memory, and the RAM is the long-term memory.

The same idea applies to computing with vectors. Operations on vectors output new vectors of the same kind, and a memory stores them for future use. Like the RAM, the memory can be made as large as needed, and large memories are necessary in systems that learn over a long life span. Furthermore, new learning should disrupt minimally what has been learned already. Such memories are called *associative* and were a topic of early neural-net research, but they are not a part of today's deep-learning nets. In deep learning, the memory function and the forming of new representations are entangled.

Memories for high-dimensional vectors have been used in two ways in the examples above. Decoding the vector **s** of three superposed variables for the value of x in Section 2.2.2 (see Figure 2.1) produces the vector **a′** that needs to be associated with its nearest neighbor among the known vectors. This is the function of the *item memory*: given a noisy vector, output its nearest neighbor among the known vectors. Using the cleaned-up **a** in further computations prevents noise from accumulating. The item memory is addressed with a bipolar vector and it outputs a bipolar vector.

The second use was to compare the profile of a test sentences to language profiles stored in memory. The profiles are (unnormalized) sum vectors with integer components, and their similarity is measured with the cosine. The profiles can be normalized to bipolar (and ultimately to binary) *after* having been accumulated, and then the memory task is identical to that of the item memory. Some information is lost to normalization [Frady et al., 2018], but the loss can be offset in part by higher vector dimensionality [Rahimi et al., 2017].

Postponing the normalization of a sum vector facilitates incremental learning – simply keep adding vectors to the sum. To make it practical, however, the range of sum-vector components needs to be limited and overflow and underflow ignored. Truncation to 8 bits per component has a minor effect on classification tasks such as language identification.

Arrays of H-dimensional vectors can be used as the memory in computer simulations. Searching through them is time-consuming, however, but the simplicity of vector algorithms can still make them practical even when simulated on standard hardware.

Associative memories at large are yet to be fully integrated into the high-dimensional computing architecture. Such memories were studied in the 1970s and 1980s [Hinton & Anderson, 1981], with the cerebellum proposed as their realization in the brain [Marr, 1969; Albus, 1971; Kanerva, 1988]. Subsequent lack of interest has a plausible explanation: a versatile algebra for computing with high-dimensional vectors was unknown, and an associative memory by itself isn't particularly useful. This has changed starting with Holographic Reduced Representation in the 1990s [Plate, 1994], and memories for high-dimensional vectors are now included in our models [Karunaratne et al., 2021]. The very large circuits that the memories require are only now becoming practical.

2.9 Outline of Systems for Autonomous Learning

Systems for computing with vectors derive their power from the remarkable properties of high-(hyper)dimensional spaces. For example, high-dimensional representation is robust and noise-tolerant in ways that human perception, learning, and memory are. It also allows data structures to be encoded, manipulated, and decoded explicitly, as in traditional computing, setting it apart from artificial neural nets trained with gradient descent. It allows data to be represented and manipulated in superposition, which sets it apart from traditional computing. The resulting system of computing combines properties that traditional computing and neural nets individually lack. Its properties seem particularly appropriate for modeling of functions controlled by brains. In this section, we outline a computing architecture for autonomous learning based on high-dimensional vectors.

Ideas for autonomous learning come from observing the animal world. For an animal to survive and prosper in an environment, it must recognize some situations as favorable and life-sustaining, and others as unfavorable and dangerous, and then act so as to favor the former and avoid the latter: seek reward and avoid punishment. To that effect, animals perceive the environment through a multitude of senses – sight, sound, smell, taste, temperature, pressure, acceleration, vibration, hunger, pain, and proprioception – and they move about and act upon the environment by controlling their muscles. The brain's job is to coordinate it all. This can be thought of as a *classification* problem where favorable motor commands serve as the classes to which sensory states are mapped, bearing in mind that classification is a particular strength of computing with high-dimensional vectors [Ge & Parhi, 2020]. What the favorable motor commands are in any sensory state can be learned in any number of ways: by explicit design, following an example, supervised, reinforced, and trial-and-error.

Coordinating a variety of sensors and actuators is a major challenge for artificial systems. Different kinds of information need to be represented in a common mathematical space that is not overly confining. As a counterexample, the number line is appropriate for representing temperature but not odor, much less combinations of the two. However, when the dimensionality of the space is high enough, all manner of things can be represented in it – i.e. *embedded* – without them unduly interfering with each other. Generic algorithms can then be used to discover relations among them and to find paths to favorable actions. That brings about the need to map raw sensory signals to high-dimensional vectors, and to map such vectors for action to signals that control actuators.

Each sensory system responds to the environment in its own peculiar way, and so the designing of the interfaces can take considerable engineering expertise. In working with speech, for example, the power spectrum is more useful than the sound wave and is universally used, and there are simple ways to turn spectra into high-dimensional vectors. Mapping a signal onto high-dimensional vectors corresponds to and is no more difficult than designing and selecting features for traditional classification algorithms [Burrello et al., 2020; Moin et al., 2021].

Designing of the interface can also be automated, at least in part, by employing deep learning or genetic algorithms. If we can define an appropriate objective function or measure of fitness, we can let the computer search for a useful mapping of signals to high-dimensional vectors, and from the vectors to commands to actuators. For the system to remain stable, however, the mappings need to remain fixed once they have been adopted, with further learning taking place with vectors in the high-dimensional space. Traditional computing can be used to program such a system and to interface it with the environment.

2.10 Energy-Efficiency

Ideas about computing with vectors need ultimately to be built into hardware [Semiconductor Research Corporation, 2021]. The requirements are fundamentally different from those for computing with numbers. The traditional model assumes determinism – identical inputs produce identical outputs – at a considerable cost in energy. Tasks that are uniquely suited for computing with vectors involve large collections of sensors and actuators of various kind where no single measurement is critical. They matter in the aggregate. When information is distributed equally over all components of a vector, individual components need not be 100% reliable. That makes it possible to use circuits with ever smaller elements and to operate them at very low voltages. Exploiting analog properties of materials also becomes possible, with further gains in energy efficiency [Wu et al., 2018]. In contrast, the requirements of the deterministic model become hard to meet when circuits get ever larger and their elements ever smaller.

The operations on vectors offer further opportunity to reduce the demand for energy. Addition and multiplication happen coordinatewise and thus can be done in parallel, meaning that the total system can be fast without its components needing to be much faster. Furthermore, the simplicity of addition, and particularly of multiplication (e.g. XOR), makes it possible to build them into the memory, reducing the need to bus data between a central processor and the memory [Gupta, Imani & Rosing, 2018]. In traditional computing, both speed and the accessing of memory are paid for in energy.

2.11 Discussion and Future Directions

Computing with vectors has grown out of attempts to understand how brains "compute," ideas for which have come from varied directions. Early evidence was qualitative and was based on observations of behavior and thought experiments, mainly by philosophers and psychologists. After the invention of the digital computer, the models became computer-like. However, their ability to explain brains has fallen far short of expectations. Meanwhile, information from neuroscience has been accumulating, starting with neuroanatomy. However, the detailed drawings of neurons and circuits by Cajal in the late 1800s and early 1900s continue to challenge our ability to explain. With advances in neurophysiology, we are able to demonstrate learning in synapses in stereotypical tasks, but the workings of entire circuits still wait to be explained.

Many things suggest that computing with vectors will help us understand brains. Dimensionality in the thousands agrees with the size of neural circuits. High-dimensional distributed representation is extremely robust, and so the failure of a single component is no more consequential than the death of a single neuron. It matters greatly that simple operations on vectors can be made into efficient algorithms for learning. Sufficiently high dimensionality allows many kinds of things to be represented in a common mathematical space without unduly interfering with each other. For example sight and sound can be both kept apart and combined. A high-capacity associative memory is an essential part of computing with vectors. That agrees with the brain's ability to learn quickly and to retain large amounts of information over long periods of time. Among the brain's circuits, the cerebellum's looks remarkably like an associative memory, and it contains over half the brain's neurons. Its importance for motor learning has been known since the 1800s, and it appears to be involved in mental functions as well. Its interpretation as an associative memory goes back half a century [Marr, 1969], and its design can instruct the engineering of high-capacity associative memories for artificial systems.

Other evidence is experiential. The human brain receives infinitely varied input through a 100 million or more sensory neurons, and from it builds a mental world of specific, nameable, repeating, more-or-less permanent colors, sounds, shapes, objects, people, events, stories, histories, and so on. The specificity can be explained by the distribution of distances – and similarity – in a high-dimensional space, and by the tendency of some operations to cluster the inputs. The representation is robust and tolerant of variation and "noise."

Computing with vectors bridges the gap between traditional computing with numbers and symbols on the one hand, and artificial neural nets and deep learning on the other. We can expect it to become an established technology for machine learning within a decade, applied widely to multimodal monitoring and control. Its ability to deal with symbols and structure makes it also a candidate technology for logic-based reasoning and language.

Wide-ranging exploration and large-scale experiments are needed, meanwhile. New algorithms are easily simulated on standard hardware. Not necessarily in the scale as ultimately desired, but the underlying math makes it possible to see whether an algorithm scales. Semantic vectors provide an example. When made with Latent Semantic Analysis [Landauer & Dumais, 1997], which uses singular-value decomposition, runtime grows with the square of the number of documents, whereas Random Indexing [Kanerva, Kristoferson & Holst, 2000] achieves comparable results in linear time.

Algorithm development needs to proceed on two fronts: *high-dimensional core* and *interfaces*. The core algorithms are generic and are what we think of as computing with vectors – and what this chapter is about. The algorithms at the core integrate input from a multitude of sensors and generate vectors that control the system's output. They also account for learning. The interfaces are specific to sensor and motor modalities, and they translate between low-dimensional signal spaces and the high-dimensional representation space as discussed in the section on autonomous learning (Section 2.9). The interface in the language-identifying experiment (Section 2.2.3 on Sequences with Permutation) is extremely simple because languages are already encoded with letters, and representing letters with high-dimensional random seed vectors was all that is needed. Designing an interface usually requires more domain knowledge but is not necessarily difficult.

Against this backdrop, we can try to see what lies ahead. We are far from understanding how brains compute, or from building artificial systems that behave like systems controlled by brains.

That would be a monumental achievement, both technological and as a source of insight into human and animal minds. Computing with high-dimensional vectors seems like a necessary step in that direction, even if only a single step of how many, we do not know.

The disparity between brains and our models can give us a clue. If we were given 100 billion neuronlike circuit elements with 100 trillion points of contact between them, akin to synapses, we would not know how to connect them into a system that works. Neither does exact copying of neural circuits yield the understanding we need, unfortunately. But the *large numbers* surely are meaningful and need to be included in our theories and models – and in brain-inspired computing.

Massive feedback is an essential feature throughout the brain, and only rarely is its role apparent. Our theories and models need to come to terms with massive feedback. Activation of neurons in the brain is *sparse*, whereas the models discussed in this chapter compute with dense vectors. Brains *learn continuously* with *minuscule energy* compared to neural-net models trained on computer clusters. The fluidity of *human language* will challenge our modeling for years to come.

Much work is needed to fully develop the idea of computing with vectors, but even incremental advances can lead to significant insight and applications. For example, adaptive robotics is likely to benefit early on, where sensor fusion and *sensor–motor integration* are absolute requirements [Räsänen & Saarinen, 2016; Mitrokhin et al., 2019; Neubert, Schubert & Protzel, 2019]. Taking our cues from the animal world, every motor action includes a proprioceptive component that reports on the execution of the action and its outcome. High-dimensional space is natural for dealing with the feedback and incorporating it into future action. From the (nervous)system's point of view, proprioception is just another set of sensory signals, to be integrated with the rest.

Signals of every kind need to be studied from the point of view of mapping them into high-dimensional vectors for further processing. Here again, we can look to nature for clues. For example, the cochlea of the inner ear breaks sound into its frequency components before passing the signal on to the brain – it is a Fourier analyzer. No doubt the frequency spectrum tells us more about the sound source than the raw sound wave. The design of an *associative memory* for perhaps thousands-to-millions of vectors is a major engineering challenge. The size and structure of the cerebellum can provide ideas for meeting it.

The statistical nature of the operations means that computing with vectors will not replace traditional computing with numbers. Instead, it allows new algorithms that benefit from high dimensionality, for example by making it possible to learn continuously from streaming data.

Computing with vectors can benefit from hardware trends to the fullest. Because the representation is extremely redundant, circuits need not be 100% reliable. The manufacture of very large circuits that operate with very little energy becomes possible, and it will be possible to compute using the analog properties of materials.

Computing with vectors has been demonstrated here with bipolar vectors, or equivalently with dense binary vectors. To extend it to vectors of other kind, and to other mathematical objects, we need to identify operations on the objects that form a useful computational algebra, and that are also suited for realization in a physical medium. Today, that medium is overwhelmingly silicon because of its success at meeting the needs of digital logic. However, the discovery of new materials and physical phenomena will widen our choices and offer new opportunities. Thinking of computing in terms of an algebra of operations on elements of a mathematical space will help us to recognize opportunities as they arise and to develop them to practical systems for computing.

Resources. Right now there are several websites that track progress in computing with high-dimensional vectors.

- **Vector Symbolic Architectures aka Hyperdimensional Computing**
  https://www.hd-computing.com/home

- **Collection of Hyperdimensional Computing Projects**
  https://github.com/HyperdimensionalComputing/collection
- **Online Speakers' Corner on Vector Symbolic Architectures and Hyperdimensional Computing**
  https://sites.google.com/ltu.se/vsaonline/home

Computing with vectors of different kind in various contexts are summarized in several papers [Neubert, Schubert & Protzel, 2019; Ge & Parhi, 2020; Hassan et al., 2021; Kleyko et al., 2021].

Acknowledgments. This chapter was written with the support of Defense Advanced Research Projects Agency (DARPA DSO program on Virtual Intelligence Processing, grant #HR00111990072) and Air Force Office of Scientific Research (AFOSR program on Cognitive and Computational Neuroscience, award number FA9550-19-1-0241). Special thanks are due Professor Bruno Olshausen, Director of UC Berkeley's Redwood Center for Theoretical Neuroscience, for fostering an environment supportive of essential research.

References

Aerts, D., Czachor, M., and De Moor, B., "Geometric analogue of holographic reduced representation," *Journal of Mathematical Psychology*, vol. 53, no. 5, pp. 389–398, 2009. doi: https://doi.org/10.1016/j.jmp.2009.02.005.

Albus, J. S., "A theory of cerebellar functions," *Mathematical Biosciences*, vol. 10, pp. 25–61, 1971.

Burrello, A., Schindler, K., Benini, L., and Rahimi, A., "Hyperdimensional computing with local binary patterns: one-shot learning of seizure onset and identification of ictogenic brain regions using short-time iEEG recordings," *IEEE Transactions on Biomedical Engineering*, vol. 67, no. 2, pp. 601–613, 2020.

Eliasmith, C., *How to Build a Brain: A Neural Architecture for Biological Cognition*, Oxford University Press, 2013.

Frady, E. P., Kent, S. J., Olshausen, B. A., and Sommer, F. T., "Resonator networks, 1: An efficient solution for factoring high-dimensional, distributed representations of data structures," *Neural Computation*, vol. 32, no. 12, pp. 2311–2331, 2020.

Frady, E. P., Kleyko, D., and Sommer, F. T., "A theory of sequence indexing and working memory in recurrent neural networks," *Neural Computation*, vol. 30, no. 6, pp. 1449–1513, 2018.

Gallant, S. I. and Okaywe, T. W., "Representing objects, relations, and sequences," *Neural Computation*, vol. 25, no. 8, pp. 2038–2078, 2013.

Gayler, R., "Multiplicative binding, representation operators, and analogy," in K. Holyoak, D. Gentner, and B. Kokinov (eds.), *Advances in Analogy Research: Integration of Theory and Data from the Cognitive, Computational, and Neural Sciences*, p. 405. Sofia, Bulgaria: New Bulgarian University Press, 1998.

Gayler, R. W., "Vector symbolic architectures answer Jackendoff's challenges for cognitive neuroscience," in P. Slezak (ed.), *ICCS/ASCS International Conference on Cognitive Science*, pp. 133–138. Sydney, Australia: University of New South Wales, 2003.

Gayler, R. W. and Levy, S. D., "A distributed basis for analogical mapping," in B. Kokinov and K. Holyoak (eds.), *New Frontiers in Analogy Research, Proceedings of the Second International Analogy Conference*, pp. 165–174. Sofia: New Bulgarian University Press, 2009.

Ge, L. and Parhi, K. K., "Classification using hyperdimensional computing: a review," *IEEE Circuits and Systems Magazine*, vol. 20, no. 2, pp. 30–47, 2020.

Gupta, S., Imani, M., and Rosing, T., "FELIX: Fast and energy-efficient logic in memory," in *2018 IEEE/ACM International Conference on Computer-Aided Design (ICCAD)*, pp. 1–7, 2018. doi: https://doi.org/10.1145/3240765.3240811.

Hassan, E., Halawani, Y., Mohammad, B., and Saleh, H., "Hyper-dimensional computing challenges and opportunities for AI applications," *IEEE Access*, 2021. doi: https://doi.org/10.1109/ACCESS.2021.3059762.

Hinton, G. E., "Mapping part–whole hierarchies into connectionist networks," *Artificial Intelligence*, vol. 46, no. 1–2, pp. 47–75, 1990.

Hinton, G. E. and Anderson, J. A. (eds.), *Parallel Models of Associative Memory*, (updated edition 1989). Hillsdale, N.J.: Lawrence Erlbaum, 1981.

Joshi, A., Halseth, J. T., and Kanerva, P., "Language geometry using random indexing," in J. A. de Barros, B. Coecke, and E. Pothos (eds.), *Quantum Interaction: 10th International Conference*, QI 2016, San Francisco, CA, July 20–22, 2016, Revised Selected Papers, pp. 265–274. Cham, Switzerland: Springer International Publishing, 2017.

Kanerva, P., *Sparse Distributed Memory*, Cambridge, MA: MIT Press, 1988.

Kanerva, P., "Binary spatter-coding of ordered K-tuples," in C. von der Malsburg, W. von Seelen, J. C. Vorbruggen, and B. Sendhoff (eds.), *Artificial Neural Networks – ICANN 96 Proceedings* (Lecture Notes in Computer Science, vol. 1112), pp. 869–873. Berlin: Springer-Verlag, 1996.

Kanerva, P., "Hyperdimensional computing: an introduction to computing in distributed representation with high-dimensional random vectors," *Cognitive Computation*, vol. 1, no. 2, pp. 139–159, 2009.

Kanerva, P., Kristoferson, J., and Holst, A., "Random indexing of text samples for latent semantic analysis," in L. R. Gleitman and A. K. Josh (eds.), *Proceedings of the 22nd Annual Meeting of the Cognitive Science Society* (CogSci'00, Philadelphia), p. 1036. Mahwah, New Jersey: Erlbaum, 2000.

Karunaratne, G., Schmuck, M., Le Gallo, M., Cherubini, G., Benini, L., Sebastian, A., and Rahimi, A., "Robust high-dimensional memory-augmented neural networks," *Nature Communications*, vol. 12, Article number 2468, 2021, 12 pp. doi: https://doi.org/10.1038/S41467-021-22364-0.

Kent, S. J., Frady, E. P., Sommer, F. T., and Olshausen, B. A., "Resonator networks, 2: Factorization performance and capacity compared to optimization-based methods," *Neural Computation*, vol. 32, no. 12, pp. 2332–2388, 2020.

Kleyko, D., Davies, M., Frady, E. P., Kanerva, P., Kent, S. J., Olshausen, B. A., Osipov, E., Rabaey, J. M., Rachkovskij, D. A., Rahimi, A., and Sommer, F. T., "Vector Symbolic Architectures as a computing framework for nanoscale hardware," arXiv:2106.05268v1 [cs.AR] 9 June 2021.

Landauer, T. and Dumais, S., "A solution to Plato's problem: the Latent Semantic Analysis theory of acquisition, induction and representation of knowledge," *Psychological Review*, vol. 104, no. 2, pp. 211–240, 1997.

Levy, S. D. and Gayler, R., "Vector symbolic architectures: a new building material for artificial general intelligence," in *Proceedings of the First Conference on Artificial General Intelligence* (AGI-08). IOS Press, 2008.

Marr, D., "A theory of cerebellar cortex," *Journal of Physiology, (London)*, vol. 202, 437–470, 1969.

Mitchell, M., *Artificial Intelligence: A Guide to Thinking Humans*, New York: Farrar, Straus and Giroux, 2019.

Mitrokhin, A., Sutor, P., Fermüller, C., and Aloimonos, Y., "Learning sensorimotor control with neuromorphic sensors: toward hyperdimensional active perception," *Science Robotics*, vol. 4, eaaw6736 2019 (15 May 2019), 10 pp. doi: https://doi.org/10.1126/scirobotics.aaw6736.

Moin, A., Zhou, A., Rahimi, A., Menon, A., Benatti, S., Alexandrov, G., Tamakloe, S., Ting, J., Yamamoto, N., Khan, Y., Burghardt, F., Benini, L., Arias, A. C., and Rabaey, J. M., "A wearable biosensing system with in-sensor adaptive machine learning for hand gesture recognition," *Nature Electronics*, vol. 4, no. 1, pp. 54–63, 2021.

Murdock, B. B., "A theory for the storage and retrieval of item and associative information," *Psychological Review*, vol. 89, no. 6, pp. 609–626, 1982. doi: https://doi.org/10.1037/0033-295X.89.6.609.

Neubert, P., Schubert, S., and Protzel, P., "An introduction to hyperdimensional computing for robotics," *Künstliche Intelligenz volume*, vol. 33, pp. 319–330, 2019. doi: https://doi.org/10.1007/s13218-019-00623-z.

Osipov, E., Kleyko, D., and Legalov, A., "Associative synthesis of finite state automata model of a controlled object with hyperdimensional computing," in *IECON 2017—43rd Annual Conference of the IEEE Industrial Electronics Society*, pp 3276–3281, 2017. doi: https://doi.org/10.1109/IECON.2017.8216554.

Plate, T. A., "Distributed representations and nested compositional structure," Doctoral dissertation, University of Toronto, 1994.

Plate, T. A., *Holographic Reduced Representation: Distributed Representation of Cognitive Structure*, Stanford, CA: CSLI Publications, 2003.

Rachkovskij, D. A. and Kussul, E. M., "Binding and normalization of binary sparse distributed representations by context-dependent thinning," *Neural Computation*, vol. 13, no. 2, pp. 411–452, 2001.

Rahimi, A., Datta, S., Kleyko, D., Frady, E. P., Olshausen, B., Kanerva, P., and Rabaey, J. M., "High-dimensional computing as a nanoscalable paradigm," *IEEE Transactions on Circuits and Systems I: Regular Papers*, vol. 64, no. 9, pp. 2508–2521, 2017. doi: https://doi.org/10.1109/tcsi.2017.2705051.

Räsänen, O. J. and Saarinen, J. P., "Sequence prediction with sparse distributed hyperdimensional coding applied to the analysis of mobile phone use patterns," *IEEE Transactions on Neural Networks and Learning Systems*, vol. 27, no. 9, pp. 1878–1889, 2016.

Semiconductor Research Corporation, "New compute trajectories for energy-efficient computing," *Decadal Plan for Semiconductors, Full Report*, pp. 122–146, 2021. https://www.src.org/about/decadal-plan.

Smolensky, P., "Tensor product variable binding and the representation of symbolic structures in connectionist networks," *Artificial Intelligence*, vol. 46, no. 1–2, pp. 159–216, 1990.

Widdows, D. and Cohen, T., "Reasoning with vectors: a continuous model for fast robust inference," *Logic Journal of the IGPL*, vol. 23, no. 2, pp. 141–173, 2015. doi: https://doi.org/10.1093/jigpal/jzu028.

Wu, T. F., Li, H., Huang, P.-C., Rahimi, A., Hills, G., Hodson, B., Hwang, W., Rabaey, J. M., Wong, H.-S. P., Shulaker, M. M., and Mitra, S., "Hyperdimensional computing exploiting carbon nanotube FETs, resistive RAM, and their monolithic 3D integration," *IEEE Journal of Solid-State Circuits*, vol. 53, no. 11, pp. 3183–3196, 2018.

3

CAD for Analog/Mixed-Signal Integrated Circuits

Ahmet F. Budak, David Z. Pan, Hao Chen, Keren Zhu, Mingjie Liu, Mohamed B. Alawieh,
Shuhan Zhang, Wei Shi, and Xiyuan Tang*

Department of Electrical and Computer Engineering, The University of Texas at Austin, Austin, TX, USA

3.1 Introduction

Integrated circuits (ICs), the backbone of modern electronics, have played a critical role in the significant advancement of modern information technology in the past decades. The ever-increasing level of integration complexity, now featuring multibillion-transistor ICs, has powered numerous complicated tasks covering cloud computing, artificial intelligence, and smart manufacturing. Such complete systems now have been integrated on a few chips or even one single chip. One example of such Systems on Chip (SoC) can be found in the wireless communication chips in smartphones, including analog, digital, and radio-frequency (RF) sections on a single chip.

As technology nodes scale into the nanometer era, digital circuits have benefited the most, with significantly boosted computing power. Despite the trend to replace analog functions with digital circuits, e.g. shifting the traditional analog channel selection into digital signal processing, a few typical functions will always remain analog, including the following:

(1) **Interfaces between the digital chip and the real world:** The physical world is analog and relies on analog/mixed-signal (AMS) components to interact. On the input side of a system, the specialized sensor captures physical signals (e.g. sound, temperature, and magnetic) and converts it into electrical representations such as voltage and current. A typical signal chain includes a low-noise amplifier, a filter, and an analog-to-digital converter (ADC). High-quality designs require a sufficient signal-to-noise-and-distortion ratio (SNDR). On the output side, the digital signal needs to be converted back to analog form to interact with the environment.

(2) **Power management components:** With the increased integration level of modern electronic devices, a single system usually requires multiple power rails. For instance, LED drivers, logic units, and power amplifiers operate under drastically different voltage supplies. Power management units, required in all modern electronics, offer high-efficiency and compact voltage conversion for various functional blocks.

(3) **Clock generation circuits:** Such as supply voltages, a modern electronic system requires multiple clock frequencies for subblocks. Derived from a standard reference block, commonly a crystal oscillator, the frequency synthesizer generates a range of frequencies required in SoC. Standard blocks include oscillators, charge pumps, and phase-lock loops.

---

*Corresponding author.

Advances in Semiconductor Technologies: Selected Topics Beyond Conventional CMOS, First Edition. Edited by An Chen.
© 2023 The Institute of Electrical and Electronics Engineers, Inc. Published 2023 by John Wiley & Sons, Inc.

(4) **Communication blocks:** The colossal data demand in the Internet of things (IoT) era poses new challenges to communication blocks. The emerging communication standards, e.g. 5G, machine-type communication, keep pushing boundaries of circuits' speed and energy efficiency. Standard blocks include transmitters and receivers. They are often referred to as RF circuits.

(5) **Highly customized mixed-signal chips targeting cutting-edge performance:** For example the state-of-the-art memory cells are mainly custom-designed like analog circuits to push speed or power limit.

Clearly, analog, mixed-signal, and RF circuits are indispensable in modern electronics systems, especially those interacting with the real world. The emerging applications, including IoT, 5G networks, advanced computing, and healthcare electronics, pose sharply increasing demand for analog, mixed-signal, RF circuits. For example, as forecasted by Semiconductor Research Corporation (SRC), more than 45 trillion sensors will appear worldwide by 2032 [1]. Therefore, a short turnaround time of analog, mixed-signal, and RF IC design is highly desired.

Despite the continuous efforts in analog layout automation, those accomplishments have not been well adopted in current industrial flows. The main reason is rooted in the characteristics of AMS circuits, which are complex and sensitive. Compared to its digital counterpart, AMS design deals with a wide range of specific circuit classes, various device types, and often requires customization for each circuit type. Besides, they are sensitive to signal couplings, layout effects, and fabrication variations. Furthermore, the lack of effective ways to model the layout effects on analog performance imposes significant challenges for automation tools. Therefore, implementing analog circuits is still mainly a manual, time-consuming, and error-prone task. Typical analog design procedure includes the following:

(1) **System architecture definition:** In a top-down design methodology, the designers first need to translate a system-level requirement into detailed subblock specifications. For example, in an RF receiver design, the system architect should break down the sensitivity requirement into low noise amplifier (LNA) noise figure, amplifier gain, and ADC signal-to-noise ratio.

(2) **Circuit topology selection:** Based on the subblock specifications, a specific circuit topology is chosen. For instance, if a wide-band design is required in an operational transconductance amplifier (OTA) design, the feed-forward compensated structure is preferred. If a high gain is needed, the telescopic amplifier is preferred.

(3) **Device sizing:** With the selected topology, devices, including transistors, capacitors, and resistors, need to be appropriately sized for the targeted performance. Due to the extensive exploration space, this progress is extremely challenging. For instance, in a simple two-stage miller-compensated OTA design, over 40 parameters need to be determined, such as transistor widths, lengths, biasing currents, voltages, and capacitor values.

(4) **Layout design:** Unlike digital circuits, which only need to satisfy design rule check (DRC), AMS layout generation faces extra challenges due to its high sensitivity. Constraints such as symmetry and critical nets directly impact the layout quality. Various layout-dependent effects, such as length of diffusion (LOD) and well proximity effect (WPE), significantly increase layout generation complexity.

(5) **Postlayout verification:** Postlayout extraction captures parasitics and layout effects. To ensure the layout quality, comprehensive verifications, including process-voltage-temperature (PVT) simulations and Monte-Carlo (MC) simulations, are performed.

Note that due to the nature of analog design, trade-offs are considered in every single step. For instance, one can hardly size an OTA to realize low-power, high-gain, and high BW simultaneously.

Hence, AMS design depends highly on the application requirement and requires customization. Besides, multiple iterations are expected to meet the final specification.

This chapter aims to review the Computer Aided Design (CAD) developments dedicated to each step of the analog design flow mentioned above and provide a helpful guide to CAD designers who are interested in improving the design efficiency of analog, mixed-signal, and RF circuits.

3.2 Front-End CAD

AMS IC design is a complex process involving multiple steps. Billions of nanoscale transistor devices are fabricated on a silicon die and connected via intricate metal layers during those steps. The final product is an IC, which powers much of our life today. Today, many systems require both digital functional units and analog subsystems to be implemented on the same chip. Thanks to advances in electronic design automation (EDA) tools, most of the digital units can be efficiently and reliably synthesized. On the other hand, AMS design lacks these tools and is mostly handcrafted by analog specialists. Therefore, AMS design suffers from long design cycles and high-design complexity and often constitutes a bottleneck in mixed analog/digital design [2]. Front-end CAD for AMS design includes topology selection and device sizing. Both these phases are considered costly since topology selection relies on the designer's experience on various architectures and device sizing is manual and conducted in an iterative manner. Therefore, front-end design consumes a significant portion of the total design budget. In a typical AMS design cycle, a designer starts with a schematic design of a circuit topology. The designer either selects a preexisting topology based on circuit performance specifications or designs an entirely new topology based on domain knowledge. Usually, a circuit is expected to meet specifications (e.g. phase, delay, gain, power, bandwidth) at tens of operation corners. If specifications are not met at any specific corner, circuit designers modify the topology or resize circuits. Circuit resizing involves modification of device parameters like widths and lengths of MOS transistors used in the circuit. Due to the complex nature of circuits, it may not be obvious which devices impact the specifications. This usually leads to trial and error, along with over design, in circuit sizing to meet performance metrics. After performance specifications are met at all corners, variation/yield/aging/reliability simulations are performed to ensure design functionality in the presence of process and usage scenarios. If a circuit fails to meet specifications at this stage, the designer has to resize the circuit again. Hence, AMS front-end is highly iterative and manual in nature. In order to tackle this labor-intensive nature and reduce time-to-market requirements, front-end automation for AMS design has attracted high interest in recent years.

3.2.1 Circuit Architecture and Topology Design Space Exploration

As it was mentioned in introduction, current practice in circuit topology selection relies on designer's expertise. After getting system requirements, designers use their understanding of circuit behaviors, in-depth analytic equations, and technology performance to choose the most appropriate topology. Automation in topology exploration is desired due to various reasons. It is an important determinant of the final circuit performance and the first part of the whole flow; hence, automating this part is necessary to have a complete automated design flow since topology selection and device sizing are generally considered jointly. Despite the hard nature of topology selection problem, there have been some efforts to automate this phase. Some methods realized the nature of human design flow and have been developed based on heuristics and design knowledge

[3–5]. On the other hand, other methods targeted topology selection problem with machine learning solutions [6, 7]. In [3], a DRC clean layout of an OPAMP is designed from a system-level specification. The topology selection is pruned via a decision tree whose branching is determined by the application purpose and performance requirements. Similarly, in [4], a method using fuzzy-logic based decision rules to determine circuit topology is adapted. The fuzzy-logic translates the expert defined design rules into a decision mechanism. Then the topology decision process is conducted via a defuzzification interface. However, as the technology advances and circuits become more complex, design knowledge is difficult to be extracted, collected, and reused. Data-driven techniques are investigated to automate the topology selection with simulation results. In [6], a Convolutional Neural Network (CNN) is used as a classifier with training dataset containing circuit specifications as the input and topology index as the output. After being trained, the CNN can select the best topology given the specifications. The main challenge is the time-consuming process of obtaining performance data of different topologies. Also, topology data-base is usually limited in prior works which make it a challenge to easily adapt it for larger domains.

It is worth noting that aforementioned studies choose one topology within a fixed database. In addition to these, some efforts to generate circuit topology from the performance requirements is shown. These studies are generally restricted to certain type/purpose of circuits, and generalizing topology generation to various applications is currently untouched research area. Koza et al. [5] presents one of the first attempts on the area where genetic programming is used as the optimization engine for topology generation. A recent study introduces a hypernetwork scheme specialized to determine alignment, type, and value of the components of two-port networks [7]. Since this method is proposed for specific type of circuits (two-port networks), adapting it for other AMS design applications may not be direct and could require further customization.

3.2.2 Device Sizing

AMS device sizing task stands for finding optimal values of design parameters to maximize the circuit performance metric. Although the majority of digital design flow is automated by EDA tools, AMS design practice is still heavily manual. In general, AMS design variables, such as transistor lengths and widths, capacitor values, biasing parameters, are adjusted based on designer's knowledge and experience [8]. However, due to the downscaling in the technology node and tightening time-to-market requirements, it gets more challenging for designers to develop the state-of-the-art, robust AMS blocks while accounting for more complex device models and severe process variations. Therefore, the need to bring automation solutions for AMS circuit sizing is obvious.

3.2.2.1 AMS Circuit Sizing: Problem Formulation
In general, AMS circuit sizing problem can be formulated in two ways. The first way is to formulate it as a constrained optimization problem succinctly as below.

$$\text{minimize } f_0(\mathbf{x})$$
$$\text{subject to } f_i(\mathbf{x}) \leq 0 \quad \text{for } i = 1, \ldots, m$$

where, $\mathbf{x} \in \mathbb{D}^d$ is the parameter vector and d is the number of design variables of sizing task. Thus, $\mathbb{D}^d$ is the design space. $f_0(\mathbf{x})$ is the objective performance metric we aim to minimize and $f_i(\mathbf{x})$ is the ith performance constraint in the design. The second way of formulating sizing problem is to transform it into an unconstrained problem by defining a Figure of Merit (FoM) and optimizing for it. In this way, circuit performance values are lumped into a single equation by using normalization constants, *w*.

$$\text{minimize FoM} = \sum_{i=0}^{m} w_i \times f_i(\mathbf{x})$$

3.2.2.2 Methods for AMS Circuit Sizing

Prior work on analog circuit sizing automation can be divided into two categories: *knowledge-based* and *optimization-based methods*. In the knowledge-based approach, design experts transcribe their domain knowledge into algorithms and equations [9, 10]. However, such methods create dependency on expert human-designers, circuit topology, and technology nodes. Thus, these methods are highly time-consuming and not scalable.

Optimization-based methods are further categorized into two classes: *equation-based* and *simulation-based* methods [11]. Equation-based methods try to express circuit performance via polynomial equations or regression models using simulation data. Then the equation-based optimization methods such as Geometric Programming (GP) [12, 13] or Semidefinite Programming (SDP) relaxations [14] are applied to convex or nonconvex formulated problems to find an optimal solution. These methods are generally fast but due to scaling in technology and advances in circuit topologies developing accurate expressions for circuit performances is not easy and may deviate largely from the actual values. Therefore, the most recent trend is to employ simulation-based methods to tackle the sizing problem [15–29]. These methods make guided exploration in the search space and target a global minimum using the real evaluations from circuit simulators. In general, black-box or learning-based optimization techniques are applied to explore design space.

Traditionally, there have existed various model-free optimization methods such as particle swarm optimization (PSO) [30] and advanced differential evolution. Liu et al. [31] is proposed for analog circuits and Afacan and Dundar [32] is proposed for RF where evolutionary strategies and simulating annealing are used as the search mechanisms, respectively. Although these methods have good convergence behavior, they are known to be sample-inefficient (i.e. SPICE simulation intensive). Recently surrogate model-based and learning-based methods are becoming increasingly popular due to their efficiency in exploring solution space. In surrogate model-based methods, Gaussian Process Regression (GPR) [33] or Artificial Neural Networks (ANN) [34] are generally used for design space modeling, and the next design point is determined through model predictions. For example, MMLDE [35] method is proposed for synthesizing passive components of high-frequency RF ICs. MMLDE is a hybrid optimization scheme, where the global search is conducted via evolutionary steps and surrogate-model is built to make selections among candidate solutions. In MMLDE, authors used both GPR and ANN as the surrogate. GASPAD method followed this work and is introduced into RF IC synthesis where GPR predictions guide evolutionary search [36]. WEIBO method proposed a GPR-based Bayesian optimization [37] algorithm, where a blended version of weighted Expected Improvement (wEI) and the probability of feasibility is selected as acquisition function to handle constrained nature of analog sizing [21, 38]. The main drawback of Bayesian Optimization methods is its scalability as GP modeling has cubic complexity in the number of samples, $\mathcal{O}(N^3)$. An attempt to overcome this is given in [39], where authors utilized a neural network to make Bayesian inference which eliminated the cubic dependency on the number of samples.

Other model-based optimization methods utilized ANN for circuit performance modeling. In [40], ANN is used as a proxy for the circuit simulator and performance estimates obtained by the trained model are used to efficiently guide an evolutionary optimization framework.

Hakhamaneshi et al. [41] is also an ANN-boosted evolutionary method, and they used the trained ANN as a performance classifier between the parent population and child population, and they modified the exploration procedure for only allowing promising child solutions to survive into next generation. In [42], authors proposed to enhance the efficiency of a Genetic Algorithm based optimizer where the local minimum search is conducted via the help of trained ANN. Another method that utilizes neural nets is ESSAB [26], where ANN training is combined with a data augmentation method to reduce the need for required number of samples for accurate training. A different approach is adapted for building their ANN model in [43], where they use circuit specifications at the input nodes and design variables are taken from the output nodes. To train and work with such a model, a large dataset is generated for offline training of the ANN model.

Recently, reinforcement learning (RL) algorithms are applied in the area as learning-based methods. GCN-RL [27] leverages Graph Neural Networks (GNN) and proposes a transferable framework. They employed actor-critic algorithms to optimize circuit performance and circuit FoM is used as the reward value for training. They also demonstrated a transfer mechanism between topologies and technology nodes by modifying the state vectors. It reports superior results over various methods and human-designer and also shows successful transfer of learning between technology nodes. Drawbacks of this method is that it requires thousands of simulations for convergence (without transfer learning) and it suffers from engineering effort to determine observation vector, architecture selection, and reward engineering. AutoCkt [28] is a sparse subsampling RL technique optimizing the circuit parameters by taking discrete actions in the solution space. AutoCkt shows better efficiency over random RL agents and differential evolution. Since it adapts an offline training scheme, it requires to be trained with thousands of SPICE simulations before deployment, which is a sink cost for using their algorithm. Another RL method is DNN-Opt [29], where the core algorithm is inspired from deep RL actor-critic methods but customized for AMS sizing task. It offers an efficient online training methodology and shows strong performance convergence both in terms of number of required samples and total optimization time. Further, it introduces a recipe to apply their method to large-scale industrial circuits.

3.3 Layout Automation

In current AMS and RF IC design flows, the physical layout implementation stage is still heavily manual, thus time-consuming and error-prone, setting limits on the turnaround time. The main reason is originated in the high complexity of the AMS/RF IC design problem itself. Compared to its digital counterpart, AMS design is much more complicated even for simple modules as it considers various primitive devices and a wide range of diverse circuit classes that require customized tuning. Sensitive signal couplings, layout-dependent effects, and process variations of AMS/RF layouts also impose further challenges toward high-quality layout design.

Though the design automation techniques for AMS and RF circuits are not yet mature enough for standardized commercial applications, recent research has achieved promising advancements. AMS layout automation techniques can be categorized into two paradigms: procedural layout generation and optimization-based layout synthesis [44]. Procedural layout generators utilizing predesigned parameterized layout templates to migrate layouts for various manufacturing technologies and device sizings have been demonstrated. Optimization-based techniques using place-and-route (P&R) algorithms to optimize the area, power, and certain performance metrics have also been developed. These approaches possess various degrees of generality and are designed for different usage scenarios. Similar paradigms are also being utilized in automating RF layout synthesis. This section introduces the two paradigms and reviews some techniques utilizing them.

3.3.1 Procedural Layout Generation

Procedural-based layout generation refers to using predesigned procedures to generate the layouts. It follows the procedures or templates to the layout structures. The manually designed templates enable sophisticated and high-quality layouts at the cost of extensive human efforts. Figure 3.1 shows a typical flow.

The idea of leveraging manually optimized templates to generate new layout traces back to VLSI design automation's early days. In the early efforts in automating analog layout generation, such as ILAC [45] and SLAM [46], procedural block generators are proposed to generate critical circuit substructures, such as differential pair and capacitors, with parameterizable templates. Later literature extends the procedural layout generation to support more sophisticated functionalities and reduce human efforts. Youssef et al. [47] developed a Python-based layout generation tool for generating primitive blocks. Han et al. [48] developed a graphical user interface (GUI)-based template engine to assist the development of layout templates. Some studies propose applications leveraging procedural analog layout generation. Stefanovic et al. [49] and Castro-Lopez et al. [50] apply procedure-based analog layout generation for parasitic estimation and device sizing. Ding et al. [51] combine template-based layout tool with digital P&R flow to automating successive approximation registers ADC (SAR ADC) layout synthesis. Wulff and Ytterdal [52] also demonstrates success in automatically generated SAR ADC layout with a Perl script-based layout compiler. Procedural layout generator also apply to RF IC demonstrated in [53].

To create a sustainable ecosystem and enable broader reuse of design templates, people are proposing comprehensive standards. A notable attempt is the series of Berkeley Analog Generator (BAG). BAG [54] and BAG2 [55] propose a Python-based interface to enable programming and parameterizing the procedures for generating layouts. Leveraging the flexibility of programming language, BAG2 provides enhanced capability to contain well-optimized layout generation procedures. BAG2 have demonstrated its effectiveness in silicon with generating a SAR ADC [56]

Figure 3.1 Procedural analog circuits layout design flow. Source: Chen et al. [44]/IOP Publishing.

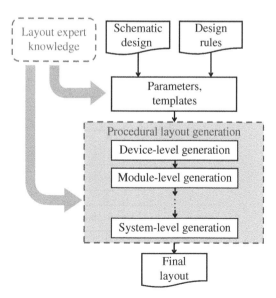

and analog parts in some other designs [57, 58]. The standardized procedural layout generation can benefits the parameterization and technology mitigation for the AMS circuit design by reusing the layout templates. It still requires manual tuning in designing and programming the layout procedures for each schematic.

3.3.2 Optimization-Based Layout Synthesis

Optimization-based layout synthesis aims to provide highly automated layout solutions for AMS circuits. Similar to the digital IC physical design flow, it usually divides the AMS layout synthesis into several stages. Each stage is then formulated as a constrained optimization problem. The sensitive analog layout considerations are indirectly handled by formulating them into objectives and constraints.

Figure 3.2 shows a typical optimization-based layout synthesis flow. As shown in the flow chart, the inputs of an optimization-based synthesis flow require minimum human effort. Only circuit netlists and technology specifications are necessary. The core of an optimization-based layout flow consists of four main subtasks: constraint extraction, module generation, placement, and routing. The computation flow is widely adopted in existing optimization-based frameworks like ALIGN [59] and MAGICAL [60, 61].

The constraint extraction step comprehends the layout design conventions by analyzing the prelayout circuit information. For instance, the sensitiveness of pairing nets, interconnections between devices and building blocks, and the geometry structure of each module. As the extracted constraints can largely impact the final layout and the postlayout performance, generating accurate constraints for optimization-based layout design methodology is crucial [62]. Typically, design conventions are realized by two major types of constraints: geometrical constraints and electrical constraints. Geometrical constraints implicitly handle some specific design aspects such as differential net current balancing by restricting the relative positions of some modules and nets. Common geometrical constraints include symmetry matching, common-centroid, proximity, etc. [62–64]. The survey paper [65] gives an overview on the recent advancements in constraint extraction for geometrical constraints. Electrical constraints prevent circuit performance degradation caused by IR-drop, signal coupling, and parasitic coupling. Some widely adopted electrical constraints are minimum metal width, minimum via cuts, and net shielding [66].

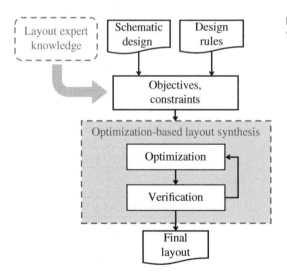

Figure 3.2 Optimization-based layout design flow. Source: Chen et al. [44]/IOP Publishing.

Module generators provide the layout for the low-level modules, such as transistors and resistors. Primary primitive cell generators are widely available in commercial tools and process design kits (PDKs). On the other hand, some academic frameworks have customized device generators for open-source purpose [59, 61].

Placement determines the physical location of each generated module. Analog placement is usually formulated as a constrained optimization problem, while essential layout considerations are handled through imposing constraints. Symmetric constraint, which restricts some modules to be placed along the symmetric axis to reduce mismatch, is widely used [67–98]. Common-centroid is another widely adopted constraint [78, 80, 81]. Other constraints have also been proposed, including array structure in layouts [82], monotonic power current flow [90, 91, 96], thermal effects [76], system signal flow [99], and WPEs [100]. In addition to AMS circuits, symmetric constraint is also applied in RF IC layout synthesis [101].

The routing stage implements the interconnection with metal wires and VIAs. Analog routing is often formulated into optimizing wirelength while satisfying the symmetry between net pairs [102–110]. Ou et al. [105] relaxes the restricted symmetric constraints into different levels, namely symmetry, common-centroid, topology, and proximity. The exact matching constraint is also proposed as an alternative, where the length of wires on each metal layer is constrained to be matched [111, 112]. There are other works that forbid routing over the active regions of transistors [107, 113], optimize power routing [114, 115], and propose shielding critical nets [110]. Recently, Chen et al. [116] propose optimizing the total symmetry across the design. They present a new algorithm to match the layouts' pins and route the nets in symmetry instead of relying on given symmetric constraints. A higher degree of symmetry is found beneficial to circuit performance.

A rising trend in optimization-based AMS circuit layout automation is to create end-to-end frameworks [117]. ALIGN [59] adopts a grid-based methodology. The primitive generation, placement, and routing are all aligned with the grid, and design rules are handled from correct-by-construction. On the other hand, MAGICAL [61] allows more flexibility to target human-like layout style. Equips with end-to-end flow consisting of module generator, analog placer, and detailed router, both ALIGN and MAGICAL have demonstrated success in automating building block-level analog circuits with low or no human efforts. MAGICAL further adopts the mixed-signal placement techniques [99] and state-of-the-art analog detailed router [116]. As a result, MAGICAL recently proves its effectiveness in silicon with a high-performance $\Delta\Sigma$ ADC [61].

In summary, optimization-based layout synthesis requires the least human effort in the three approaches. However, it does not directly generate the layout from human expertise and requires a carefully formulated optimization problem. On the other hand, constraints to the automated tools still need manual input. To further improve the quality and reduce human efforts for the optimization-based AMS circuit layout automation, recent studies explore various emerging techniques.

Several studies focus on optimizing the postlayout performance beyond conventional constrained optimization settings. Xu et al. [118] propose to generate the wells similar to human practice. The proposed framework uses a generative adversarial network (GAN) to automatically learn how experienced engineer draws the wells and apply the learned pattern in automated flow. It is also extended to become part of the placement process in [100]. Zhu et al. [119] apply a similar idea to analog routing. Instead of relying on conventional constraints, the proposed router uses a variational autoencoder (VAE) to learn where human engineers route the special nets and use the machine learning model to guide the automated detailed router. In addition to mimicking the human implementations, people are proposing techniques to directly optimizing the performance. Liu et al. [120] explores the layout-to-performance modeling problem. The proposed machine learning model applies a CNN to predict the postlayout performance from the placement results. Li et al. [121] integrate the performance prediction model in automated placement. The proposed framework treats the performance prediction as to the placement problem's objective to enable direct optimization on performance.

3.4 Post-Layout Extraction and Verification

A clean DRC report shows that the layout abides by all the design rules for the technology node such as metal width, minimum spacing. However, these checks do not guarantee that the layout corresponds to the desired schematic level design. Hence, a critical step at this stage is to extract a postlayout netlist to be compared against the schematic level netlist. Using commercial tools, a post-layout extraction step is performed where all components in the design, including connectivities and parasitic, are extracted to form a postlayout netlist. In general, resistance and capacitance parasitic are the main focus of the extraction step. However, in analog/RF designs inductance extraction for interconnect is also needed.

Given both netlists at the schematic and postlayout levels, Layout vs. Schematic (LVS) check is performed to ensure that the layout corresponds to the schematic-level design. A clean LVS check indicates a correct correspondence between layout and schematic. At this stage, the layout corresponds to the circuit intent at the schematic level while abiding by all design rules.

Despite these checks, the circuit may not be design ready at this point. In fact, the layout quality has a significant impact on the circuit performance especially in AMS designs. The layout extraction step generates a new netlist that includes layout-dependent parasitic. Hence, it is imperative that the new netlist is verified through simulation to make sure that, given the layout induced parasitic, the design still meets the requirements.

A final step in the design process is to assess the reliability of the AMS design against variation in the manufacturing process. Two major assessments are necessary [122]. The first is evaluating the performance under different operating conditions; mainly PVT corners that the chip may face after fabrications. This analysis is application-dependent since the fields of use are different for different chips, and so are the requirements. Therefore, PVT analysis is performed to ensure the correct operation of the design under the conditions it is meant to encounter.

While PVT corners can be determined in most cases, the design should also be simulated under random process variations. Examples of such variations include change in oxide thickness (Δ_{tox}) and random dopant fluctuations (RDF), among others. In fact, with the continuous scaling, process variations manifest itself among the most prominent factors limiting the yield of AMS circuits [122, 123]. To quantify the impact of this variation, the design is simulated under different realizations of this variation [122, 123]. Using MC analysis, a set of points are sampled from the variation distribution and circuit performance is performed. With such analysis, it is possible to evaluate the expected parametric yield of the manufactured chips, i.e. the percentage of manufactured chips that meet the design requirements. It is important to note though that MC analysis is computationally expensive, and achieving high-confidence estimation for modern designs may require thousands of simulations. In literature, different techniques have been proposed to reduce the computational cost of MC analysis, e.g. using importance sampling and scaled sigma sampling [124, 125].

3.5 Conclusion

Although most of the digital circuits can be efficiently and reliably designed with advanced electronics design automation tools, AMS circuits still heavily rely on manual design. This book chapter reviews the recent advancements in CAD techniques and covers key steps in AMS circuit designs flow, including system architecture definition, circuit topology selection, device sizing, layout design, and postlayout verification. Large efforts have been made to apply the most advanced machine learning techniques to reduce the burden of human in the loop and further automate the AMS circuit design process. Due to the rapid advancements of machine learning, future AMS IC designs will significantly leverage both human and machine intelligence for quality of results and design productivity.

Acknowledgments

This work is supported in part by the NSF under Grant No. 1704758, and the DARPA IDEA program.

References

1 "The Decadal Plan for Semiconductors," https://www.src.org/about/decadal-plan/, accessed: 2021-01-30.

2 P. R. Gray, "Analog IC's in the submicron era: trends and perspectives," in *1987 International Electron Devices Meeting*, 1987, pp. 5–9.

3 H. Y. Koh, C. H. Sequin, and P. R. Gray, "OPASYN: A compiler for CMOS operational amplifiers," *IEEE Transactions on Computer-Aided Design of Integrated Circuits and Systems*, vol. 9, no. 2, pp. 113–125, 1990.

4 J. Chavez, A. Torralba, and L. G. Franquelo, "A fuzzy-logic based tool for topology selection in analog synthesis," in *Proceedings of IEEE International Symposium on Circuits and Systems - ISCAS '94*, vol. 1, 1994, pp. 367–370.

5 J. R. Koza, F. H. Bennett, D. Andre, and M. A. Keane, *Automated Design of Both the Topology and Sizing of Analog Electrical Circuits Using Genetic Programming*, Dordrecht: Springer Netherlands, 1996, pp. 151–170.

6 T. Matsuba, N. Takai, M. Fukuda, and Y. Kubo, "Inference of suitable for required specification analog circuit topology using deep learning," in *2018 International Symposium on Intelligent Signal Processing and Communication Systems (ISPACS)*, 2018, pp. 131–134.

7 M. Rotman and L. Wolf, "Electric analog circuit design with hypernetworks and a differential simulator," in *ICASSP 2020–2020 IEEE International Conference on Acoustics, Speech and Signal Processing (ICASSP)*, 2020, pp. 4157–4161.

8 B. Razavi, *Design of Analog CMOS Integrated Circuits*, 1st Edition. New York, NY: McGraw-Hill, Inc., 2001.

9 N. Horta, "Analogue and mixed-signal systems topologies exploration using symbolic methods," *Analog Integrated Circuits and Signal Processing*, vol. 31, pp. 161–176, 2002.

10 N. Jangkrajarng, S. Bhattacharya, R. Hartono, and C.-J. Shi, "Iprail–intellectual property reuse-based analog IC layout automation," *Integration, the VLSI Journal*, 2003, analog and Mixed-signal IC Design and Design Methodologies.

11 R. Rutenbar, "Analog design automation: where are we? Where are we going?" in *Proceedings of CICC*, 1993.

12 W. Daems, G. Gielen, and W. Sansen, "Simulation-based generation of posynomial performance models for the sizing of analog integrated circuits," in *IEEE TCAD*, 2003.

13 M. d. Hershenson, S. P. Boyd, and T. H. Lee, "Optimal design of a CMOS Op-Amp via geometric programming," in *IEEE TCAD*, 2001.

14 Y. Wang, M. Orshansky, and C. Caramanis, "Enabling efficient analog synthesis by coupling sparse regression and polynomial optimization," in *Proceedings of DAC*, 2014.

15 W. Nye, D. Riley, A. Sangiovanni-Vincentelli, and A. Tits, "DELIGHT.Spice: An optimization-based system for the design of integrated circuits," *IEEE Transactions on Computer-Aided Design of Integrated Circuits and Systems*, vol. 7, no. 4, pp. 501–519, 1988.

16 B. Liu, Y. Wang, Z. Yu, L. Liu, M. Li, Z. Wang, J. Lu, and F. V. Fernández, "Analog circuit optimization system based on hybrid evolutionary algorithms," *Integration, The VLSI Journal*, vol. 42, no. 2, pp. 137–148, Feb. 2009. [Online]. Available: https://doi.org/10.1016/j.vlsi.2008.04.003.

17 J. Koza, F. Bennett, D. Andre, M. Keane, and F. Dunlap, "Automated synthesis of analog electrical circuits by means of genetic programming," *IEEE Transactions on Evolutionary Computation*, vol. 1, no. 2, pp. 109–128, 1997.

18 G. Alpaydin, S. Balkir, and G. Dundar, "An evolutionary approach to automatic synthesis of high-performance analog integrated circuits," *IEEE Transactions on Evolutionary Computation*, vol. 7, no. 3, pp. 240–252, 2003.

19 R. A. Vural and T. Yildirim, "Swarm intelligence based sizing methodology for CMOS operational amplifier," in *2011 IEEE 12th International Symposium on Computational Intelligence and Informatics (CINTI)*, 2011, pp. 525–528.

20 W. Lyu, P. Xue, F. Yang, C. Yan, Z. Hong, X. Zeng, and D. Zhou, "An efficient Bayesian optimization approach for automated optimization of analog circuits," *IEEE Transactions on Circuits and Systems I: Regular Papers*, vol. 65, no. 6, pp. 1954–1967, 2017.

21 W. Lyu, F. Yang, C. Yan, D. Zhou, and X. Zeng, "Batch Bayesian optimization via multi-objective acquisition ensemble for automated analog circuit design," in *International Conference on Machine Learning*. PMLR, 2018, pp. 3306–3314.

22 S. Zhang, F. Yang, D. Zhou, and X. Zeng, "An efficient asynchronous batch Bayesian optimization approach for analog circuit synthesis," in *2020 57th ACM/IEEE Design Automation Conference (DAC)*. IEEE, 2020, pp. 1–6.

23 S. Zhang, W. Lyu, F. Yang, C. Yan, D. Zhou, X. Zeng, and X. Hu, "An efficient multi-fidelity Bayesian optimization approach for analog circuit synthesis," in *2019 56th ACM/IEEE Design Automation Conference (DAC)*. IEEE, 2019, pp. 1–6.

24 S. Zhang, F. Yang, C. Yan, D. Zhou, and X. Zeng, "An efficient batch constrained Bayesian optimization approach for analog circuit synthesis via multi-objective acquisition ensemble," *IEEE Transactions on Computer-Aided Design of Integrated Circuits and Systems*, vol. 41, no. 1, pp. 1–14, 2021.

25 B. He, S. Zhang, F. Yang, C. Yan, D. Zhou, and X. Zeng, "An efficient Bayesian optimization approach for analog circuit synthesis via sparse Gaussian process modeling," in *2020 Design, Automation & Test in Europe Conference & Exhibition (DATE)*. IEEE, 2020, pp. 67–72.

26 A. Budak, M. Gandara, W. Shi, D. Pan, N. Sun, and B. Liu, "An efficient analog circuit sizing method based on machine learning assisted global optimization," *IEEE Transactions on Computer-Aided Design of Integrated Circuits and Systems*, p. 1, vol. 41, no. 5, pp. 1209–1221, 2021.

27 H. Wang, K. Wang, J. Yang, L. Shen, N. Sun, H. Lee, and S. Han, "GCN-RL circuit designer: transferable transistor sizing with graph neural networks and reinforcement learning," in *Proceedings of DAC*, 2020.

28 K. Settaluri, A. Haj-Ali, Q. Huang, K. Hakhamaneshi, and B. Nikolić, "AutoCkt: Deep reinforcement learning of analog circuit designs," in *Proceedings of DATE*, 2020.

29 A. F. Budak, P. Bhansali, B. Liu, N. Sun, D. Z. Pan, and C. V. Kashyap, "DNN-Opt An RL inspired optimization for analog circuit sizing using deep neural networks," in *Proceedings of the 58th ACM/EDAC/IEEE Design Automation Conference*, ser. DAC '21, 2021.

30 R. A. Vural and T. Yildirim, "Analog circuit sizing via swarm intelligence," *AEU - International Journal of Electronics and Communications*, vol. 66, no. 9, pp. 732–740, 2012.

31 B. Liu, G. Gielen, and F. V. Fernndez, *Automated Design of Analog and High-frequency Circuits: A Computational Intelligence Approach*, Springer, 2013.

32 E. Afacan and G. Dundar, "A mixed domain sizing approach for RF circuit synthesis," 04 2016.

33 C. E. Rasmussen and C. K. I. Williams, *Gaussian Processes for Machine Learning (Adaptive Computation and Machine Learning)*, The MIT Press, 2005.

34 A. Jain, J. Mao, and K. Mohiuddin, "Artificial neural networks: a tutorial," *Computer*, vol. 29, no. 3, pp. 31–44, 1996.

35 B. Liu, D. Zhao, P. Reynaert, and G. Gielen, "Synthesis of integrated passive components for high-frequency RF ICs based on evolutionary computation and machine learning techniques," *IEEE Transactions on Computer-Aided Design of Integrated Circuits and Systems*, vol. 30, no. 10, pp. 1458–1468, 2011.

36 B. Liu, D. Zhao, P. Reynaert, and G. G. E. Gielen, "GASPAD: A general and efficient mm-wave integrated circuit synthesis method based on surrogate model assisted evolutionary algorithm," in *IEEE TCAD*, Feb. 2014.

37 J. Snoek, H. Larochelle, and R. P. Adams, "Practical Bayesian optimization of machine learning algorithms," in F. Pereira, C. J. C. Burges, L. Bottou, and K. Q. Weinberger (eds.), *Proceedings of NIPS*, vol. 25. Lake Tahoe, CA: Curran Associates, Inc., 2012.

38 W. Lyu, F. Yang, C. Yan, D. Zhou, and X. Zeng, "Multi-objective Bayesian optimization for analog/RF circuit synthesis," in *Proceedings of DAC*.

39 S. Zhang, W. Lyu, F. Yang, C. Yan, D. Zhou, and X. Zeng, "Bayesian optimization approach for analog circuit synthesis using neural network," in *Proceedings of DATE*, 2019, pp. 1463–1468.

40 G. İslamoğlu, T. O. Çakici, E. Afacan, and G. Dündar, "Artificial neural network assisted analog IC sizing tool," in *2019 16th International Conference on Synthesis, Modeling, Analysis and Simulation Methods and Applications to Circuit Design (SMACD)*, 2019, pp. 9–12.

41 K. Hakhamaneshi, N. Werblun, P. Abbeel, and V. Stojanovi?, "BagNet: Berkeley analog generator with layout optimizer boosted with deep neural networks," in *2019 IEEE/ACM International Conference on Computer-Aided Design (ICCAD)*, 2019, pp. 1–8.

42 Y. Li, Y. Wang, Y. Li, R. Zhou, and Z. Lin, "An artificial neural network assisted optimization system for analog design space exploration," *IEEE Transactions on Computer-Aided Design of Integrated Circuits and Systems*, vol. 39, no. 10, pp. 2640–2653, 2020.

43 N. Lourenço, J. Rosa, R. Martins, H. Aidos, A. Canelas, R. Póvoa, and N. Horta, "On the exploration of promising analog IC designs via artificial neural networks," in *2018 15th International Conference on Synthesis, Modeling, Analysis and Simulation Methods and Applications to Circuit Design (SMACD)*, 2018, pp. 133–136.

44 H. Chen, M. Liu, X. Tang, K. Zhu, N. Sun, and D. Z. Pan, "Challenges and opportunities toward fully automated analog layout design," *Journal of Semiconductors*, vol. 41, no. 11, p. 111407, Nov. 2020. [Online]. Available: https://doi.org/10.1088/1674-4926/41/11/111407.

45 J. Rijmenants, J. B. Litsios, T. R. Schwarz, and M. G. R. Degrauwe, "ILAC: An automated layout tool for analog CMOS circuits," *IEEE Journal Solid-State Circuits*, vol. 24, no. 2, pp. 417–425, 1989.

46 D. J. Chen, J. Lee, and B. J. Sheu, "SLAM: A smart analog module layout generator for mixed analog-digital VLSI design," in *Proceedings of ICCD*, 1989, pp. 24–27.

47 S. Youssef, F. Javid, D. Dupuis, R. Iskander, and M. Louerat, "A Python-based layout-aware analog design methodology for nanometric technologies," in *IEEE International Design and Test Workshop (IDT)*, 2011, pp. 62–67.

48 S. Han, S. Jeong, C. Kim, H.-J. Park, and B. Kim, "GUI-Enhanced layout generation of FFE SST TXs for fast high-speed serial link design," in *Proceedings of DAC*, 2020.

49 D. Stefanovic, M. Kayal, M. Pastre, and V. B. Litovski, "Procedural analog design (PAD) tool," in *Proceedings of ISQED*, 2003, pp. 313–318.

50 R. Castro-Lopez, O. Guerra, E. Roca, and F. V. Fernandez, "An integrated layout-synthesis approach for analog ICs," *IEEE TCAD*, vol. 27, no. 7, pp. 1179–1189, 2008.

51 M. Ding, P. Harpe, G. Chen, B. Busze, Y. Liu, C. Bachmann, K. Philips, and A. van Roermund, "A hybrid design automation tool for SAR ADCs in IoT," *IEEE TVLSI*, vol. 26, no. 12, pp. 2853–2862, 2018.

52 C. Wulff and T. Ytterdal, "A compiled 9-bit 20-MS/s 3.5-fJ/conv. step SAR ADC in 28-nm FDSOI for bluetooth low energy receivers," *IEEE Journal Solid-State Circuits*, vol. 52, no. 7, pp. 1915–1926, 2017.

53 R. Martins, N. Lourenço, F. Passos, R. Póvoa, A. Canelas, E. Roca, R. Castro-López, J. Sieiro, F. V. Fernandez, and N. Horta, "Two-step RF IC block synthesis with preoptimized inductors and full layout generation in-the-loop," *IEEE TCAD*, vol. 38, no. 6, pp. 989–1002, 2019.

54 J. Crossley, A. Puggelli, H. Le, B. Yang, R. Nancollas, K. Jung, L. Kong, N. Narevsky, Y. Lu, N. Sutardja, E. J. An, A. L. Sangiovanni-Vincentelli, and E. Alon, "BAG: A designer-oriented integrated framework for the development of AMS circuit generators," in *Proceedings of ICCAD*, 2013, pp. 74–81.

55 E. Chang, J. Han, W. Bae, Z. Wang, N. Narevsky, B. NikoliC, and E. Alon, "BAG2: A process-portable framework for generator-based AMS circuit design," in *Proceedings of CICC*, 2018, pp. 1–8.

56 J. Han, E. Chang, S. Bailey, Z. Wang, W. Bae, A. Wang, N. Narevsky, A. Whitcombe, P. Lu, B. Nikoli?, and E. Alon, "A generated 7GS/s 8b time-interleaved SAR ADC with 38.2dB SNDR at nyquist in 16nm CMOS FinFET," in *Proceedings of CICC*, 2019.

57 E. Chang, N. Narevsky, J. Han, and E. Alon, "An automated SerDes frontend generator verified with a 16-nm instance achieving 15 Gb/s at 1.96 pJ/bit," *IEEE Solid-State Circuits Letters*, vol. 1, no. 12, pp. 245–248, 2018.

58 N. Mehta, S. Lin, B. Yin, S. Moazeni, and V. Stojanovi?, "A laser-forwarded coherent transceiver in 45-nm SOI CMOS using monolithic microring resonators," *IEEE Journal Solid-State Circuits*, vol. 55, no. 4, pp. 1096–1107, 2020.

59 K. Kunal, M. Madhusudan, A. K. Sharma, W. Xu, S. M. Burns, R. Harjani, J. Hu, D. A. Kirkpatrick, and S. S. Sapatnekar, "ALIGN: Open-source analog layout automation from the ground up," in *Proceedings of DAC*, 2019, pp. 1–4.

60 B. Xu, K. Zhu, M. Liu, Y. Lin, S. Li, X. Tang, N. Sun, and D. Z. Pan, "MAGICAL: Toward fully automated analog IC layout leveraging human and machine intelligence: invited paper," in *Proceedings of ICCAD*, 2019, pp. 1–8.

61 H. Chen, M. Liu, X. Tang, K. Zhu, A. Mukherjee, S. Nan, and D. Z. Pan, "MAGICAL 1.0: An open-source fully-automated AMS layout synthesis framework verified with a 40-nm 1 Gs/s $\delta\sigma$ ADC," in *Proceedings of CICC*, 2021.

62 H. Chen, K. Zhu, M. Liu, X. Tang, N. Sun, and D. Z. Pan, "Universal symmetry constraint extraction for analog and mixed-signal circuits with graph neural networks," in *Proceedings of DAC*, 2021, pp. 1–6.

63 I.-P. Wu, H.-C. Ou, and Y.-W. Chang, "QB-trees: Towards an optimal topological representation and its applications to analog layout designs," in *Proceedings of DAC*, 2016, pp. 1–6.

64 M. Liu, W. Li, K. Zhu, B. Xu, Y. Lin, L. Shen, X. Tang, N. Sun, and D. Z. Pan, "S3DET: Detecting system symmetry constraints for analog circuits with graph similarity," in *Proceedings of ASPDAC*, 2020, pp. 193–198.

65 K. Zhu, H. Chen, M. Liu, and D. Z. Pan, "Automating analog constraint extraction: from heuristics to learning," in *Proceedings of ASPDAC*, 2022.

66 A. Hastings and R. A. Hastings, *The Art of Analog Layout*, 1st Edition. Saddle River, New Jersey, USA: Prentice Hall, 2001.

67 J. Rijmenants, J. B. Litsios, T. R. Schwarz, and M. G. R. Degrauwe, "ILAC: An automated layout tool for analog CMOS circuits," *IEEE Journal Solid-State Circuits*, vol. 24, no. 2, pp. 417–425, 1989.

68 J. M. Cohn, D. J. Garrod, R. A. Rutenbar, and L. R. Carley, "KOAN/ANAGRAM II: New tools for device-level analog placement and routing," *IEEE Journal Solid-State Circuits*, vol. 26, no. 3, pp. 330–342, 1991.

69 F. Balasa and K. Lampaert, "Module placement for analog layout using the sequence-pair representation," in *Proceedings of DAC*, 1999, pp. 274–279.

70 Y. Pang, F. Balasa, K. Lampaert, and C.-K. Cheng, "Block placement with symmetry constraints based on the O-tree non-slicing representation," in *Proceedings of DAC*, 2000, pp. 464–467.

71 F. Balasa, S. C. Maruvada, and K. Krishnamoorthy, "Efficient solution space exploration based on segment trees in analog placement with symmetry constraints," in *Proceedings of ICCAD*, 2002, pp. 497–502.

72 F. Balasa, S. C. Maruvada, and K. Krishnamoorthy, "Using red-black interval trees in device-level analog placement with symmetry constraints," in *Proceedings of ASPDAC*, 2003, pp. 777–782.

73 J.-M. Lin, G.-M. Wu, Y.-W. Chang, and J.-H. Chuang, "Placement with symmetry constraints for analog layout design using TCG-S," in *Proceedings of ASPDAC*, 2005, pp. 1135–1137.

74 F. Balasa, S. C. Maruvada, and K. Krishnamoorthy, "On the exploration of the solution space in analog placement with symmetry constraints," *IEEE TCAD*, vol. 23, no. 2, pp. 177–191, Nov. 2006.

75 D. Long, X. Hong, and S. Dong, "Signal-path driven partition and placement for analog circuit," in *Proceedings of ASPDAC*, 2006.

76 J. Liu, S. Dong, Y. Ma, D. Long, and X. Hong, "Thermal-driven symmetry constraint for analog layout with CBL representation," in *Proceedings of ASPDAC*, 2007, pp. 191–196.

77 P.-H. Lin and S.-C. Lin, "Analog placement based on hierarchical module clustering," in *Proceedings of DAC*, 2008, pp. 50–55.

78 M. Strasser, M. Eick, H. Gräb, U. Schlichtmann, and F. M. Johannes, "Deterministic analog circuit placement using hierarchically bounded enumeration and enhanced shape functions," in *Proceedings of ICCAD*, 2008, pp. 306–313.

79 P.-H. Lin, Y.-W. Chang, and S.-C. Lin, "Analog placement based on symmetry-island formulation," *IEEE TCAD*, vol. 28, no. 6, pp. 791–804, 2009.

80 L. Xiao and E. F. Y. Young, "Analog placement with common centroid and 1-D symmetry constraints," in *Proceedings of ASPDAC*, 2009, pp. 353–360.

81 P.-H. Lin, H. Zhang, M. D. F. Wong, and Y.-W. Chang, "Thermal-driven analog placement considering device matching," in *Proceedings of DAC*, 2009, pp. 593–598.

82 S. Nakatake, M. Kawakita, T. Ito, M. Kojima, M. Kojima, K. Izumi, and T. Habasaki, "Regularity-oriented analog placement with diffusion sharing and well island generation," in *Proceedings of ASPDAC*, 2010, pp. 305–311.

83 C.-W. Lin, J.-M. Lin, C.-P. Huang, and S.-J. Chang, "Performance-driven analog placement considering boundary constraint," in *Proceedings of DAC*, 2010, pp. 292–297.

84 Q. Ma, L. Xiao, Y.-C. Tam, and E. F. Young, "Simultaneous handling of symmetry, common centroid, and general placement constraints," in *IEEE TCAD*, pp. 85–95, Jan. 2011.

85 H.-F. Tsao, P.-Y. Chou, S.-L. Huang, Y.-W. Chang, M. P.-H. Lin, D.-P. Chen, and D. Liu, "A corner stitching compliant B*-tree representation and its applications to analog placement," in *Proceedings of ICCAD*, 2011, pp. 507–511.

86 P.-Y. Chou, H.-C. Ou, and Y.-W. Chang, "Heterogeneous B*-trees for analog placement with symmetry and regularity considerations," in *Proceedings of ICCAD*, 2011, pp. 512–516.

87 C.-W. Lin, C.-C. Lu, J.-M. Lin, and S.-J. Chang, "Routability-driven placement algorithm for analog integrated circuits," in *Proceedings of ISPD*, 2012, pp. 71–78.

88 P.-H. Wu, M. P.-H. Lin, Y.-R. Chen, B.-S. Chou, T.-C. Chen, T.-Y. Ho, and B.-D. Liu, "Performance-driven analog placement considering monotonic current paths," in *Proceedings of ICCAD*, 2012.

89 H.-C. C. Chien, H.-C. Ou, T.-C. Chen, T.-Y. Kuan, and Y.-W. Chang, "Double patterning lithography-aware analog placement," in *Proceedings of DAC*, 2013.

90 H.-C. Ou, H.-C. C. Chien, and Y.-W. Chang, "Simultaneous analog placement and routing with current flow and current density considerations," in *Proceedings of DAC*, 2013.

91 P.-H. Wu, M. P.-H. Lin, T.-C. Chen, C.-F. Yeh, T.-Y. Ho, and B.-D. Liu, "Exploring feasibilities of symmetry islands and monotonic current paths in slicing trees for analog placement," *IEEE TCAD*, vol. 33, no. 6, pp. 879–892, 2014.

92 I.-P. Wu, H.-C. Ou, and Y.-W. Chang, "QB-trees: towards an optimal topological representation and its applications to analog layout designs," in *Proceedings of DAC*, 2016.

93 H.-C. Ou, K.-H. Tseng, J.-Y. Liu, I.-P. Wu, and Y.-W. Chang, "Layout-dependent effects-aware analytical analog placement," *IEEE TCAD*, vol. 35, no. 8, pp. 1243–1254, 2016.

94 B. Xu, S. Li, X. Xu, N. Sun, and D. Z. Pan, "Hierarchical and analytical placement techniques for high-performance analog circuits," in *Proceedings of ISPD*, 2017.

95 Y.-S. Lu, Y.-H. Chang, and Y.-W. Chang, "WB-Trees: A meshed tree representation for finfet analog layout designs," in *Proceedings of DAC*, 2018.

96 A. Patyal, P.-C. Pan, K. A. Asha, H.-M. Chen, H.-Y. Chi, and C.-N. Liu, "Analog placement with current flow and symmetry constraints using PCP-SP," in *Proceedings of DAC*, 2018.

97 B. Xu, B. Basaran, M. Su, and D. Z. Pan, "Analog placement constraint extraction and exploration with the application to layout retargeting," in *Proceedings of ISPD*, 2018.

98 B. Xu, S. Li, C.-W. Pui, D. Liu, L. Shen, Y. Lin, N. Sun, and D. Z. Pan, "Device layer-aware analytical placement for analog circuits," in *Proceedings of ISPD*, 2019.

99 K. Zhu, H. Chen, M. Liu, X. Tang, N. Sun, and D. Z. Pan, "Effective analog/mixed-signal circuit placement considering system signal flow," in *Proceedings of ICCAD*, 2020.

100 K. Zhu, H. Chen, M. Liu, X. Tang, W. Shi, N. Sun, and D. Z. Pan, "Generative-adversarial-network-guided well-aware placement for analog circuits," in *Proceedings of ASPDAC*, 2022.

101 N. Jangkrajarng, L. Zhang, S. Bhattacharya, N. Kohagen, and C.-J. R. Shi, "Template-based parasitic-aware optimization and retargeting of analog and RF integrated circuit layouts," in *Proceedings of ICCAD*, 2006, p. 342–348.

102 U. Choudhury and A. Sangiovanni-Vincentelli, "Constraint-based channel routing for analog and mixed analog/digital circuits," *IEEE TCAD*, vol. 12, no. 4, pp. 497–510, 1993.

103 J. M. Cohn, D. J. Garrod, R. A. Rutenbar, and L. R. Carley, "KOAN/ANAGRAM II: New tools for device-level analog placement and routing," *IEEE Journal Solid-State Circuits*, vol. 26, no. 3, pp. 330–342, March 1991.

104 P. Lin, H. Yu, T. Tsai, and S. Lin, "A matching-based placement and routing system for analog design," in *Proceedings of VLSI-DAT*, April 2007, pp. 1–4.

105 H. Ou, H. C. Chien, and Y. Chang, "Non-uniform multilevel analog routing with matching constraints," in *Proceedings of DAC*, June 2012, pp. 549–554.

106 P. Pan, H. Chen, Y. Cheng, J. Liu, and W. Hu, "Configurable analog routing methodology via technology and design constraint unification," in *Proceedings of ICCAD*, November 2012, pp. 620–626.

107 L. Xiao, E. F. Y. Young, X. He, and K. P. Pun, "Practical placement and routing techniques for analog circuit designs," in *Proceedings of ICCAD*, 2010, pp. 675–679.

108 C. Wu, H. Graeb, and J. Hu, "A pre-search assisted ILP approach to analog integrated circuit routing," in *Proceedings of ICCD*, October 2015, pp. 244–250.

109 H. Chi, H. Tseng, C. J. Liu, and H. Chen, "Performance-preserved analog routing methodology via wire load reduction," in *Proceedings of ASPDAC*, January 2018, pp. 482–487.

110 Q. Gao, Y. Shen, Y. Cai, and H. Yao, "Analog circuit shielding routing algorithm based on net classification," in *Proceedings of ISLPED*, August 2010, pp. 123–128.

111 M. M. Ozdal and R. F. Hentschke, "Exact route matching algorithms for analog and mixed signal integrated circuits," in *Proceedings of DAC*, November 2009, pp. 231–238.

112 M. M. Ozdal and R. F. Hentschke, "Maze routing algorithms with exact matching constraints for analog and mixed signal designs," in *Proceedings of ICCAD*, November 2012, pp. 130–136.

113 H. Ou, H. C. Chien, and Y. Chang, "Simultaneous analog placement and routing with current flow and current density considerations," in *Proceedings of DAC*, May 2013, pp. 1–6.

114 J.-W. Lin, T.-Y. Ho, and I. H.-R. Jiang, "Reliability-driven power/ground routing for analog ICs," in *ACM TODAES*, vol. 17, no. 1, pp. 6:1–6:26, 2012.

115 R. Martins, N. Lourenço, A. Canelas, and N. Horta, "Electromigration-aware and IR-drop avoidance routing in analog multiport terminal structures," in *Proceedings of DATE*, March 2014, pp. 1–6.

116 H. Chen, K. Zhu, M. Liu, X. Tang, N. Sun, and D. Z. Pan, "Toward silicon-proven detailed routing for analog and mixed-signal circuits," in *Proceedings of ICCAD*, 2020.

117 H. Chen, M. Liu, B. Xu, K. Zhu, X. Tang, S. Li, Y. Lin, N. Sun, and D. Z. Pan, "MAGICAL: An open-source fully automated analog ic layout system from netlist to GDSII," *IEEE Design & Test*, vol. 38, no. 2, pp. 19–26, 2020.

118 B. Xu, Y. Lin, X. Tang, S. Li, L. Shen, N. Sun, and D. Z. Pan, "WellGAN: Generative-adversarial-network-guided well generation for analog/mixed-signal circuit layout," in *Proceedings of DAC*, 2019, pp. 1–6.

119 K. Zhu, M. Liu, Y. Lin, B. Xu, S. Li, X. Tang, N. Sun, and D. Z. Pan, "GeniusRoute: A new analog routing paradigm using generative neural network guidance," in *Proceedings of ICCAD*, 2019, pp. 1–8.

120 M. Liu, K. Zhu, J. Gu, L. Shen, X. Tang, N. Sun, and D. Z. Pan, "Towards decrypting the art of analog layout: placement quality prediction via transfer learning," in *Proceedings of DATE*, 2020, pp. 1–6.

121 Y. Li, Y. Lin, M. Madhusudan, A. Sharma, W. Xu, S. Sapatnekar, R. Harjani, and J. Hu, "A customized graph neural network model for guiding analog IC placement," in *Proceedings of ICCAD*, 2020.

122 R. Reis, Y. Cao, and G. Wirth, *Circuit Design for Reliability*, Springer, 2015.

123 X. Li and L. Pileggi, *Statistical Performance Modeling and Optimization*, Now Publishers, 2007.

124 R. Kanj, R. Joshi, and S. Nassif, "Mixture importance sampling and its application to the analysis of SRAM designs in the presence of rare failure events," in *2006 43rd ACM/IEEE Design Automation Conference*. IEEE, 2006, pp. 69–72.

125 S. Sun, X. Li, H. Liu, K. Luo, and B. Gu, "Fast statistical analysis of rare circuit failure events via scaled-sigma sampling for high-dimensional variation space," *IEEE Transactions on Computer-Aided Design of Integrated Circuits and Systems*, vol. 34, no. 7, pp. 1096–1109, 2015.

4

Magnetoelectric Transistor Devices and Circuits with Steering Logic

Andrew Marshall[1] and Peter A. Dowben[2]

[1]*Department of Electrical Engineering, The Erik Johnson School of Engineering and Computer Science, University of Texas at Dallas, Richardson, TX, USA*
[2]*Department of Physics and Astronomy, Jorgensen Hall, University of Nebraska, Lincoln, NE, USA*

4.1 Introduction

The magnetoelectric field effect transistor (MEFET) concept has been given increasing consideration as a versatile option for beyond CMOS circuitry [1–6], and there are now some benchmarking efforts [4–7], where there has been an effort to compare the MEFET with CMOS [5]. The MEFET schemes are based on the polarization of the semiconductor channel, by the boundary polarization of the magnetoelectric gate [1–5] and thus are inherently nonvolatile spintronic devices. The advantage of the MEFET is that such schemes avoid the complexity and detrimental switching energetics associated with magnetoelectric exchange-coupled ferromagnetic devices. In addition, the canting of the interface magnetization and low barriers to domain switching can be assets to device utilization and may no longer be an issue to device implementation. In fact, an applied magnetic field is no longer essential for symmetry breaking, and deterministic switching by voltage alone is now possible [8, 9]. Spintronic MEFET devices based solely on the switching of magnetoelectric are expected to have a switching speed limited only by the switching dynamics of the magnetoelectric material, and above all are voltage-controlled spintronic devices. The basic device is shown in Figure 4.1. Such magnetoelectric spintronic devices are considered a versatile option for beyond CMOS circuitry. This device is an enhancement to the spin transistor proposed by Datta and Das [10]. In the original spin transistor, the applied gate voltage controls the channel spin precession, through electric field generated due to the spin–orbit dependent Rashba effect. This results in a change in the source-drain (IDS) current, which represents the state variable of the device.

This MEFET device could work well on gate voltages of around 100 mV or less, possesses inherent memory due to the nonvolatile AFM order of the magnetoelectric and has a sharp turn-on voltage [11, 12]. A particular advantage is that the device also has a potential on–off ratio of $\sim 10^6$ [13], which is comparable to CMOS and advantageous for implementing logic functions. It also features an extremely low switching delay of around 6 ps (or less), if the antiferromagnetic domain reversal mechanism is coherent rotation [14]. Variants of the device have been demonstrated to create conventional-style logic such as inverters, NAND, and NOR, recognizable from CMOS logic [2–5].

Advances in Semiconductor Technologies: Selected Topics Beyond Conventional CMOS, First Edition. Edited by An Chen.
© 2023 The Institute of Electrical and Electronics Engineers, Inc. Published 2023 by John Wiley & Sons, Inc.

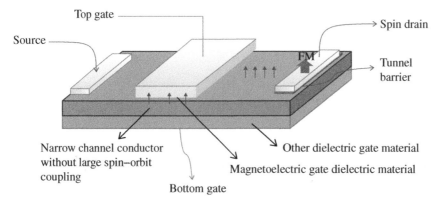

Figure 4.1 Simple MEFET gate, used in prior logic schemes [1–3, 5, 7]. Here the channel is polarized depending on the chromia spin vector orientation. The basic top-gated magnetoelectric spin-FET with a ferromagnetic (FM) source and drain. The thin channel conductor/semiconductor could be anything really, suitable to the task (graphene, TiS_3, In_4Se_3, InP, GaSb, PbS, MoS_2, WS_2, HfS_3, etc.). Source: Sharma et al. [5]/IOP Publishing.

In addition, they can be used to form very area-efficient designs for what in CMOS are compound gates, such as XOR, AND, and OR gates [4–6]. While efficient, the problem with the prior evolution of devices, based on the MEFET, is that they have aimed to optimize the logic operations at the device level [5–7].

Introduced here is a single-source, double-drain magnetoelectric FET [1], such that each gate has two outputs that allow logic to be configured with a steering function, through an anomalous Hall effect mechanism and spin–orbit coupling [15], directing the electron flux to one of the two outputs depending on the polarization of the spin current. Recent work suggests this one source, two drain device appears to be very close to realization [16–22], and it is only a matter of time before such a device is constructed with an antiferromagnetic magnetoelectric as the gate dielectric. Indeed, there are already detailed theoretical efforts to understand how such a device might work [23, 24]. The circuit architecture obtained from this type of component is termed "steering logic," and the results are a highly efficient and simple scheme for logic implementation, at the circuit level. The new magnetoelectric "steering logic" approach to circuit architecture optimizes gate level logic, resulting in greater area efficiency, performance, and leakage over conventional designs.

4.2 Simple Logic Functions with the MEFET "Steering Logic"

Steering logic in the MEFET case steers the electron flux to one or the other of two outputs, based on the local field introduced by the magnetoelectric gate on the semiconductor channel, which should have spin–orbit coupling, to make the device operational [15, 23–25]. Thus, in this steering logic MEFET, any gate should have two output ports (Out+ and Out−) to which the electrons are steered. These correspond to the logic "1" and "0" voltage or current levels in the usual transistor logic. It is important to note that when carriers enter either drains, only the charge current matters at the outputs. Using a directional spin–orbit state at the source voltage, voltage applied to the gate steers current to the left or right drain (Figure 4.2). This occurs as a result of changes to

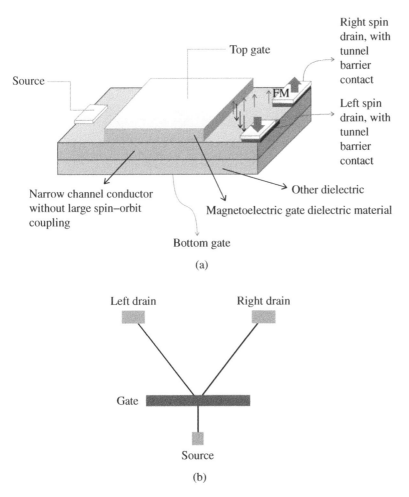

Figure 4.2 (a) 3D view and detail of operation for steering MEFET (b) Schematic view of MEFET steering device depicted in (a). Source: (a) Sharma et al. [5]/IOP Publishing and (b) Andrew Marshall.

the boundary polarization of the magnetoelectric gate, that in turn perturbs, through exchange interactions, the spin polarization of the current in the semiconductor channel [5, 15]. This can be used for transparent logic schemes.

Combining steering MEFET components allows for construction of all input combinations that impact all outputs, to potentially provide eight output possibilities for the three inputs, that is to say, with three tiers of gates, as in Figure 4.3. Depending on the voltages applied at each tier of gates (blocks labeled A, B, and C in Figure 4.3), encoding logic inputs 0 or 1, the current will be steered into one of the eight outputs. To implement a certain logic gate, the outputs need to be merged according to the truth table of the corresponding logic function. Outputs which produce the function value of 1 are merged to the "left drain" output, and the ones to the right drain have the function value of 0. As noted above, when we switch into the metal level (unlabeled blocks in Figure 4.3), we quickly lose all spin coherence. This is actually desirable, as we can then regenerate both spins as we go into the next device.

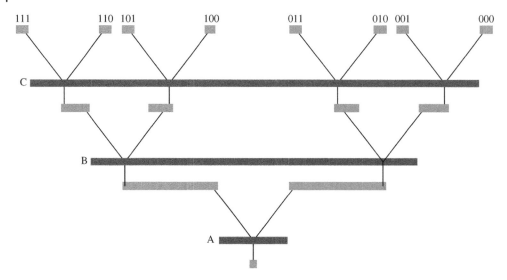

Figure 4.3 Schematic of a steering logic gate with three input gates (A, B, C). This configuration allows any output from any input. It can be stacked for any number of inputs, just by adding extra rows of devices. Source: Andrew Marshall.

AND, XOR, or NOR logic gates, are possible from two-input gates, that is to say two tiers of gates, as seen Figure 4.4. Obviously, there are different ways to merge the four outputs into the left and right drains that produce a variety of gates. This has been noted previously in the context of binary-decision logic [26–28], with a stacking of device similar to that outline in Figure 4.3 [28].

4.3 Logic Functions – Majority Gate

We can also generate logic functions such as the Majority/Minority gate (Figure 4.5 showing majority gate functions in both current and voltage output configurations). The majority function typically requires at least four conventional logic gates, but here is constructed with the same three-tier steering stack from which other three input functions are made.

It is clear that the basic majority gate of Figure 4.5 has several unused steering one source, two drain devices. We can form a gate-optimized version of the majority gate of Figure 4.5 by reviewing the stack for unused gates as well as removing any of the superfluous gates, as in Figure 4.6.

Further simplifications can be made by combining outputs that have the same function. Here, the two steering gates connected to the "C" input are doing the same logical function, so the outputs can be merged to form the resultant, much simplified, majority logic function circuit of Figure 4.7.

We can compare this new MEFET steering logic gate to prior magnetoelectric based transistor logic gate designs. Figure 4.8 shows the prior MEFET majority gate design, as described in Sharma et al. [5], based on the device illustrated in Figure 4.1, although with two device elements having their gates "split," again as described in detail elsewhere [5]. This ME-FET majority gate concept has three components (two multi-input, one single input), five gates, two leakage paths, and a single clock cycle is required for the function. The new design, based on the one source, two drain steering ME-FET majority logic function of Figure 4.7, uses four components (four single inputs), four gates, and only one leakage path. Once again, only a single clock cycle is needed for the function.

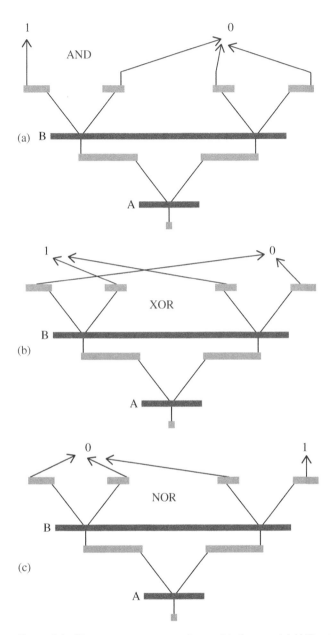

Figure 4.4 The same structure can be used to form an (a) AND gate function, (b) XOR gate function, or (c) a NOR gate function, as well as the inverse, NAND, OR, and XNOR can be similarly formed. Source: Andrew Marshall.

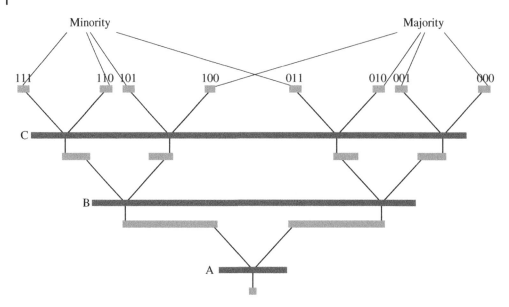

Figure 4.5 A schematic of a majority gate using the same three tier input stack of one source, two drain devices. The combination of outputs, indicated as the left drain, would be 1 (on) and the combination of outputs, indicated as the right drain, would be 0 (off). Source: Andrew Marshall.

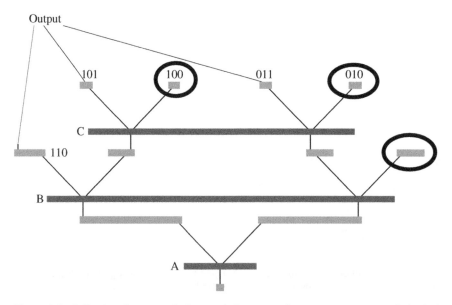

Figure 4.6 Following the removal of unneeded ore superfluous one source, two drain devices, which in this case is "on" whichever state is on the "C" input. Source: Andrew Marshall.

Figure 4.7 The schematic of the condensed majority logic function based on the steering, one source, two drain ME-FET devices, of Figure 4.2. Source: Andrew Marshall.

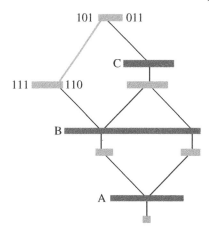

Figure 4.8 Prior version of the MEFET majority gate. This uses an AND gate, a transmission gate, and an exclusive OR in order to achieve the required function. Source: [5]. Reproduced with permission of IoP Publishing.

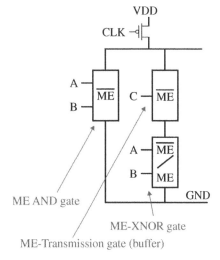

4.4 The Full Adder and the Dual XOR (Sum) Gates

The full adder is the most important digital benchmarking circuit. In CMOS, using NAND gates, the full adder requires nine gates (as in Figure 4.9a). This can be simplified if nonvolatile majority logic is available, as is possible with magnetoelectric-based transistor logic gate designs (Figure 4.9b) [5]. While the magnetoelectric devices are nonvolatile because the all-majority version of the MEFET configuration requires multiple clocks (we need to use the Carry Out function), the magnetoelectric majority gate and inverter full adder (Figure 4.9b) is a slower cell than the majority gate and XOR option, previously described for the ME-FET based on Figure 4.1, as was described in [5].

While we can, in principle, create two separate XOR devices, and cascade them, as described in [5], this is not as efficient as would be the case exploiting the one source, two drain ME-FET steering transistor of Figure 4.2. Since we only use each of the three inputs (Ain, Bin, Carry In) once for the Sum function, we can achieve the sum function in a single logical cycle using steering logic (Figure 4.10).

The original MEFET dual XOR version of the Sum function, described in Sharma et al. [5], is efficient. This MEFET dual XOR (Figure 4.11) requires just two components, based on the split

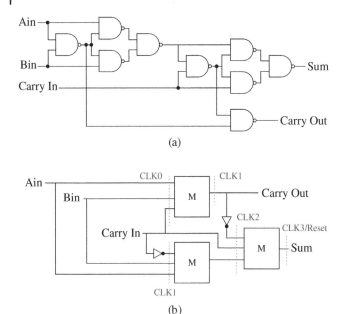

(a)

(b)

Figure 4.9 (a) is a conventional CMOS-based full adder, based on NAND gates, (b) is a majority and inverter-based full adder, where the "M" is ME-FET nonvolatile majority gates, as described in [5]. Source: Andrew Marshall.

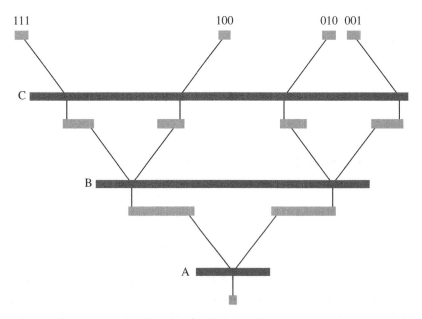

Figure 4.10 The double XOR MEFET function, requiring only one clock cycle to achieve the Sum function, based on the ME-FET of Figure 4.2. Source: Andrew Marshall.

gate magnetoelectric transistor version [3] of Figure 4.1, but this ME-FET XOR gate takes two clock cycles: two components (two multi-input), four gates, two leakage paths, and a dual clock cycle (i.e. 2°clock cycles required). The new MEFET steering logic based dual XOR by contrast requires seven components (seven single inputs), seven gates, one leakage path, but only a single clock cycle (Figures 4.10 and 4.12).

Figure 4.11 The MEFET double XOR to achieve the sum function, previously described in Sharma et al. [5]. This requires two clock cycles to complete the function and obtain an output. Source: Andrew Marshall.

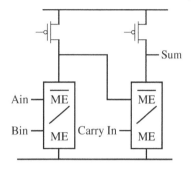

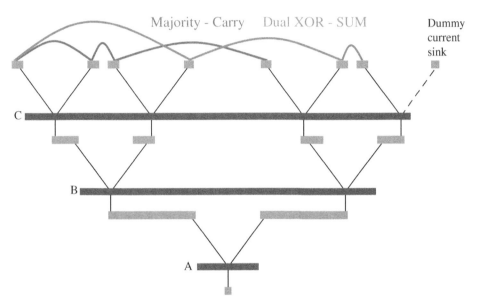

Figure 4.12 Full adder using combined functions from the steering logic. This same functionality can be obtained from simply combining the circuits in Figures 4.6 and 4.12; however, such a combination is much less efficient in terms of area, speed, and both active and passive power. Source: Andrew Marshall.

A full adder from the rendered majority and double XOR gates is inefficient when considering the entire circuit and the best way to obtain both functions is as shown in Figure 4.12. Using the ME-FET majority gate and cascaded XOR, described in Sharma et al. [5], we can build the full adder circuit using a combination of five MEFET components (four multi-input, one single input), for a total of nine gates, but this [3] has four leakage paths, and requires two clock cycles.

The one source, two drain MEFET steering logic full adder configuration (Figure 4.12), again based on the device of Figure 4.2, has seven MEFET components compared to the ME-FET full adder of five components (as in the case of Figure 4.9b) but offers other advantages. All seven components are single input devices, for a total of seven gates, but there is just one leakage path, and this full adder configuration only needs a single clock cycle to complete the full add sum and carry functionality. So while there are more components in the steering logic full adder, compared to prior ME-FET full adder concepts, there are expected improvements of the steering logic system over the prior ME-FET logic of Sharma et al. [5]. The area of these ME-FET full adders is similar, but the delay for the entire cell, based on the steering logic MEFET device concept, is now 50% of the earlier ME-FET version described in [3]. The delay-to-carry is now 20% lower because of a capacitance reduction, with the steering logic MEFET device full adder concept, and there

Table 4.1 The performance comparison of the full adder in CMOS [10], the more conventional prior MEFET logic [3] and MEFET (this work).

Technology	Energy (fJ)	Delay (ps)
High-performance CMOS	0.6	5.4
Low-power CMOS	0.07	65
High-performance MEFET (conventional)	0.16	30
Low-power MEFET (conventional)	0.085	20
High-performance MEFET (steering)	0.16	25
Low-power MEFET (steering)	0.08	16

is a reduction of leakage by 25% compared to prior ME-FET logic described in Sharma et al. [5]. A performance comparison of the full adder in CMOS, MEFET (using conventional logic configurations), and MEFET (using a steering logic configuration) assuming a 15 nm equivalent process node, as used in prior benchmarking exercises with beyond CMOS technologies [12] are shown in Table 4.1.

4.5 Latch and Memory

While the MEFET device has inherent memory applications because of nonvolatility, it has been described above in terms of conventional logic performance. A basic latch can be formed from a single device (Figure 4.13), with functionality as given in Table 4.2 and Figure 4.14. With the back-gates of CMOS, which are created by dopant diffusion in the semiconductor, and hence are difficult to electrically isolate in an area-efficient manner. In comparison, the back gate in MEFET devices can readily be isolated from adjacent devices, making an independently top and bottom gated MEFET a viable option.

In contrast, the NAND latch, used in CMOS designs, requires two NAND gates, each of which uses four transistors. The circuit for the CMOS latch is shown in Figure 4.15, with associated truth table, provided in Table 4.3.

Figure 4.13 Memory latch using a single MEFET. The top gate and back gate are the inputs and outputs in the drain. Source: Andrew Marshall.

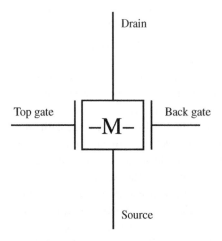

Drain

Top gate —||— –M– —||— Back gate

Source

Table 4.2 The timing diagram for operation of the MEFET when used as a memory latch is given here.

TG (A)	BG (A)	Drain (Q)
1	0	0
1/0	1/0	0
0	1	1
1/0	1/0	1

Operation is very similar to a conventional CMOS bistable latch.

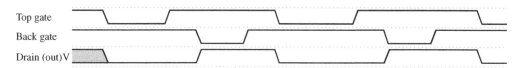

Top gate

Back gate

Drain (out) V

Figure 4.14 The timing diagram for operation of the MEFET when used as a memory latch is presented here. While this diagram shows logic level "1" as the standby state, the device can work equally with a "0" as the standby state. Source: Andrew Marshall.

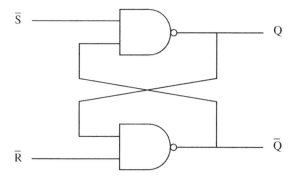

Figure 4.15 Memory latch using conventional CMOS. The latch uses a cross coupled pair of NAND gates. Its two inputs are typically referred to as S (set) and R (reset), with outputs Q and Q' or Q-bar.

Table 4.3 The timing diagram for operation of the MEFET when used as a memory latch is given here.

S	R	Q	Q'	
1	0	0	1	
1	1	0	1	After S = 1, R = 0
0	1	1	0	
1	1	1	0	After S = 0, R = 1
0	0	1	1	

Operation is very similar to a conventional CMOS bistable latch.

4.6 The JK Master–Slave Flip-Flop

By adding clocked gating to the latch inputs, we create a flip-flop. This is the basic circuit used for sequential logic. An enhancement is the two input JK master–slave flip-flop, which is two gated SR flip-flops connected in a series combination. This is schematically shown as a CMOS circuit in Figure 4.16, with the timing diagram in Figure 4.17. Here when master flip-flop level is triggered and then slave flip-flop level is triggered, so the master responds before the slave. If J = 0 and K = 1, the high Q′ output of the master goes to the K input of the slave and the clock forces the slave to reset, thus the slave copies the master. The slave has an inverted clock pulse, designed to eliminate the race condition observed in simple flip-flop devices. During the positive clock cycle the master flip-flop gives the intermediate output but the slave flip-flop does not give the final output. During the negative part of the clock cycle the slave is enabled to give the final output.

A JK master-slave flip-flop is a reliable counter module. It is readily observed that the Q output is half the frequency of the clock input. J and K act as the set and reset, and if not required can be tied to supply. As is seen, the master and slave latches are the same as shown above in Figure 4.15, and so it is very evident that we can substitute a MEFET latch in place of the CMOS latches, modifying the circuit because the MEFET only has a "Q" output, not a complimentary pair of outputs. The resultant schematic is shown in Figure 4.18.

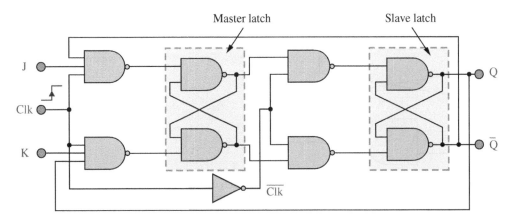

Figure 4.16 Schematic of the JK master–slave flip-flop in CMOS.

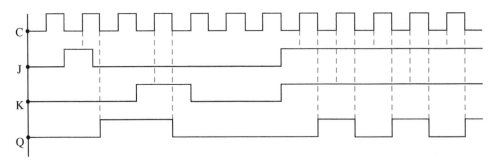

Figure 4.17 The timing diagram for operation of the JK master–slave flip-flop.

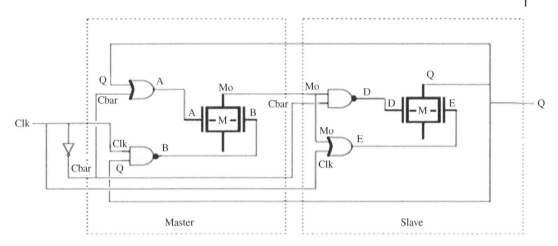

Figure 4.18 Schematic of the MEFET flip-flop. Because we do not have a Qbar output from the MEFET latch, some minor logic changes provide a more efficient scheme. Source: Andrew Marshall.

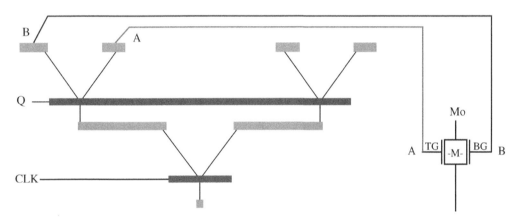

Figure 4.19 Schematic of the MEFET master section of the flip-flop, using steering logic for the support gates. Source: Andrew Marshall.

Using steering logic together with the MEFET latch gives us the schematic of Figure 4.19. Similarly, the slave section of the flip-flop is as shown in Figure 4.20.

If the circuits of Figures 4.19 and 4.20 are combined and simplified, then the result is a formation of a flip-flop, as in Figure 4.21, but out of nonvolatile magnetoelectric transistors. Figure 4.21 shows the design of the master and slave sections of the magnetoelectric transistor flip-flop, showing that significant efficiency is achieved by the use of MEFET latches and steering logic. The truth table for the flip-flop illustrated in Figure 4.21 is shown in Table 4.4.

The resulting magnetoelectric transistor JK master–slave flip-flop has only five active components, and six pull-ups, compared to the CMOS version which requires 34 components. This represents an approximately 60% area reduction, and only has to pass through six devices to switch states. In contrast, the complete path for a CMOS JK flip-flop takes seven devices. Hence, switching speed of the MEFET JK master–slave flip-flop is anticipated to be approximately 14% faster than CMOS, for equivalent performance gates.

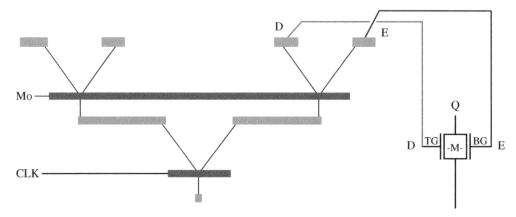

Figure 4.20 Schematic of the MEFET slave section of the flip-flop, using steering logic for the support gates. Source: Andrew Marshall.

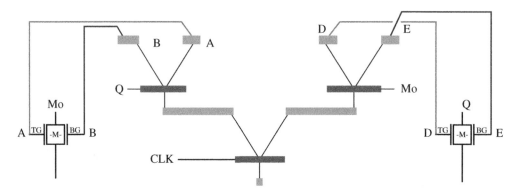

Figure 4.21 Schematic of the complete MEFET flip-flop, using steering logic for the support gates. Source: Andrew Marshall.

Table 4.4 Truth table of the complete MEFET flip-flop of Figure 4.21, showing expected frequency halving between Clk and Q.

Clk	A	B	D	E	Mo	Q
0	1	1	1	0	0	0
1	0	1	1	1	1	0
0	1	1	0	1	1	1
1	1	0	1	1	0	1
0	1	1	1	0	0	0
1	0	1	1	1	1	0
0	1	1	0	1	1	1

4.7 Conclusion

We have presented a new magnetoelectric transistor concept that lends itself to efficient logic architecture, based on MEFET anomalous Hall effect steering logic, where all possible outputs, from a set of static inputs, are available. These outputs can be monitored to generate a variety of logical output functions. We find that in the case of the full adder benchmarking circuit, we achieve a reduction in delay time from between 20% and 50%, while at the same time reducing leakage power by 75%. These improvements are achieved in a similar die area. This use of the MEFET anomalous Hall effect steering logic also permits multiple logical functions to be generated from the same circuitry, by tapping off the different outputs.

We have also demonstrated how the MEFET can be used as a latch, and how an efficient JK master–slave flip-flop can be created from magnetoelectric transistor latches and steering logic, reducing the component count from the CMOS required 34 MOS devices down to five active magnetoelectric transistor devices and six pull-up devices. Together these components form the basics of logical systems, thus any required logic function can be efficiently made using magnetoelectric FET technology.

Acknowledgments

Supported by nCORE, a wholly owned subsidiary of the Semiconductor Research Corporation (SRC), through the Center on Antiferromagnetic Magnetoelectric Memory and Logic task ID 2760.001, 2760.002, and 2760.003 and by the National Science Foundation (NSF), through Grant No. NSF-ECCS 1740136. The authors are grateful to Dmitri Nikonov for some seminal contributions to this work and Ian Young for pointing out pertinent background to this work.

References

1 P. A. Dowben, C. Binek, and D. E. Nikonov, Chapter 11: Potential of nonvolatile magnetoelectric devices for spintronic applications. In: S. Oda and D. Ferry (eds.), *Nanoscale Silicon Devices*, Taylor and Francis, London, pp. 255–278, 2016.

2 P. A. Dowben, C. Binek, K. Zhang, L. Wang, W.-N. Mei, J. P. Bird, U. Singisetti, X. Hong, K. L. Wang, and D. Nikonov, "Towards a strong spin–orbit coupling magnetoelectric transistor," *IEEE Journal on Exploratory Solid-State Computational Devices and Circuits*, vol. 4, no. 1, pp. 1–9, 2018.

3 P. A. Dowben, D. E. Nikonov, A. Marshall, and Ch. Binek, "Magneto-electric antiferromagnetic spin-orbit logic devices," *Applied Physics Letters*, vol. 116, 080502, 2020.

4 C. Pan and A. Naeemi, "Complementary logic implementation for antiferromagnet field-effect transistors," *IEEE Journal on Exploratory Solid-State Computational Devices and Circuits*, vol. 4, no. 2, pp. 69–75, 2018.

5 N. Sharma, J. P. Bird, Ch. Binek, P. A. Dowben, D. Nikonov, and A. Marshall, "Evolving magneto-electric device technologies," *Semiconductor Science and Technology*, vol. 35, 073001, 2020.

6 N. Sharma, A. Marshall, J. Bird, and P. A. Dowben, "Verilog-A Based Compact Modeling of the ME-FET device", in *2017 Fifth Berkeley Symposium on Symposium & Steep Transistors Workshop (E3S)*, IEEE Xplore INSPEC Accession Number: 17453928, doi: 10.1109/E3S.2017.8246186, 2018.

7 Sharma, N., Binek, C., Marshall, A., Bird, J. P., Dowben, P. A., and Nikonov, D., "Compact Modeling and Design of Magneto-Electric Transistor Devices and Circuits, in *2018 31st IEEE International System-on-Chip Conference (SOCC)*, Arlington, VA, USA, IEEE Explore, pp. 146–151, 2018.

8 A. Mahmood, W. Echtenkamp, M. Street, J. -L. Wang, S. Cao, T. Komesu, P. A. Dowben, P. Buragohain, H. Lu, A. Gruverman, A. Parthasaraty, S. Rakheja, and C. Binek, "Voltage controlled Néel vector rotation in zero magnetic field at CMOS-compatible temperatures," *Nature Communications*, vol. 12, 1674, 2021. doi: 10.21203/rs.3.rs-38435/v1, 2021.

9 E. Y. Vedmedenko, R. K. Kawakami, D. Sheka, P. Gambardella, A. Kirilyuk, A. Hirohata, C. Binek, O. Chubykalo-Fesenko, S. Sanvito, B. J. Kirby, J. Grollier, K. Everschor-Sitte, T. Kampfrath, C.-Y. You, and A. Berger, "The 2020 magnetism roadmap," *Journal of Physics D: Applied Physics* vol. 53, 453001, 2020.

10 S. Datta and B. Das, "Electronic analog of the electro-optic modulator," *Applied Physics Letters* vol. 56, pp. 665–667, 1990.

11 W. Echtenkamp and C. Binek, "Electric control of exchange bias training," *Physical Review Letters,* vol. 111, 187204, 2013.

12 A. Iyama and T. Kimura, "Magnetoelectric hysteresis loops in Cr_2O_3 at room temperature," *Physical Review B*, vol. 87, 180408, 2013.

13 H. -J. Chuang, B. Chamlagain, M. Koehler, M. Madusanka Perera, J. Yan, D. Mandrus, D. Tománek, and Z. Zhou, "Low-resistance 2D/2D Ohmic contacts: a universal approach to high-performance WSe_2, MoS_2, and $MoSe_2$ transistors," *NanoLetters* vol. 16, pp. 1896–1902.

14 A. Parthasarathy and S. Rakheja "Dynamics of magnetoelectric reversal of an antiferromagnetic domain," *Physics Review Applied,* vol. 11, 034051, 2019.

15 J. Qi, X. Li, Q. Niu, and J. Feng, "Giant and tunable valley degeneracy splitting in $MnTe_2$," *Physical Review B,* vol. 92, 121403(R), 2015.

16 A. M. Afzal, K. H. Min, B. M. Ko, and J. Eom "Observation of giant spin–orbit interaction in graphene and heavy metal heterostructures," *RSC Advances,* vol. 9, pp. 31797–31805, 2019.

17 A. Avsar, J. Y. Tan, T. Taychatanapat, J. Balakrishnan, G. K. W. Koon, Y. Yeo, J. Lahiri, A. Carvalho, A. S. Rodin, E. C. T. O'Farrell, G. Eda, A. H. Castro Neto, and Özyilmaz, B., "Spin–orbit proximity effect in graphene," *Nature Communications,* vol. 5, 4875, 2014.

18 J. Balakrishnan, G. K. W. Koon, A. Avsar, Y. Ho, J. H. Lee, M. Jaiswal, S. -J. Baeck, J. -H. Ahn, A. Ferreira, M. A. Cazalilla, A. H. Castro Neto, and B. Özyilmaz "Giant spin Hall effect in graphene grown by chemical vapour deposition," *Nature Communications*, vol. 5, 4748, 2014.

19 X. Cui, G. -H. Lee, Y. D. Kim, G. Arefe, P. Y. Huang, C. -H. Lee, D. A. Chenet, X. Zhang, L. Wang, F. Ye, F. Pizzocchero, B. S. Jessen, K. Watanabe, T. Taniguchi, D. A. Muller, T. Low, P. Kim, and J. Hone, "Multi-terminal transport measurements of MoS_2 using a van der Waals heterostructure device platform," *Nature Nanotechnology* vol. 10, 534, 2015.

20 J. Park, H. Duk Yun, M. -J. Jin, J. Jo, I. Oh, V. Modepalli, S. -Y. Kwon, and J. -W. Yoo, "Gate-dependent spin Hall induced nonlocal resistance and the symmetry of spin–orbit scattering in Au-clustered graphene," *Physical Review B,* vol. 95, 245414, 2017.

21 Z. Wang, D.-K. Ki, H. Chen, H. Berger, A. H. MacDonald, and A. F. Morpurgo, "Strong interface-induced spin–orbit interaction in graphene on WS_2," *Nature Communications*, vol. 6, 8339, 2015.

22 Z. Wang, C. Tang, R. Sachs, Y. Barlas, and J. Shi, "Proximity-induced ferromagnetism in graphene revealed by the anomalous Hall effect," *Physical Review Letters* vol. 114, 016603, 2015.

23 Z. Qiao, W. Ren, H. Chen, L. Bellaiche, Z. Zhang, A. H. MacDonald, and Q. Niu, "Quantum anomalous Hall effect in graphene proximity coupled to an antiferromagnetic insulator," *Physical Review Letters* vol. 112, 116404, 2014.

24 H. Takenaka, S. Sandhoefner, A. A. Kovalev, and E. Y. Tsymbal, "Magnetoelectric control of topological phases in graphene," *Physical Review B* vol. 100, 125156, 2019.

25 L. -H. Wu and X. Hu, "Scheme for achieving a topological photonic crystal by using dielectric material," *Physical Review Letters*, vol. 114, 223901, 2015.

26 N. Asahi, M. Akazawa, and Y. Amemiya "Binary-decision-diagram device," *IEEE Transactions on Electron Devices*, vol. 42, no. 11, pp. 1999–2003, 1995.

27 N. Asahi, M. Akazawa, and Y. Amemiya "Single-electron logic device based on the binary decision diagram," *IEEE Transactions on Electron Devices*, vol. 44, no. 7, pp. 1109–1116, 1997.

28 S. Kasai and H. Hasegawa "A single electron binary-decision-diagram quantum logic circuit based on Schottky wrap gate control of a GaAs nanowire hexagon," *IEEE Electron Device Letters* vol. 23, no. 8, pp. 446–448, 2002.

5

Nonvolatile Memory Based Architectures Using Magnetoelectric FETs

Shaahin Angizi[1], Deliang Fan[2], Andrew Marshall[3], and Peter A. Dowben[4]

[1]*Department of Electrical and Computer Engineering, New Jersey Institute of Technology, Newark, NJ, USA*
[2]*Department of Electrical, Computer and Energy Engineering, Arizona State University, Tempe, AZ, USA*
[3]*Department of Electrical Engineering, The Erik Johnson School of Engineering and Computer Science, University of Texas at Dallas, Richardson, TX, USA*
[4]*Department of Physics and Astronomy, Jorgensen Hall, University of Nebraska, Lincoln, NE, USA*

5.1 Introduction

Magnetoelectric field effect transistor (MEFET) devices have been proposed for logic applications [1–8]. The anticipated operating range, which is expected to be higher than 125 °C, covers most application ranges. Here we investigate the use of MEFET in one transistor plus one MEFET (1T-1M) configuration, which has the advantage of a simple, two-component cell for reduced area arrays. Nonvolatile memories (NVM) are being actively explored due to their robustness, low stand-by leakage, and high integration density [9–13]. By substituting the MEFET structure (Figure 5.1), in place of the capacitor in a DRAM cell, we may achieve the benefits of nonvolatile memory (Figure 5.2), in approximately the same layout footprint that would normally be occupied by DRAM volatile memory. This forms a hybrid cell made of a ME-FET and an n-channel metal-oxide semiconductor (NMOS) device. The use of a nonvolatile element reduces the complexity of the read and write circuit, while retaining the read performance, though at some expense in write time. Prior work on logic design based on ME devices, such as ME–MTJ [14, 15] and MEFET [1–8], has shown that ME-based devices have utility as logic devices. In this context, there are good reasons to investigate various memory functions based on the MEFET.

5.2 Magnetoelectric Field Effect Transistor (MEFET)

In its most common form, the MEFET is a four-terminal device, which superficially is similar to the conventional complementary metal-oxide semiconductor (CMOS) FET device. Figure 5.1 shows the basic single source version of MEFET as a four-terminal device with gate, source, drain, and back gate terminals [2, 3, 8]. The device is a stacked structure of a narrow semiconductor channel sandwiched by two dielectrics, i.e. a magnetoelectric (ME) material (typically chromia) and an insulator such as alumina (Al_2O_3). There are two electrodes contacting the structure, to the top gate via the ME layer and at the bottom via a back gate alumina layer, although reversing the top

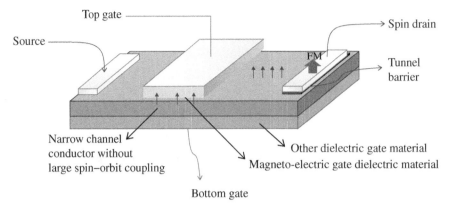

Figure 5.1 (a) Simple MEFET gate, used in prior logic schemes [6–8]. Here the channel is polarized depending on the chromia spin vector orientation. The basic top-gated magnetoelectric spin-FET with a ferromagnetic (FM) source and drain. The thin channel conductor/semiconductor (dark gray) could be any suitable material, e.g. (graphene, TiS_3, In_4Se_3, InP, GaSb, PbS, MoS_2, WS_2, HfS_3, etc.) [2, 3, 8], and (b) Magnetoelectric spin-FET symbol. Source: (a) Reproduced from Sharma et al. [8], with permission, copyright 2020 Institute of Physics Publishing.

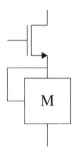

Figure 5.2 The proposed MEFET 1T-1M memory circuit scheme. This architecture is similar to DRAM, but is nonvolatile.

and bottom gate dielectric results in an equally realistic device. The channel is a 2D semiconductor with large spin–orbit coupling [2, 3, 8], like tungsten diselenide (WSe_2), or HfS_3, which potentially provides an on–off ratio of 10^6, and has high hole mobility (WSe_2) or electron mobility (HfS_3) compared to silicon FETs. The drain is a fixed magnetization ferromagnet, for improved spin fidelity (but which may not be essential [2, 3, 8]). This forms a hybrid cell, of ME-FET and NMOS device, which is acceptable, since ME-FET devices can be constructed as a back-end process adder onto a processed silicon CMOS wafer.

The MEFET operates based on the programming of the semiconductor channel polarization, by the boundary polarization of the ME gate, through the proximity effect. As a result, the channel can be polarized by the ME layer with extremely low voltages of around $100\,mV$ [6–8] at the ME layer gate, while the bottom gate is grounded. Depending on whether a positive or negative voltage is applied to the gate, a vertical field across the gate is created that changes the direction of orientation of chromia boundary spin polarization through magnetoelectric coupling. The ME boundary polarization can have an exchange interaction with the semiconductor channel to generate a spin current, especially if the semiconductor channel is very thin, e.g. a 2D material [2, 3, 8].

5.3   1T-1M Memory Design Based on the MEFET

A MEFET device model has been developed [6, 7]. While this model was principally intended for logical element simulations, it is also effective in-memory applications [9]. The proposed non-volatile 1T-1M random-access memory (RAM) bit-cell, consists of one MEFET as the main storage element and one access transistor, as shown in Figure 5.2. The memory bit cell has a combination read/write path which reduces cell area.

Each cell (Figure 5.3) is controlled by three controlling signals: the Word line (WL), which controls the gates of the cells; the Bit Line (BL), which controls the supply and sense to the cell; and the Source Line (SL), which on a conventional DRAM typically connects to ground.

5.3.1   Read Operation

The concept behind 1T-1M's read operation is to sense the resistance of the selected memory cell and compare it to a reference resistance using a sense amplifier (SA), as shown in Figure 5.4. At the array level, as shown in Figure 5.3, the row and column decoders activate the Word and Bit paths, respectively, while SL is grounded. When a memory cell is selected, by applying a very small sense voltage to the BL, a current is generated through the cell. Owing to the low- or high-resistance state of the selected 1T-1M RAM bit-cell, the sense current is lower or higher than the reference, thus, through setting the appropriate reference the SA outputs binary "1"or "0."

Figure 5.3   A MEFET 1T-1M array configuration, showing the cell implementation in the array.

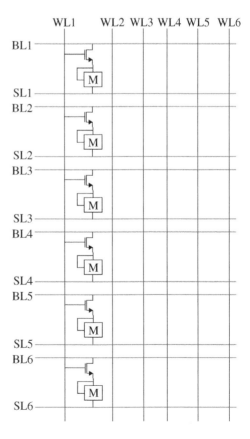

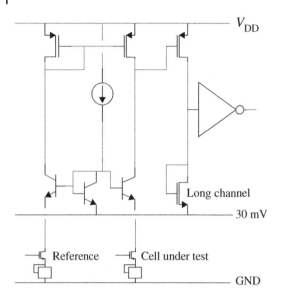

Figure 5.4 The simplified schematic of the sense amplifier for the MEFET 1T-1M memory.

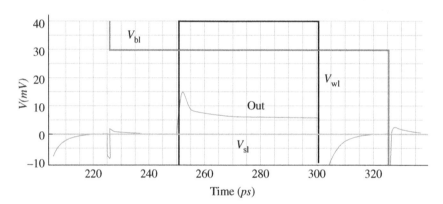

Figure 5.5 The simulation results showing data read time is about 10 ps. This depends on the access transistor.

For the access transistor, the most efficient device would appear to be a ME-FET compatible device, suited to a sub 0.2 V region of operation. The design proposed here uses an NMOS device. Simulation results show that the read time is about 10 ps (Figure 5.5), with a read power of just 0.15 aJ. The reference can be either a stack of "on" state MEFET cells (typically about five), along with a dummy access transistor, or, since there is a large difference between on and off state MEFET resistance, an appropriate resistance may be substituted.

An alternative sense amplifier design is shown in Figure 5.6. This has the advantage of not requiring a 30 mV reference supply but is somewhat larger in area.

5.3.2 Write Operation

The write operation, shown in Figure 5.7, is accomplished by activating the WL at the same time while making the SL grounded and asserting appropriate write voltage through voltage driver

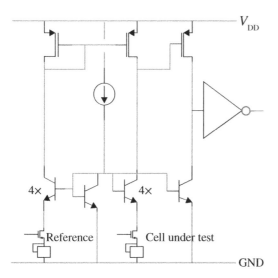

Figure 5.6 The simplified schematic of an alternative sense amplifier design for the MEFET 1T-1M memory, for when a 30 mV reference supply is not available.

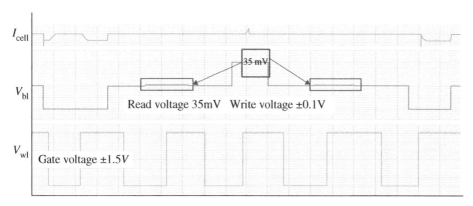

Figure 5.7 Biasing for program and read of the double-diffused metal-oxide-semiconductor (DMOS)-like MEFET cell.

on BL. The programming voltage provides sufficient vertical electrical field across the gate to switch the spin vectors of the underlying chromia layer, thus flipping the boundary polarization and thus the spins in the channel of MEFET. The SL is grounded for the entire write period. The biasing condition for the write operation is shown in Figure 5.7, with write times shown in Figure 5.8.

Since generating a negative programming voltage is complex on chip, an option is to use a positive voltage on the SL, while the BL is held more negative. Exact conditions vary depending on the transistor technology and are not considered in detail here.

Simulations were made using the Cadence Spectre simulator with a commercial Spice library used for MOS devices, while for the MEFET, a custom Verilog-A model was used, adapted from use as a nonvolatile component used for logic simulation [6–8].

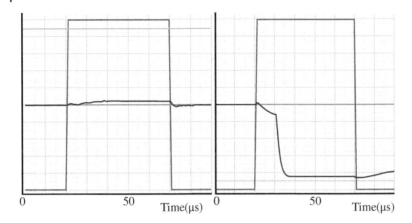

0   50   Time(µs)   0   50   Time(µs)

Figure 5.8 Program (write) time to program high/low. Programming time is around 12 µs (this is partially dependent upon the access device), with an energy for programing being 5.5 aJ.

5.4 2T-1M Memory Design Based on the MEFET

The presented nonvolatile 2T-1M RAM bit-cell [9] consists of one MEFET as the main storage element and two access transistors, as shown in Figure 5.9a. By virtue of the three-terminal structure of MEFET, depicted in Figure 5.1b, we designed the memory bit-cell to have separate read and write paths which facilitates independent optimization of both operations, as well as

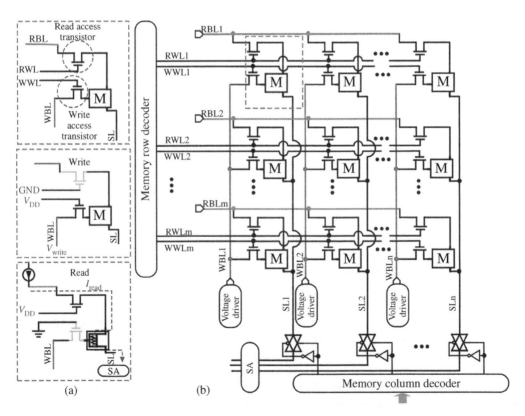

Figure 5.9 (a) The 2T-1M bit-cell with Read and Write signals, (b) an m × n 2T-1M array with peripheral circuitry.

avoiding common read-write conflicts in many 1T-1R resistive nonvolatile memory designs. Each cell is controlled by five controlling signals, i.e. Write Word Line (WWL), Write Bit Line (WBL), Read Word Line (RWL), Read Bit Line (RBL), and SL. The read/write access transistor is controlled by a RWL/WWL enabling selective read/write operation on the cells located within one row. An $m \times n$ array, developed based on the 2T-1M bit-cell, is shown in Figure 5.9b. The WLs and RBLs are shared among cells within the same row and WBLs, while SLs are shared between cells in the same column. The WLs are controlled by the memory row decoder (active-high output). The column decoder (active-high output) controls the activation of the read current path through the SL. The voltage driver component is designed to set the proper voltage on the WBL. We explain the detailed read and write operations, respectively, below (*vide infra*).

5.4.1 Read Operation

The concept behind 2T-1M's read operation is to sense the resistance of the selected memory cell and compare it by a reference resistor using a sense amplifier (SA), as shown in Figure 5.9a. At the array level, shown in Figure 5.10a, the row and column decoders activate the RWL and SL paths, respectively. When a memory cell is selected, by applying a very small sense current (sub-micro) to RBL, a voltage (V_{sense}) is generated on the corresponding SL, which is taken as the input of the sense circuit, as shown in Figure 5.10b. Owing to the low- or high-resistance state of the selected 2T-1M bit-cell (R_{M1}), the sense voltage is V_{low} or V_{high} ($V_{low} < V_{high}$), respectively. Thus, through setting the reference voltage at ($V_{low} + V_{high}$)/2, the SA outputs binary "1" when $V_{sense} > V_{ref}$, whereas the output is "0." We designed and tuned the sense circuit based on StrongARM latch shown in Figure 5.10c. Each read operation requires two clock phases: pre-charge (CLK "high") and sensing (CLK "low"). During the precharge phase, both SA's outputs are reset to ground potential. Then, in

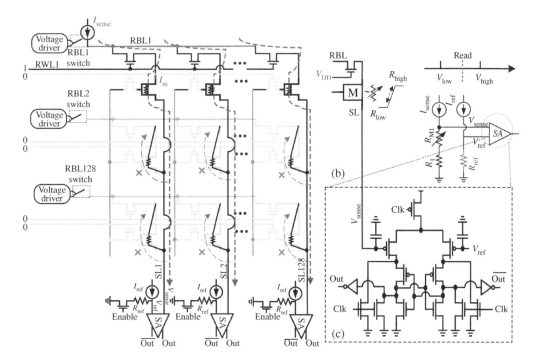

Figure 5.10 (a) The array-level read operation. Read access transistors isolate the unaccessed cells avoiding sneak paths. (b) The idea of voltage comparison between V_{sense} and V_{ref} for memory read, (c) The sense amplifier (SA).

the sensing phase, the input transistors provide various charging currents based on the gate biasing voltages (V_{sense} and V_{ref}), leading to various switching speeds for the latch's cross-coupled inverters.

5.4.2 Write Operation

The write operation, shown in Figure 5.9a, is accomplished by activating WWL and asserting appropriate bipolar write voltage ($V_{write}/-V_{write}$) through voltage driver on WBL. As detailed above, the voltage provides a vertical electrical field magnitude across the gate to switch the spin vectors of the underlying chromia layer, thus causing 180° rotation of the boundary polarization which would then flip the spins within the channel of MEFET through exchange coupling. The RBL and SL do not require grounding for the write period. Unaccessed rows are isolated by driving the corresponding WLs to GND through the row decoder. This is necessary to avoid unwanted current paths that can cause false write/read states.

Figure 5.11 shows the transient simulation results of a 2T-1M cell located in a 256×256 subarray, based on the architecture shown in Figure 5.9b. Here, we consider two experimental scenarios for the write operation, as indicated by the solid and the dotted lines in Figure 5.11. For clarity, we assumed a 3 ns period clock that synchronizes the write and read operation. However, a <1 ns

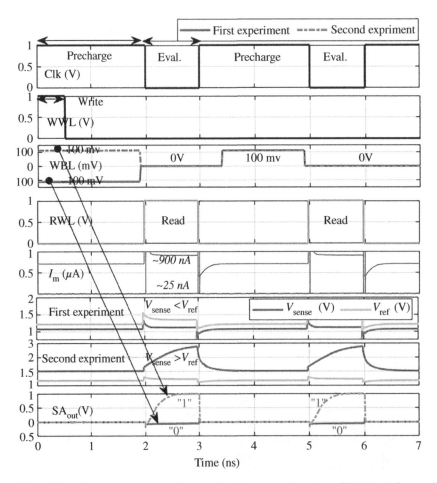

Figure 5.11 The transient simulation results of two experiments on 2T-1M cell. Source: Reproduced from Angizi et al. [9], with permission, copyright 2021 Association for Computing Machinery.

period can be used for a reliable read operation. During the precharge phase of SA (Clk = 1), the V_{write} voltage is set (=−100 mV in the first experiment or +100 mV in the second experiment) and applied to the WBL to change the MEFET resistance to $R_{low} = 1.05\,k\Omega$ or $R_{high} = 63.4\,M\Omega$. Before the evaluation phase (Eval.) of SA, WWL, and WBL is grounded, while RBL is fed by the very small sense current, $I_{sense} = 900\,nA$. In the evaluation phase, RWL goes high, and depending on the resistance state of 2T-M bit-cell and accordingly SL, V_{sense} is generated at the first input of SA, when V_{ref} is generated at the second input of SA. The comparison between V_{sense} and V_{ref} for both experiments is plotted in Figure 5.11.

We observe that when $V_{sense} < V_{ref}$ (first experiment), the SA generates the binary output "0," whereas the output is "1" (second experiment). As can be seen, the same value can be sensed again in the next sensing cycle (3–6 ns) regardless of WBL voltage since WWL is deactivated, so the 2T-1M bit-cell remains unchanged.

5.5 MEFET Steering Memory

In addition to the relatively conventional circuit options for memory, we also consider here a new memory option, based on the steering version of the MEFET [4]. The steering MEFET is based on the conventional MEFET of Figure 5.1 but has a dual drain. The gate boundary spin polarization steers the charge either to the right or left, though the channel spin orbit coupling, as seen in Figure 5.12a, with the schematic view of the device shown in Figure 5.12b. A similar device can be created that achieves the inverse function, utilizing two source inputs to give one drain output, i.e. a multiplexer [8].

Combining steering MEFETs, as in Figure 5.13, allows all input combinations to impact all outputs, to potentially give eight outputs for the three inputs. It is important to note that when we switch into the metal level, we quickly lose all spin coherence. This is actually desirable, as we can then regenerate both spins as we go into the next device. This configuration provides a discrete output from any combination of inputs. It can be stacked for any number of inputs, just by adding extra rows of devices. Since the device inherently retains a memory of the last input state, by polling each output, we can determine the input state.

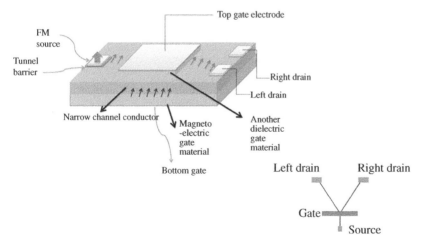

Figure 5.12 (a) Steering MEFET gate, used in prior logic schemes. Source: Modified on Marshall and Dowben [4]. The channel is polarized according to the chromia spin vector orientation, and (b) steering magnetoelectric spin-FET symbol.

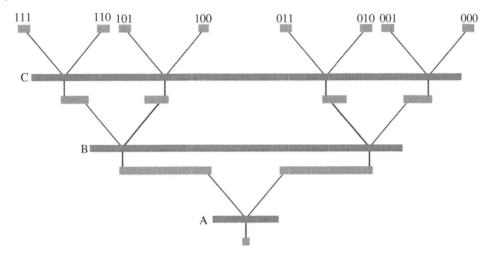

Figure 5.13 Stacked Steering MEFET gates, showing input state options and outputs. Source: From Marshall and Dowben [4] ©2020 IEE.

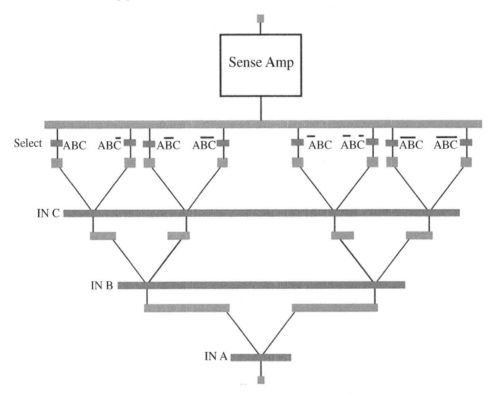

Figure 5.14 Sequential read allows only one sense-amp for better area efficiency, but this is a slow read technique.

Memory Read Options

1. **Sequencing**: In Figure 5.14, we see a stack of devices to store the data, as above in Figure 5.13. In this case, we use the example of three bits of data. These are read through a single Sense-amp, depending on the state of the select inputs. Here the inputs must sequence through the states, or home in on the correct state using a binary search. This is area efficient, but slow.

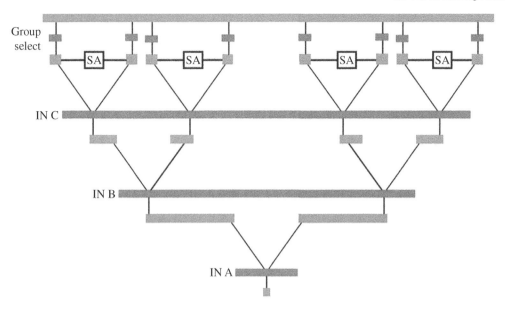

Figure 5.15 This provides faster performance than the circuit of Figure 5.11, at the expense of die area.

2. **Steering CACHE**: The cache-style steering memory, of Figure 5.15, provides faster performance. Here individual sense-amplifiers detect the logic state for each output. This provides for a fast data read, at the expense of the die area.

3. **Modified steering CACHE**: A modified version of the steering memory may provide improved performance while reducing the number of sense amplifiers needed (Figure 5.16). In this instance, the outputs are read through a modified bank of Sense-amps, depending on the state of the select inputs, reducing the read time by about 50%. This results a somewhat smaller area array than the basic steering CACHE. The Sense amplifier in this case compares adjacent states of the output, and so must not only identify when a different output is detected, but also not give a false positive.

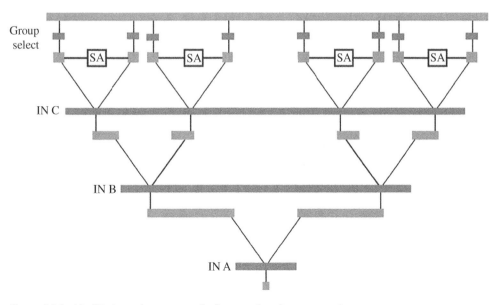

Figure 5.16 Modified steering memory for improved performance and area.

5.6 Evaluation

A comparison between MEFET 1T-1M memory, MEFET Steering memory, MEFET 2T-1M, and other memory types requires an analysis of comparative read and write times, energies and device area. This is summarized in Table 5.1. The values presented in Table 5.1 are perhaps a little more optimistic than prior efforts of this type [11], but similar nonetheless [9].

5.6.1 Comparative Read Time

The MEFET memory cells lie in an intermediate position; not as fast as static random-access memory (SRAM), which benefits from a rail-to-rail and bipolar output state, but faster than NAND flash, due to the stacked nature of the NAND-flash cell. They are also somewhat faster than conventional DRAM since DRAM is looking at a small charge on a capacitor, as opposed to a resistance change.

Speed performance varies by architecture but is approximately the same as the read time of the STT-MRAM cell. Actual read time will depend upon the array size and configuration, together with sense amplifier performance, so from that perspective, we would expect a high-speed read capability due to the small unit cell size of the MEFET cells.

5.6.2 Comparative Write Time

Write time depends significantly on the magnetoelectric device and the array resistance and capacitance. At the cell level, the MEFET cells programs in about 20 ps. In practice, the resistance-capacitance of the array dominates this timing, so we should expect programming speeds similar to DRAM. SRAM is again faster, due to the cell's inherent gain.

However, a large advantage is seen with ME devices when compared to STT-MRAM. The STT-MRAM programming time is of the order of 1 ns minimum, and more generally is longer, in the range 2–5 ns [10]. Nand-flash has the longest programming time of the major semiconductor memories. The write power required to change the state of ME devices is also relatively low for MEFET memory compared to other high-density memory.

5.6.3 Comparison of Cell Areas

Since the ME-FET memory cells are a simple, three-terminal, nonvolatile device, they have an inherent area advantage over SRAM. While not as dense as NAND-flash, they can, at best be similar to DRAM in the area.

Table 5.1 Cell level memory technology parameters at 15 nm technology equivalent.

Parameter	SRAM	DRAM	NAND-Flash	STT-MRAM	ME:1T1ME	ME:2T1ME	ME-steering
Volatile	Yes	Yes	No	No	No	No	No
Cell size (F^2)	50–100	6–10	5–6	8–20	8–12	13–18	5–10
Read time (ps)	10–20	100	1000	4–80	20–30	15–25	50–70
Write time (ps)	10–20	200	10^6	1000	20–30	20–30	20–30
Write power	Low	Medium	High	Medium	Low	Low	Low

Source: Based on Refs. [4, 8–10, 12, 13].

In addition, depending on the configuration, there may be an extra supply line which will potentially increase the footprint a little more. The SRAM is by far the largest of the cells, with flash being smaller and STT-MRAM around the same size as most ME memory. While these cell area comparisons have been tabulated at 15 nm, we expect the ME devices can be constructed down to geometries of 5 nm, and possibly smaller.

5.7 Conclusion

We have presented several nonvolatile MEFET memory bit-cell configurations. The MEFET designs are shown to be competitive with non-MEFET nonvolatile designs in die area and performance, especially read time. Write time is also faster than NAND flash and STT-MRAM. MEFET thus shows promise as a nonsilicon-based memory architecture.

Acknowledgments

This work was supported by the National Science Foundation under Grants No. 2005209 and No. 2003749, and EPSCoR RII Track-1: Emergent Quantum Materials and Technologies (EQUATE), Award OIA-2044049. We wish to thank Nishtha Sharma for developing the MEFET model used for this evaluation. The authors are also grateful to Dmitri Nikonov for some seminal contributions to this work and Ian Young for pointing out the pertinent background.

References

1 P. A. Dowben, C. Binek, and D. E. Nikonov, Chapter 11: Potential of nonvolatile magneto-electric devices for spintronic applications. In: S. Oda and D. Ferry (eds.), *Nanoscale Silicon Devices*. London: Taylor and Francis, pp. 255–278, 2016.

2 P. A. Dowben, C. Binek, K. Zhang, L. Wang, W.-N. Mei, J. P. Bird, U. Singisetti, X. Hong, K. L. Wang, and D. Nikonov, "Towards a strong spin–orbit coupling magnetoelectric transistor," *IEEE Journal on Exploratory Solid-State Computational Devices and Circuits*, vol. 4, pp. 1–9, 2018.

3 P. A. Dowben, D. E. Nikonov, A. Marshall, and Ch. Binek, "Magneto-electric antiferromagnetic spin-orbit logic devices," *Applied Physics Letters*, vol. 116, 080502, 2020.

4 A. Marshall and P. A. Dowben, "Magneto-electric transistor devices and circuits with steering logic," in *2020 IEEE 14th Dallas Circuits and Systems Conference (DCAS)*, IEEE Xplore 1–4, INSPEC Accession Number: 20384727, 2021, doi: 10.1109/DCAS51144.2020.9330662.

5 C. Pan and A. Naeemi, "Complementary logic implementation for antiferromagnet field-effect transistors," *IEEE Journal on Exploratory Solid-State Computational Devices and Circuit*, vol. 4, pp. 69–75, 2018.

6 N. Sharma, A. Marshall, J. Bird, and P. A. Dowben, "Verilog-A based compact modeling of the ME-FET device," in *2017 Fifth Berkeley Symposium on Symposium & Steep Transistors Workshop (E3S)*, IEEE Xplore (2018), INSPEC Accession Number: 17453928, 2018, doi: 10.1109/E3S.2017.8246186.

7 N. Sharma, C. Binek, A. Marshall, J. P. Bird, P. A. Dowben, and D. Nikonov, "Compact modeling and design of magneto-electric transistor devices and circuits," in: *2018 31st IEEE*

International System-on-Chip Conference (SOCC), Arlington, VA, USA, IEEE Explore, 2018, pp. 146–151.

8 N. Sharma, J. P. Bird, Ch. Binek, P. A. Dowben, D. Nikonov, and A. Marshall, "Evolving magneto-electric device technologies," *Semiconductor Science and Technology*, vol. 35, 073001, 2020.

9 S. Angizi, N. Khoshavi, A. Marshall, P. Dowben, and D. Fan, "MeF-RAM: a new non-volatile cache memory based on magneto-electric FET," *ACM Transactions on Automation of Electronic Systems*, vol. 27, p. 18, 2021.

10 K. C. Chun, H. Zhao, J. D. Harms, T. -H. Kim, J. -P. Wang, and C. H. Kim, "A scaling roadmap and performance evaluation of in-plane and perpendicular MTJ based STT-MRAMs for high-density cache memory," *IEEE Journal of Solid State Circuits*, vol. 48, pp. 598–610, 2013.

11 Y. -C. Liao, C. Pan, and A. Naeemi, "Benchmarking and optimization of spintronic memory arrays," *IEEE Journal on Exploratory Solid-State Computational Devices and Circuits*, vol. 6, pp. 9–17, 2020.

12 A. Shafaei, Y. Wang, and M. Pedram, "Low write-energy STT-MRAMs using FinFET-based access transistors," in *2014 IEEE 32nd International Conference on Computer Design (ICCD)*, IEEE EXPLORE, INSPEC Accession Number: 14806035, 2014, doi: 10.1109/ICCD.2014.6974708.

13 K. L. Wang, J. G. Alzate, and P. K. Amiri, "Low-power non-volatile spintronic memory: STT-RAM and beyond," *Journal of Physics D: Applied Physics*, vol. 46, 074003, 2013.

14 D. E. Nikonov and I. A. Young, "Benchmarking of beyond-CMOS exploratory devices for logic integrated circuits," *IEEE Journal on Exploratory Solid-State Computational Devices and Circuits*, vol. 1, pp. 3–11, 2015.

15 N. Sharma, J. P. Bird, P. A. Dowben, and A. Marshall, "Compact-device model development for the energy-delay analysis of magneto-electric magnetic tunnel junction structures," *Semiconductor Science and Technology*, vol. 31, 065022, 2016.

6

Organic Electronics

Hagen Klauk

Max Planck Institute for Solid State Research, Stuttgart, Germany

6.1 Introduction

Organic semiconductors are solid-state materials that are composed of conjugated organic molecules and which display semiconducting properties, such as variable electrical conductivity [1], electroluminescence [2], charge excitation [3], and dopability [4]. Like inorganic semiconductors, organic semiconductors can be produced as single-crystalline, polycrystalline, or amorphous materials and can be employed for the fabrication of a wide range of electronic and optoelectronic devices, including light-emitting diodes [5, 6], photovoltaic cells [7, 8], photodetectors [9], field-effect transistors [10, 11], and chemical and biological sensors [12].

A notable difference between organic and inorganic semiconductors is that the conjugated molecules that constitute organic semiconductors are held together by intermolecular van der Waals forces, which are substantially weaker than the covalent bonds by which the atomic constituents of inorganic semiconductors are held together [13]. As a result, organic semiconductors tend to be mechanically soft materials, which makes the fabrication of large, free-standing, mechanically rigid, and robust single-crystals based on organic semiconductors substantially more difficult and time-consuming compared to, for example, silicon. Many organic semiconductors can, however, be deposited over large areas in the form of thin amorphous or polycrystalline films. Depending on the choice of materials, organic-semiconductor films can often be produced at or near room temperature and thus on a wide range of substrates, including glass, plastics, fabrics, and paper. This makes organic semiconductors particularly attractive for large-area applications, such as flat-panel displays, general lighting, photovoltaics, and sensor arrays. The intrinsic mechanical softness of organic semiconductors greatly simplifies the fabrication of flexible and stretchable devices and systems [14].

A myriad of conjugated organic molecules have been employed in (or explored for) the fabrication of electronic and optoelectronic devices [15–17]. The majority of these materials do not occur naturally but are produced by more or less efficient synthetic protocols. Conjugated organic semiconductors can be broadly categorized into small-molecule materials (oligomers) and macromolecules (polymers). An important advantage of small-molecule semiconductors over polymers is that small-molecule materials can often be deposited from the vapor phase and thus without the need for liquid organic solvents. This is a significant advantage, since organic solvents are often toxic, carcinogenic, and environmentally harmful. Increasing global awareness of public health and environmental protection thus makes the availability of all-dry manufacturing approaches that

Advances in Semiconductor Technologies: Selected Topics Beyond Conventional CMOS, First Edition. Edited by An Chen.
© 2023 The Institute of Electrical and Electronics Engineers, Inc. Published 2023 by John Wiley & Sons, Inc.

do not require liquid organic solvents a critical benefit, if not a fundamental requirement, for the industrial mass manufacturing of electronic devices and systems.

6.2 Organic Light-Emitting Diodes

The first organic-semiconductor device in commercial mass production was the organic light-emitting diode (OLED). Milestones in OLED development include the first reports of electroluminescence in organic compounds in the 1950s and 1960s [18–21], the first efficient single-layer [22] and double-layer [23] thin-film OLEDs in the 1980s, the introduction of phosphorescent emitters [24] and doped charge-transport layers [25, 26] in 1998, the first white OLEDs with fluorescent-tube efficiency in 2009 [27], and the introduction of delayed-fluorescence emitters in 2012 [28].

The emission color of OLEDs is determined mainly by the chemical structure of the organic emissive molecules [15] and can thus be anywhere between infrared [29] and ultraviolet [30]. High-efficiency, modern-day OLEDs are heterostructures of at least one emission layer sandwiched between stacks of dedicated charge-injection, charge-transport, and charge-blocking layers, each with a specific material composition and thickness (usually 10–50 nm). This organic heterostructure is sandwiched between a hole-injecting anode and an electron-injecting cathode, at least one of which must be optically transparent (Figure 6.1a) [31]. White OLEDs are fabricated either by mixing two or three emitters of different color within a single emission layer (using one or multiple host materials) or by stacking two or three dedicated emission layers separated by charge-generation layers [6]. High-performance white OLEDs often consist of as many as 12 or more organic layers (Figure 6.1b) [32].

Advantages of OLEDs over traditional light sources include superior luminous efficiency (>200 lm/W for monochrome OLEDs [33, 34]; >160 lm/W for white OLEDs [35, 36]), uniform large-area emission characteristics (Figure 6.2a) [37], extremely small thickness of the

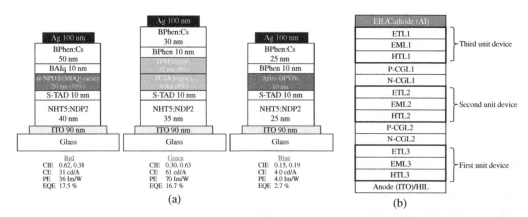

(a)   (b)

Figure 6.1 (a) Schematic cross sections of monochrome OLEDs with a red, green, or blue emission layer sandwiched between charge-injection, charge-transport, and charge-blocking layers. Source: Reproduced with permission Meerheim et al. [31]. Copyright 2008, American Institute of Physics. (b) Schematic cross section of a white OLED in the three-stack, three-color configuration. EIL, electron injection layer; ETL, electron-transport layer; EML, emission layer; HTL, hole transport layer; P-CGL, p-type charge-generation layer; N-CGL, n-type charge-generation layer; ITO, indium tin oxide; HIL, hole-injection layer. Source: Reproduced with permission Han et al. [32]. Copyright 2017, Wiley.

(a)

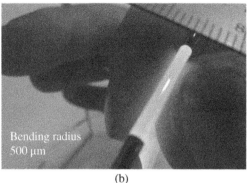
(b)

Figure 6.2 (a) Large-area white OLED panel developed by General Electric. Source: Reproduced with permission [37]. Copyright 2005, John Wiley & Sons. (b) Flexible OLED developed by TNO/Holst Center. Source: Reproduced with permission [38]. Copyright 2018, John Wiley & Sons.

devices (usually below 100 nm), and the option of fabricating flexible or stretchable OLEDs (Figure 6.2b) [34, 36, 38, 39].

The first commercially successful and mass-produced OLED application is the active-matrix organic light-emitting diode (AMOLED) display, first introduced in 2008 for mobile phones and later for television sets, smartwatches, notebook and tablet computers and automotive dashboards. Compared with active-matrix liquid-crystal displays (AMLCDs), AMOLED displays typically provide higher contrast, larger viewing angles, and shorter response times. In 2019, a total of 700 million AMOLED displays were manufactured, composed of a total of about 10^{15} OLEDs covering an area of about 10^7 m^2. While the market share of TV sets and notebook computers with AMOLED displays is still below 10%, the majority of high-end smartphones and smartwatches are currently equipped with AMOLED displays. Emerging OLED applications include general lighting, especially for buildings and automotive indoor and rear lighting, and clinical pulse oximetry, especially for preemies.

Currently, the most common approach for OLED manufacturing in mass-produced AMOLED displays is the vacuum deposition of small-molecule materials. Currently, the most common substrates sizes for AMOLED-display mass manufacturing are 1500 mm × 1800 mm (Gen-6) for smartphone and smartwatch displays and 2200 mm × 2500 mm (Gen-8.5) and 2940 mm × 3370 mm (Gen-10.5) for TV displays.

AMOLED displays for smartphones and smartwatches usually consist of an array of top-emitting red, green, and blue (RGB) OLEDs patterned using fine-metal mask (FMM) technology, which provides a pattern accuracy of about 3 μm and a display resolution up to 600 ppi [40, 41]. The emission of each OLED is controlled by an integrated circuit located directly below each OLED. These pixel-control circuits are comprised of five to seven thin-film transistors (TFTs) that are based on low-temperature polycrystalline silicon (LTPS) as the semiconductor and have a photolithographically defined channel length of 2–3 μm [42]. Depending on the display size and resolution, each OLED occupies an area of about 100–1000 μm^2. Samsung has been developing and perfecting this technology since the early 2000s and is currently the undisputed champion in the global market for small- and medium-sized AMOLED displays.

AMOLED displays for TV sets typically utilize large-area, bottom-emitting, white OLEDs stacked on top of an array of RGB color filters [6, 32]. The pixel-control circuits are usually fabricated using TFTs based on amorphous indium gallium zinc oxide (IGZO) as the semiconductor with a

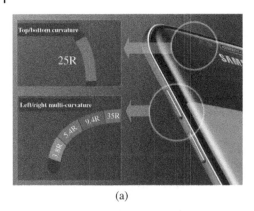 

(a)   (b)

Figure 6.3 (a) Curved AMOLED display fabricated on a flexible polyimide substrate in a Samsung Galaxy S8 mobile phone. Source: Reproduced with permission [44]. Copyright 2018, John Wiley & Sons. (b) Flexible AMOLED TV monitor developed by LG [45]. Source: LG Display.

photolithographically defined channel length of 5–8 µm [43]. LG began developing this technology in the mid-2000s and currently controls the majority of the global market for large-area AMOLED displays.

A trend that started in 2015 is to first coat the surface of the glass substrates with a thin film of polyimide, on which the displays are then fabricated. The polyimide foil (along with the entire display) is then released from the glass carrier, yielding extremely thin, extremely light-weight, and fully flexible AMOLED displays [44]. For most commercially available applications (at the time of this writing), the display is then mounted between a rigid heat sink and a rigid cover glass, resulting in a potentially curved, but no longer flexible display (Figure 6.3a) [44]. The transition from there to truly foldable or rollable AMOLED displays is obvious (Figure 6.3b) [45], and given sufficient consumer interest, the share of such devices is expected to increase.

6.3 Organic Solar Cells

The commercialization of organic photovoltaics lags behind that of OLEDs by more than a decade, due mainly to the challenges in achieving power-conversion efficiencies and operational lifetimes competitive with those of inorganic photovoltaics [8, 46]. Due to the comparatively small permittivity of organic semiconductors, excitons in organic semiconductors (including those generated by optical absorption) are typically characterized by large binding energies. The key to designing efficient organic solar cells and photodetectors is thus the formation of a heterojunction with appropriate energy-level offsets to provide an interface at which exciton dissociation can occur efficiently [3]. Organic solar cells and organic photodetectors thus virtually always consist of two different organic semiconductors, one serving as an electron donor and the other as an electron acceptor. Ideally, donor and acceptor are characterized by complementary absorption spectra, so as to cover as much of the solar spectrum as possible [47]. The two most commonly implemented device architectures are the planar heterojunction [48] and the bulk heterojunction [49, 50].

For bulk-heterojunction solar cells fabricated using solution-deposited blends of small-molecule acceptors and polymer donors, certified power-conversion efficiencies have reached and surpassed 17% for individual cells [51] and 11% for modules [52], on par with solar cells based on

Figure 6.4 Organic photovoltaic modules installed on the facade of an industrial building (Copyright 2020, Innogy SE) and on a wind turbine (Copyright 2020, Acciona S.A.). Source: Riede et al. [59], John Wiley & Sons.

hydrogenated amorphous silicon [46]. One drawback of solution-processed bulk-heterojunction solar cells is that the morphology of the bulk heterojunction and thus the power-conversion efficiency tend to degrade over time. Another drawback is that improvements in efficiency or lifetime often come at the expense of significantly higher material costs [8].

For planar-heterojunction solar cells fabricated using vacuum-deposited small-molecule semi-conductors, accelerated-lifetime tests have indicated operational lifetimes exceeding 10,000 years [53], which ought to suffice for most terrestrial application scenarios. High-throughput man-ufacturing of such cells is possible by means of continuous roll-to-roll fabrication approaches [54]. Advantages of vacuum-processed, planar-heterojunction solar cells include the high purity of the vacuum-deposited organic materials and the high level of control of the resulting mor-phology and interface quality. The principal drawback of planar-heterojunction solar cells is that the power-conversion efficiency is typically no greater than about 10% in a single-junction architecture. In multi-junction architectures, power-conversion efficiencies in excess of 13% have been reported, and there is evidence that multi-junction cells show better scalability to large-area modules than single-junction cells [55].

In order for organic photovoltaics to be able to compete with inorganic technologies in gigawatt-scale electricity generation, it will be necessary to break the existing compromises between efficiency, lifetime, and material costs and reduce the minimum sustainable price (MSP) and the levelized cost of energy (LCOE) of organic solar modules below those of inorganic pho-tovoltaics [56]. This is going to be a challenging task. On the other hand, organic solar cells hold distinct advantages over inorganic photovoltaics under low-incidence and indirect-illumination conditions, which makes them the perfect choice for indoor applications [57]. Also, given that organic solar cells can be readily fabricated on a wide range of thin, light-weight, and mechanically flexible and even stretchable substrates, organic photovoltaics hold great promise for unconven-tional off-grid applications, such as building-integrated photovoltaics [58, 59] (Figure 6.4) and wearable electronics [60]. Somewhat similar considerations hold for organic photodetectors [61].

6.4 Organic Thin-Film Transistors

The modulation of electric currents in organic-semiconductor films by a field effect was first reported in the 1980s [62–66]. Since then, the static and dynamic performance as well as the operational and shelf-life stability of organic TFTs have been greatly advanced.

Figure 6.5 Rollable AMOLED display developed by Sony, based on 311,040 top-emitting OLEDs and 630,000 organic TFTs fabricated on a polymeric substrate. Source: Reproduced with permission [70] Copyright 2011, John Wiley & Sons.

One of the potential applications envisioned for organic TFTs is in flexible or stretchable AMOLED displays and sensor arrays for mobile or wearable electronics applications [67–69]. In 2011, Sony demonstrated the visionary concept of a rollable AMOLED display in which both the pixel-control circuits and the gate-driver circuit were implemented using a total of 630,000 organic TFTs with a photolithographically defined channel length of 5 μm and a truly remarkable integration density of 13,600 TFTs/cm^2 [70]. The display was capable of showing full-color videos with a resolution of 121 ppi and a frame rate of 60 Hz while being firmly wrapped around a cylinder with a radius of 4 mm (Figure 6.5).

The pixel-control circuits in most flexible AMOLED displays currently in commercial mass production are implemented using LTPS TFTs [42, 44]. Since the fabrication of LTPS TFTs with sufficient performance and stability requires process temperatures in excess of 400 °C, the choice of the substrate material is limited essentially to polyimide, owing to its unique thermal stability. Unlike LTPS TFTs, organic TFTs are readily fabricated at temperatures below 100 °C, which permits the use of a much wider range of plastic substrates, including polyethylene naphthalate (PEN) and polyethylene terephthalate (PET). These materials are not only mechanically flexible and optically transparent, but also colorless, which allows the implementation of bottom-emitting OLEDs. In addition, PEN and especially PET tend to also be less expensive than polyimide. Organic TFTs have also been successfully fabricated on various types of paper [71] and fabric [72].

LTPS TFTs have excellent static and dynamic performance. State-of-the-art LTPS TFTs typically have channel-width-normalized on-state and off-state drain currents of about 10^{-4} and 10^{-13} A/μm, respectively, on/off current ratios of about 10^9, subthreshold swings of about 70 mV/decade and transit frequencies of about 600 MHz for n-channel and 300 MHz for p-channel TFTs (for a photolithographically defined channel length of 3.5 μm and a supply voltage of 3 V) [73]. For organic TFTs to be able to compete with LTPS TFTs, they will have to provide similar performance, irrespective of the fact that they provide the benefit of lower-temperature processing.

In terms of the static device performance, organic TFTs on clear plastic substrates are already on par with LTPS TFTs on polyimide, as exemplified by organic TFTs fabricated on PEN substrates showing an on/off current ratio of 10^{10} and a subthreshold swing of 59 mV/decade for a channel length of 8 μm and gate-source and drain-source voltages of 3 V [74].

Competing with LTPS, TFTs in terms of the dynamic performance is far more challenging. The transit frequency f_T of a field-effect transistor in the saturation regime can be written approximately as follows [74]:

$$f_T = \frac{\mu_0 \left(V_{GS} - V_{th}\right)}{2\pi \left(L + \frac{1}{2} \mu_0 R_C W C_{diel} \left(V_{GS} - V_{th}\right)\right) \left(L_{ov,t} + \frac{2}{3} L\right)} \tag{6.1}$$

where μ_0 is the intrinsic channel mobility, V_{GS} is the gate-source voltage, V_{th} is the threshold voltage, L is the channel length, R_C is the parasitic contact resistance, W is the channel width, C_{diel} is the unit-area gate-dielectric capacitance, and $L_{ov,t}$ is the sum of the parasitic gate-to-source and gate-to-drain overlaps.

To illustrate the parameter space in which organic TFTs can have transit frequencies similar to those of LTPS TFTs (i.e. several hundred megahertz), solutions to Eq. (6.1) are plotted in Figure 6.6 for various parameter combinations. For the purpose of this analysis, the gate-overdrive voltage $(V_{GS} - V_{th})$ is set to a value of 3 V, since this represents approximately the output voltage of small batteries or small energy-harvesting devices and thus approximately the upper limit of the supply voltage available in mobile or wearable electronics applications. The unit-area gate-dielectric capacitance is set to a value of 1 µF/cm², since this is approximately the minimum value required for TFTs to have a sufficiently large channel-width-normalized transconductance (>1 S/m) when operated at voltages not exceeding 3 V. For Figure 6.6a, the channel-width-normalized contact resistance $(R_C W)$ is set to a value of 10 Ωcm, since this is the smallest contact resistance reported for organic TFTs at the time of this writing [74].

Figure 6.6a shows that the influence of the intrinsic channel mobility (μ_0) on the transit frequency is rather weak once it exceeds about 1 cm²/Vs. This implies that improvements in the intrinsic mobility of organic TFTs beyond about 1 cm²/Vs would provide little or no benefit in terms of the dynamic TFT performance. For Figure 6.6b,c, the intrinsic channel mobility is thus set to a value of 1 cm²/Vs.

Figure 6.6b shows that the influence of the contact resistance on the transit frequency is somewhat more pronounced compared to that of the intrinsic mobility, but only down to a

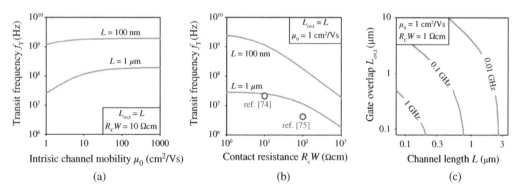

Figure 6.6 Transit frequency of field-effect transistors in the saturation regime calculated using Eq. (6.1) for various combinations of intrinsic channel mobility (μ_0), channel length (L), total gate-to-contact overlap $(L_{ov,t})$, and channel-width-normalized contact resistance $(R_C W)$. The gate-overdrive voltage $(V_{GS} - V_{th})$ and the unit-area gate-dielectric capacitance (C_{diel}) were set to values of 3 V and 1 µF/cm², respectively. The two data points in panel (b) indicate results of transit-frequency measurements performed on organic TFTs having a channel length of 0.6 or 1 µm, a total gate-to-contact overlap of 10 µm and a contact resistance of 10 or 100 Ωcm [74, 75].

channel-width-normalized contact resistance (R_CW) of about 1 Ωcm. Reducing R_CW below 1 Ωcm would yield diminishing returns in terms of the transit frequency. For Figure 6.6c, the contact resistance is thus set to a value of 1 Ωcm.

Figure 6.6c indicates that transit frequencies between 100 MHz and 1 GHz can be expected for organic TFTs with critical dimensions ($L, L_{ov,t}$) of a few hundred nanometers. Similar design guidelines were derived for flexible TFTs based on IGZO [76, 77]. The fabrication of TFTs with this range of lateral dimensions on flexible polymeric substrates is possible using a variety of techniques, including optical lithography [77, 78], nanoimprint lithography [79], and stencil-mask lithography [74] (Figure 6.7).

Figure 6.8 shows the static current–voltage characteristics of p-channel and n-channel organic TFTs with a channel length (L) of 600 nm, a total gate-to-contact overlap ($L_{ov,t}$) of 400 nm and a gate-dielectric thickness of 6 nm, confirming that the desirable long-channel-transistor characteristics, including a large on/off current ratio (10^8), a steep subthreshold swing (67 mV/decade) and drain-current saturation at large drain–source voltages, are preserved in nanoscale organic TFTs, provided the field-effect-transistor (FET) scaling laws [81, 82] are properly observed.

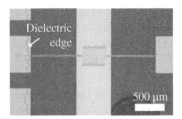

Figure 6.7 Organic TFTs with submicron channel lengths on plastic substrates. Source: J. W. Borchert et al., 2020, American Association for the Advancement of Science.

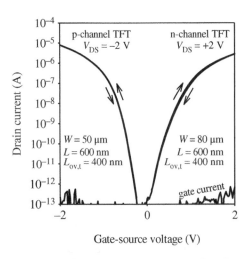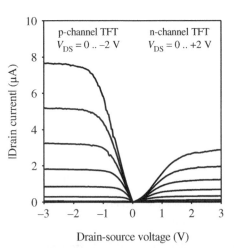

Figure 6.8 Current–voltage characteristics of p-channel and n-channel organic TFTs fabricated using the vacuum-deposited small-molecule semiconductors 2,9-diphenyl-dinaphtho[2,3-b : 2′,3′-f]thieno[3,2-b] thiophene (DPh-DNTT; for the p-channel TFT) [74] and N,N′-bis(2,2,3,3,4,4,4-heptafluorobutyl)-1,7-dicyano-perylene-(3,4:9,10)-tetracarboxylic diimide (Polyera ActivInk N1100; for the n-channel TFT) [80]. Both TFTs have a lithographically defined channel length (L) of 600 nm and a total gate-to-contact overlap ($L_{ov,t}$) of 400 nm.

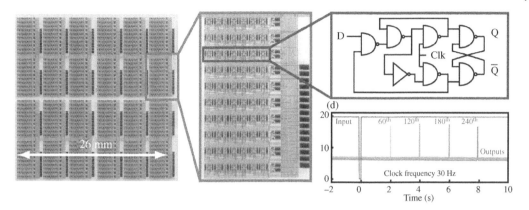

Figure 6.9 Photographs, circuit schematic, and measured output waveforms of a 240-stage shift register designed as a gate-driver circuit for a rollable electrophoretic display developed by PolymerVision [83]. The circuit consists of 13,440 p-channel organic TFTs with an integration density of 2000 TFTs/cm^2 and was fabricated on a flexible plastic substrate. Source: Reproduced with permission [84]. Copyright 2014, IEEE.

A prerequisite for the mechanical rollability of flexible displays is the integration of the gate-driver circuit on the flexible substrate [70]. In early rollable-display demonstrations by Sony [70] (Figure 6.5) and PolymerVision [83], the gate-driver circuits were designed and fabricated in unipolar circuit topologies based exclusively on p-channel organic TFTs, since the performance and stability of organic n-channel TFTs have historically been notably inferior to those of organic p-channel TFTs (Figure 6.9) [84]. However, for mobile applications powered by small batteries or energy-harvesting devices, the gate-driver circuits will necessarily have to be implemented in a low-power complementary circuit topology based on both p-channel and n-channel TFTs [85]. This will require further improvements in the performance and long-term stability of n-channel organic TFTs [86].

6.5 Outlook

The rapid commercialization of OLEDs in smartphone and TV displays in the 2000s and 2010s has been a remarkable success, especially when taking into account that OLEDs with luminous efficiencies exceeding 1 lm/W had only been first reported in 1987 [23]. It could be argued that the rapid market penetration of AMOLED displays was aided by a lack of alternative emissive display technologies at the time, and that the dominance of AMOLED displays in smartphones might come to an end as soon as an emissive, large-area, low-power display technology with viewing characteristics superior to those of AMOLED displays will become commercially viable. But even then, the organic LED will likely continue to serve as an example of a truly disruptive technology. Along the way, the unexpected commercial success of AMOLED displays forced the AMLCD manufacturers to double their efforts in developing improved liquid-crystal displays to defend market shares against AMOLED technology [87]. In a way, OLEDs have made other display technologies better.

For organic photovoltaics and organic transistors, competition from existing and emerging technologies has been far more challenging. Organic solar cells are competing against photovoltaics based on single-crystalline silicon, polycrystalline silicon, nanocrystalline silicon, hydrogenated amorphous silicon, CdTe, Cu(In,Ga)(Se,S)$_2$, Cu(Zn,Sn)(Se,S)$_2$, dye-sensitized TiO$_2$, perovskites, and various quantum dot materials [45]. Competition for organic TFTs comes mainly from LTPS, hydrogenated amorphous silicon, ZnO, IGZO, and semiconducting carbon nanotubes. Against this wall of competition, organic photovoltaics and organic TFTs have been and still are struggling to gain a commercial foothold. Part of the reason is that research on organic photovoltaics and organic TFTs has traditionally been driven mainly by molecular design and synthetic chemistry, and less by device or process engineering. As a result, public funding has been funneled largely into materials development, and research has focused mainly on aspects that have little or no influence on the relevant device properties, while the pursuit of critically necessary advances in device engineering and manufacturing has been woefully neglected. Without the necessary progress there, organic photovoltaics and organic transistors are destined to never reach commercial maturity.

References

1 C. K. Chiang, C. R. Fincher, Y. W. Park, A. J. Heeger, H. Shirakawa, E. J. Louis, S. C. Gau, and A. G. MacDiarmid, "Electrical conductivity in doped polyacetylene," *Physical Review Letters*, vol. 39, no. 17, pp. 1098–1101, 1977, doi: 10.1103/PhysRevLett.39.1098.

2 L. J. Rothberg and A. J. Lovinger, "Status of and prospects for organic electroluminescence," *Journal of Materials Research*, vol. 11, no. 12, pp. 3174–3187, 1996, doi: 10.1557/JMR.1996.0403.

3 P. Peumans, A. Yakimov, and S. R. Forrest, "Small molecular weight organic thin-film photodetectors and solar cells," *Journal of Applied Physics*, vol. 93, no. 7, pp. 3693–3723, 2003, doi: 10.1063/1.1534621.

4 K. Walzer, B. Maennig, M. Pfeiffer, and K. Leo, "Highly efficient organic devices based on electrically doped transport layers," *Chemical Reviews*, vol. 107, no. 4, pp. 1233–1271, 2007, doi: 10.1021/cr050156n.

5 S. -J. Zou, Y. Shen, F. -M. Xie, J. -D. Chen, Y. -Q. Li, and J. -X. Tang, "Recent advances in organic light-emitting diodes: toward smart lighting and displays," *Materials Chemistry Frontiers*, vol. 4, no. 3, pp. 788–820, 2020, doi: 10.1039/c9qm00716d.

6 Y. Yin, M. U. Ali, W. Xie, H. Yang, and H. Meng, "Evolution of white organic light-emitting devices: from academic research to lighting and display applications," *Materials Chemistry Frontiers*, vol. 3, no. 6, pp. 970–1031, 2019, doi: 10.1039/c9qm00042a.

7 W. Ye, Y. Yang, Z. Zhang, Y. Zhu, L. Ye, C. Miao, Y. Lin, and S. Zhang, "Nonfullerene all-small-molecule organic solar cells: prospect and limitation," *Solar RRL*, vol. 4, no. 11, 2000258, 2020, doi: 10.1002/solr.202000258.

8 C. J. Brabec, A. Distler, X. Du, H. -J. Egelhaaf, J. Hauch, T. Heumueller, and N. Li, "Material strategies to accelerate OPV technology toward a GW technology," *Advanced Energy Materials*, vol. 10, no. 43, 2001864, 2020, doi: 10.1002/aenm.202001864.

9 P. C. Y. Chow, and T. Someya, "Organic photodetectors for next-generation wearable electronics," *Advanced Materials*, vol. 32, no. 15, 1902045, 2020, doi: 10.1002/adma.201902045.

10 H. Klauk, "Organic thin-film transistors," *Chemical Society Reviews*, vol. 39, no. 7, pp. 2643–2666, 2010, doi: 10.1039/b909902f.

11 X. Guo, Y. Xu, S. Ogier, T. N. Ng, M. Caironi, A. Perinot, L. Li, J. Zhao, W. Tang, R. A. Sporea, A. Nejim, J. Carrabina, P. Cain, and F. Yan, "Current status and opportunities of organic thin-film transistor technologies," *IEEE Transactions on Electron Devices*, vol. 64, no. 5, pp. 1906–1921, 2017, doi: 10.1109/TED.2017.2677086.

12 H. Li, W. Shi, J. Song, H. -J. Jang, J. Dailey, J. Yu, and H. E. Katz, "Chemical and biomolecule sensing with organic field-effect transistors," *Chemical Reviews*, vol. 119, no. 1, pp. 3–35, 2019, doi: 10.1021/acs.chemrev.8b00016.

13 C. Ambrosch-Draxl, D. Nabok, P. Puschnig, and C. Meisenbichler, "The role of polymorphism in organic thin films: oligoacenes investigated from first principles," *New Journal of Physics*, vol. 11, no. 12, 125010, 2009, doi: 10.1088/1367-2630/11/12/125010.

14 J. A. Rogers, T. Someya, and Y. Huang, "Materials and mechanics for stretchable electronics," *Science*, vol. 327, no. 5973, pp. 1603–1607, 2010, doi: 10.1126/science.1182383.

15 J. -H. Jou, S. Kumar, A. Agrawal, T. -H. Lia, and S. Sahoo, "Approaches for fabricating high efficiency organic light emitting diodes," *Journal of Materials Chemistry*, vol. 3, no. 13, pp. 2974–3002, 2015, doi: 10.1039/c4tc02495h.

16 M. D. M. Faurea and B. H. Lessard, "Layer-by-layer fabrication of organic photovoltaic devices: material selection and processing conditions," *Journal of Materials Chemistry C*, vol. 9, no. 1, pp. 14–40, 2021, doi: 10.1039/d0tc04146g.

17 C. Wang, H. Dong, W. Hu, Y. Liu, and D. Zhu, "Semiconducting π-conjugated systems in field-effect transistors: a material odyssey of organic electronics," *Chemical Reviews*, vol. 112, no. 4, pp. 2208–2267, 2012, doi: 10.1021/cr100380z.

18 A. Bernanose, M. Comte, and P. Vouaux, "A new method of emission of light by certain organic compounds," *Journal de Chimie Physique*, vol. 50, pp. 64–68, 1953, doi: 10.1051/jcp/1953500064.

19 S. Namba, M. Yoshizawa, and H. Tamura, "Electroluminescence of organic fluorescent compounds (I)," *Journal Applied Physics (Japan)*, vol. 28, no. 8, p. 439, 1959.

20 H. P. K. M. Pope, and P. Magnante, "Electroluminescence in organic crystals," *The Journal of Chemical Physics*, vol. 38, no. 8, pp. 2042–2043, 1963, doi: 10.1063/1.1733929.

21 W. Helfrich and W. G. Schneider, "Recombination radiation in anthracene crystals," *Physical Review Letters*, vol. 14, no. 7, pp. 229–231, 1965, doi: 10.1103/PhysRevLett.14.229.

22 P. S. Vincett, W. A. Barlow, R. A. Hann, and G. G. Roberts, "Electrical conduction and low voltage blue electroluminescence in vacuum-deposited organic films," *Thin Solid Films*, vol. 94, no. 2, pp. 171–183, 1982, doi: 10.1016/0040-6090(82)90509-0.

23 C. W. Tang and S. A. VanSlyke, "Organic electroluminescent diodes," *Applied Physics Letters*, vol. 51, no. 12, pp. 913–915, 1987, doi: 10.1063/1.98799.

24 M. A. Baldo, D. F. O'Brien, Y. You, A. Shoustikov, S. Sibley, M. E. Thompson, and S. R. Forrest, "Highly efficient phosphorescent emission from organic electroluminescent devices," *Nature*, vol. 395, no. 6698, pp. 151–154, 1998, doi: 10.1038/25954.

25 J. Blochwitz, M. Pfeiffer, T. Fritz, and K. Leo, "Low voltage organic light emitting diodes featuring doped phthalocyanine as hole transport material," *Applied Physics Letters*, vol. 73, no. 6, pp. 729–731, 1998, doi: 10.1063/1.121982.

26 J. Huang, M. Pfeiffer, A. Werner, J. Blochwitz, K. Leo, and S. Liu, "Low-voltage organic electroluminescent devices using pin structures," *Applied Physics Letters*, vol. 80, no. 1, pp. 139–141, 2002, doi: 10.1063/1.1432110.

27 S. Reineke, F. Lindner, G. Schwartz, N. Seidler, K. Walzer, B. Lussem, and K. Leo, "White organic light-emitting diodes with fluorescent tube efficiency," *Nature*, vol. 459, no. 7244, pp. 234–238, 2009, doi: 10.1038/nature08003.

28 H. Uoyama, K. Goushi, K. Shizu, H. Nomura, and C. Adachi, "Highly efficient organic light-emitting diodes from delayed fluorescence," *Nature*, vol. 492, no. 7428, pp. 234–240, 2012, doi: 10.1038/nature11687.

29 Y. -C. Wei, S. F. Wang, Y. Hu, L. -S. Liao, D. -G. Chen, K. -H. Chang, C. -W. Wang, S. -H. Liu, W. -H. Chan, J. -L. Liao, W. -Y. Hung, T. -H. Wang, P. -T. Chen, H.- F. Hsu, Y. Chi, and

P. -T. Chou, "Overcoming the energy gap law in near-infrared OLEDs by exciton-vibration decoupling," *Nature Photonics*, vol. 14, no. 9, pp. 570–577, 2020, doi: 10.1038/s41566-020-0653-6.

30 Y. Luo, S. Li, Y. Zhao, C. Li, Z. Pang, Y. Huang, M. Yang, L. Zhou, X. Zheng, X. Pu, and Z. Lu, "An ultraviolet thermally activated delayed fluorescence OLED with total external quantum efficiency over 9%," *Advanced Materials*, vol. 32, no. 32, 2001248, 2020, doi: 10.1002/adma.202001248.

31 R. Meerheim, R. Nitsche, and K. Leo, "High-efficiency monochrome organic light emitting diodes employing enhanced microcavities," *Applied Physics Letters*, vol. 93, no. 4, 043310, 2008, doi: 10.1063/1.2966784.

32 C. -W. Han, M. -Y. Han, S. -R. Joung, J. -S. Park, Y. -K. Jung, J. -M. Lee, H. -S. Choi, G. -J. Cho, D. -H. Kim, M. -K. Yee, H. -G. Kim, H. -C. Choi, C. -H. Oh, I. -B. Kang, "3 Stack-3 Color White OLEDs for 4K Premium OLED TV," SID Symposium Digest of Technical Papers, vol. 48, no. 1, pp. 1–4, 2017, doi: 10.1002/sdtp.11555.

33 M. G. Helander, Z. B. Wang, J. Qiu, M. T. Greiner, D. P. Puzzo, Z. W. Liu, and Z. H. Lu, "Chlorinated indium tin oxide electrodes with high work function for organic device compatibility," *Science*, vol. 332, no. 6032, pp. 944–947, 2011, doi: 10.1126/science.1202992.

34 Z. B. Wang, M. G. Helander, J. Qiu, D. P. Puzzo, M. T. Greiner, Z. M. Hudson, S. Wang, Z. W. Liu, Z. H. Lu, "Unlocking the full potential of organic light-emitting diodes on flexible plastic," *Nature Photonics*, vol. 5, no. 12, pp. 753–757, 2011, doi: 10.1038/NPHOTON.2011.259.

35 S. Jeon, S. Lee, K. -H. Han, H. Shin, K. -H. Kim, J. -H. Jeong, and J. -J. Kim, "High-quality white OLEDs with comparable efficiencies to LEDs," *Advanced Optical Materials*, vol. 6, no. 8, 1701349, 2018, doi: 10.1002/adom.201701349.

36 H. -Y. Xiang, Y. -Q. Li, S. -S. Meng, C. -S. Lee, L. -S. Chen, and J. -X. Tang, "Extremely efficient transparent flexible organic light-emitting diodes with nanostructured composite electrodes," *Advanced Optical Materials*, vol. 6, no. 21, 1800831, 2018, doi: 10.1002/adom.201800831.

37 A. R. Duggal, J. J. Shiang, D. F. Foust, L. G. Turner, W. F. Nealon, and J. C. Bortscheller, "Large Area White OLEDs," SID Symposium Digest of Technical Papers, vol. 36, no. 1, pp. 28–31, 2005, doi: 10.1889/1.2036427.

38 H. B. Akkerman, R. K. Pendyala, P. Panditha, A. Salem, J. Shen, P. van de Weijer, P. C. P. Bouten, S. H. P. M. de Winter, K. van Diesen-Tempelaars, G. de Haas, S. Steudel, C. -Y. Huang, M. -H. Lai, Y. -Y. Huang, M. -H. Yeh, A. J. Kronemeijer, P. Poodt, and G. H. Gelinck, "High-temperature thin-film barriers for foldable AMOLED displays," *Journal of the Society for Information Display*, vol. 26, no. 4, pp. 214–222, 2018, doi: 10.1002/jsid.647.

39 D. Yin, J. Feng, R. Ma, Y. -F. Liu, Y. -L. Zhang, X. -L. Zhang, Y. -G. Bi, Q. -D. Chen, and H. -B. Sun, "Efficient and mechanically robust stretchable organic light-emitting devices by a laser-programmable buckling process," *Nature Communications*, vol. 7, 11573, 2016, doi: 10.1038/ncomms11573.

40 T. A. Mai and B. Richerzhagen, "Manufacturing of 4th Generation OLED Masks with the Laser Microjet Technology," SID Symposium Digest of Technical Papers, vol. 38, no. 1, pp. 1596–1598, 2007, doi: 10.1889/1.2785624.

41 S. -K. Hong, J. -H. Sim, I. -G. Seo, K. -C. Kim, S. -I. Bae, H. -Y. Lee, N. -Y. Lee, and J. Jang, "New pixel design on emitting area for high resolution active-matrix organic light-emitting diode displays," *Journal of Display Technology*, vol. 6, no. 12, pp. 601–606, 2010, doi: 10.1109/JDT.2010.2063694.

42 N. -H. Keum, C. -C. Chai, S. -K. Hong, and O. -K. Kwon, "A compensation method for variations in subthreshold slope and threshold voltage of thin-film transistors for AMOLED

displays," *IEEE Journal of the Electron Devices Society*, vol. 7, pp. 462–469, 2019, doi: 10.1109/JEDS.2019.2904852.

43 C. W. Han, H. -S. Choi, C. Ha, H. Shin, H. C. Choi, and I. B. Kang, Advanced technologies for large-sized OLED display. In C. Ravariu (ed.), *Green Electronics*, IntechOpen, 2018, doi: 10.5772/intechopen.74869.

44 J. Lee, T. T. Nguyen, J. Bae, G. Jo, Y. Lee, S. Yang, H. Chu, and J. Kwag, "5.8-inch QHD flexible AMOLED display with enhanced bendability of LTPS TFTs," *Journal of the Society for Information Display*, vol. 26, no. 4, pp. 200–207, 2018, doi: 10.1002/jsid.655.

45 LG invented a bendable TV that sticks to walls like a refrigerator magnet, Available: https://tinyurl.com/y7zpzb27 (accessed January 2021).

46 A. Polman, M. Knight, E. C. Garnett, B. Ehrler, and W. C. Sinke, "Photovoltaic materials: present efficiencies and future challenges," *Science*, vol. 352, no. 6283, aad4424, 2016, doi: 10.1126/science.aad4424.

47 Z. Li, W. Zhang, X. Xu, Z. Genene, D. Di Carlo Rasi, W. Mammo, A. Yartsev, M. R. Andersson, R. A. J. Janssen, and E. Wang, "High-performance and stable all-polymer solar cells using donor and acceptor polymers with complementary absorption," *Advanced Energy Materials*, vol. 7, no. 14, 1602722, 2017, doi: 10.1002/aenm.201602722.

48 C. W. Tang, "Two-layer organic photovoltaic cell," *Applied Physics Letters*, vol. 48, no. 2, pp. 183–185, 1986, doi: 10.1063/1.96937.

49 J. J. M. Halls, C. A. Walsh, N. C. Greenham, E. A. Marseglia, R. H. Friend, S. C. Moratti, and A. B. Holmes, "Efficient photodiodes from interpenetrating polymer networks," *Nature*, vol. 376, no. 6540, pp. 498–500, 1995, doi: 10.1038/376498a0.

50 G. Yu, J. Gao, J. C. Hummelen, F. Wudl, and A. J. Heeger, "Polymer photovoltaic cells: enhanced efficiencies via a network of internal donor–acceptor heterojunctions," *Science*, vol. 270, no. 5243, pp. 1789–1791, 1995, doi: 10.1126/science.270.5243.1789.

51 Q. Liu, Y. Jiang, K. Jin, J. Qin, J. Xu, W. Li, J. Xiong, J. Liu, Z. Xiao, K. Sun, S. Yang, X. Zhang, and L. Ding, "18% Efficiency organic solar cells," *Science Bulletin*, vol. 65, no. 4, pp. 272–275, 2020, doi: 10.1016/j.scib.2020.01.001.

52 A. Distler, C. J. Brabec, and H. -J. Egelhaaf, "Organic photovoltaic modules with new world record efficiencies," *Progress in Photovoltaics: Research and Applications*, vol. 29, no. 1, pp. 24–31, 2021, doi: 10.1002/pip.3336.

53 Q. Burlingame, X. Huang, X. Liu, C. Jeong, C. Coburn, and S. R. Forrest, "Intrinsically stable organic solar cells under high-intensity illumination," *Nature*, vol. 573, no. 7774, pp. 394–398, 2019, doi: 10.1038/s41586-019-1544-1.

54 B. Qu and S. R. Forrest, "Continuous roll-to-roll fabrication of organic photovoltaic cells via interconnected high-vacuum and low-pressure organic vapor phase deposition systems," *Applied Physics Letters*, vol. 113, no. 5, 053302, 2018, doi: 10.1063/1.5039701.

55 X. Xiao, K. Lee, and S. R. Forrest, "Scalability of multi-junction organic solar cells for large area organic solar modules," *Applied Physics Letters*, vol. 106, no. 21, 213301, 2015, doi: 10.1063/1.4921771.

56 J. Guo and J. Min, "A cost analysis of fully solution-processed ITO-free organic solar modules," *Advanced Energy Materials*, vol. 9, no. 3, 1802521, 2019, doi: 10.1002/aenm.201802521.

57 H. S. Ryu, S. Y. Park, T. H. Lee, J. Y. Kim, and H. Y. Woo, "Recent progress in indoor organic photovoltaics," Nanoscale, vol. 12, no. 10, pp. 5792–5804, 2020, doi: 10.1039/d0nr00816h.

58 E. Biyik, M. Araz, A. Hepbasli, M. Shahrestani, R. Yao, L. Shao, E. Essah, A. C. Oliveira, T. del Caño, E. Rico, J. L. Lechón, L. Andrade, A. Mendes, and Y. B. Atlı, "A key review of building integrated photovoltaic (BIPV) systems," *Engineering Science* and *Technology*, vol. 20, no. 3, pp. 833–858, 2017, doi: 10.1016/j.jestch.2017.01.009.

59 M. Riede, D. Spoltore, and K. Leo, "Organic solar cells – the path to commercial success," *Advanced Energy Materials*, vol. 11, no. 1, 2002653, 2021, doi: 10.1002/aenm.202002653.

60 K. Fukuda, K. Yu, and T. Someya, "The future of flexible organic solar cells," *Advanced Energy Materials*, vol. 10, no. 25, 2000765, 2020, doi: 10.1002/aenm.202000765.

61 C. Fuentes-Hernandez, W. -F. Chou, T. M. Khan, L. Diniz, J. Lukens, F. A. Larrain, V. A. Rodriguez-Toro, and B. Kippelen, "Large-area low-noise flexible organic photo-diodes for detecting faint visible light," *Science*, vol. 370, no. 6517, pp. 698–701, 2020, doi: 10.1126/science.aba2624.

62 F. Ebisawa, T. Kurokawa, and S. Nara, "Electrical properties of polyacetylene/polysiloxane inter-face," *Journal of Applied Physics*, vol. 54, no. 6, pp. 3255–3259, 1983, doi: 10.1063/1.332488.

63 K. Kudo, M. Yamashina, and T. Moriizumi, "Field effect measurement of organic dye films," *Japanese Journal of Applied Physics*, vol. 23, part 1, no. 1, p. 130, 1984, doi: 10.1143/JJAP.23.130.

64 A. Tsumura, H. Koezuka, and T. Ando, "Macromolecular electronic device: field-effect transistor with a polythiophene thin film," *Applied Physics Letters*, vol. 49, no. 18, pp. 1210–1212, 1986, doi: 10.1063/1.97417.

65 C. Clarisse, M. T. Riou, M. Gauneau, and M. Le Contellec, "Field-effect transistor with diphthalocyanine thin film," *Electronics Letters*, vol. 24, no. 11, pp. 674–675, 1988, doi: 10.1049/el:19880456.

66 J. H. Burroughes, C. A. Jones, and R. H. Friend, "New semiconductor device physics in polymer diodes and transistors," *Nature*, vol. 335, no. 6186, pp. 137–141, 1988, doi: 10.1038/335137a0.

67 L. Zhou, A. Wang, S. -C. Wu, J. Sun, S. Park, and T. N. Jackson, "All-organic active matrix flexible display," *Applied Physics Letters*, vol. 88, no. 8, 083502, 2006, doi: 10.1063/1.2178213.

68 T. Sekitani, H. Nakajima, H. Maeda, T. Fukushima, T. Aida, K. Hata, and T. Someya, "Stretch-able active-matrix organic light-emitting diode display using printable elastic conductors," *Nature Materials*, vol. 8, no. 6, pp. 494–499, 2009, doi: 10.1038/nmat2459.

69 S. Elsaegh, C. Veit, U. Zschieschang, M. Amayreh, F. Letzkus, H. Sailer, M. Jurisch, J. N. Burghartz, U. Würfel, H. Klauk, H. Zappe, and Y. Manoli, "Low-power organic light sensor array based on active-matrix common-gate transimpedance amplifier on foil for imag-ing applications," *IEEE Journal of Solid-State Circuits*, vol. 55, no. 9, pp. 2553–2566, 2020, doi: 10.1109/JSSC.2020.2993732.

70 M. Noda, N. Kobayashi, M. Katsuhara, A. Yumoto, S. Ushikura, R. Yasuda, N. Hirai, G. Yukawa, I. Yagi, K. Nomoto, and T. Urabe, "An OTFT-driven rollable OLED display," *Journal of the Soci-ety for Information Display*, vol. 19, no. 4, pp. 316–322, 2011, doi: 10.1889/JSID19.4.316.

71 U. Zschieschang and H. Klauk, "Organic transistors on paper: a brief review," *Journal of Mate-rials Chemistry C*, vol. 7, no. 19, pp. 5522–5533, 2019, doi: 10.1039/c9tc00793.

72 J. S. Kim and C. K. Song, "AMOLED panel driven by OTFTs on polyethylene fabric substrate," *Organic Electronics*, vol. 30, pp. 45–51, 2016, doi: 10.1016/j.orgel.2015.12.007.

73 Y. Kubota, T. Matsumoto, H. Tsuji, N. Suzuki, S. Imai, and H. Kobayashi, "1.5-V-operation ultralow power circuit of poly-Si TFTs fabricated using the NAOS method," *IEEE Transactions on Electron Devices*, vol. 59, no. 2, pp. 385–392, 2012, doi: 10.1109/TED.2011.2175395.

74 J. W. Borchert, U. Zschieschang, F. Letzkus, M. Giorgio, R. T. Weitz, M. Caironi, J. N. Burghartz, S. Ludwigs, and H. Klauk, "Flexible low-voltage high-frequency organic thin-film transistors," *Science Advances*, vol. 6, no. 21, eaaz5156, 2020, doi: 10.1126/sciadv.aaz5156.

75 T. Zaki, R. Rödel, F. Letzkus, H. Richter, U. Zschieschang, H. Klauk, and J. N. Burghartz, "AC characterization of organic thin-film transistors with asymmetric gate-to-source and gate-to-drain overlaps," *Organic Electronics*, vol. 14, no. 5, pp. 1318–1322, 2013, doi: 10.1016/j.orgel.2013.02.014.

76 N. Münzenrieder, G. A. Salvatore, L. Petti, C. Zysset, L. Büthe, C. Vogt, G. Cantarella, and G. Tröster, "Contact resistance and overlapping capacitance in flexible sub-micron long oxide thin-film transistors for above 100 MHz operation," *Applied Physics Letters*, vol. 105, no. 26, 263504, 2014, doi: 10.1063/1.4905015.

77 C. Tückmantel, U. Kalita, T. Haeger, M. Theisen, U. Pfeiffer, and T. Riedl, "Amorphous indium-gallium-zinc-oxide TFTs patterned by self-aligned photolithography overcoming the GHz threshold," *IEEE Electron Device Letters*, vol. 41, no. 12, pp. 1786–1789, 2020, doi: 10.1109/LED.2020.3029956.

78 N. Münzenrieder, K. Ishida, T. Meister, G. Cantarella, L. Petti, C. Carta, F. Ellinger, and G. Tröster, "Flexible InGaZnO TFTs with f_{max} above 300 MHz," *IEEE Electron Device Letters*, vol. 39, no. 9, pp. 1310–1313, 2018, doi: 10.1109/LED.2018.2854362.

79 S. G. Higgins, B. V. O. Muir, G. Dell'Erba, A. Perinot, M. Caironi, and A. J. Campbell, "Complementary organic logic gates on plastic formed by self-aligned transistors with gravure and inkjet printed dielectric and semiconductors," *Advanced Electronic Materials*, vol. 2, no. 2, 1500272, 2016, doi: 10.1002/aelm.201500272.

80 B. A. Jones, M. J. Ahrens, M.-H. Yoon, A. Facchetti, T. J. Marks, and M. R. Wasielewski, "High-mobility air-stable n-type semiconductors with processing versatility: dicyanoperylene-3,4:9,10-bis(dicarboximides)," *Angewandte Chemie, International Edition*, vol. 43, no. 46, pp. 6363–6366, 2004, doi: 10.1002/anie.200461324.

81 R. H. Dennard, F. H. Gaensslen, H. -N. Yu, V. L. Rideout, E. Bassous, and A. R. LeBlanc, "Design of ion-implanted MOSFET's with very small physical dimensions," *IEEE Journal of Solid-State Circuits*, vol. 9, no. 5, pp. 256–268, 1974, doi: 10.1109/JSSC.1974.1050511.

82 G. Baccarani, M. R. Wordeman, and R. H. Dennard, "Generalized scaling theory and its application to a ¼ micrometer MOSFET design," *IEEE Transactions on Electron Devices*, vol. 31, no. 4, pp. 452–462, 1984, doi: 10.1109/T-ED.1984.21550.

83 H. E. A. Huitema, G. H. Gelinck, P. J. G. van Lieshout, E. van Veenendaal, and F. J. Touwslager, "Flexible electronic-paper active-matrix displays," *Journal of the Society for Information Display*, vol. 14, no. 8, pp. 729–733, 2006, doi: 10.1889/1.2336100.

84 D. Raiteri, P. van Lieshout, A. van Roermund, and E. Cantatore, "Positive-feedback level shifter logic for large-area electronics," *IEEE Journal of Solid-State Circuits*, vol. 49, no. 2, pp. 524–535, 2014, doi: 10.1109/JSSC.2013.2295980.

85 S. Abdinia, T.-H. Ke, M. Ameys, J. Li, S. Steudel, J. L. Vandersteen, B. Cobb, F. Torricelli, A. van Roermund, and E. Cantatore, "Organic CMOS line drivers on foil," *Journal of Display Technology*, vol. 11, no. 6, pp. 564–569, 2015, doi: 10.1109/JDT.2015.2421344.

86 J. T. E. Quinn, J. Zhu, X. Li, J. Wang, and Y. Li, "Recent progress in the development of n-type organic semiconductors for organic field effect transistors," *Journal of Materials Chemistry C*, vol. 5, no. 34, pp. 8654–8681, 2017, doi: 10.1039/c7tc01680h.

87 H. Chen and S. -T. Wu, "Advanced liquid crystal displays with supreme image qualities," *Liquid Crystal Today*, vol. 28, no. 1, pp. 4–11, 2019, doi: 10.1080/1358314X.2019.1625138.

7

Active-Matrix Electroluminescent Displays

Xiaojun Guo[1], Li'ang Deng[1], and Arokia Nathan[2]

[1]*Department of Electronic Engineering, Shanghai Jiao Tong University, China*
[2]*Darwin College, University of Cambridge, Cambridge, UK*

7.1 Introduction

Following the growing popularity of Internet-based communications coupled with the advent of digital multi-media, large-area, high information-content flat-panel displays (FPDs) are becoming the central feature of many consumer products, ranging from smart phones and tablets to note-book personal computers and televisions. The FPD industry has been dominated by liquid crystal displays (LCDs), which continue to advance with newly emerging component technologies. On the other hand, with continuous advances in materials, device architectures, and manufacturing technologies, the organic light-emitting diode (OLED) display is rapidly expanding its market share in small-sized, mobile applications. Since the launch of mass production in 2007, OLED displays are well into the phase of economic scale-up to large areas for large-screen TVs [1, 2]. The OLED display provides a set of attractive attributes for deployment in high-quality video applications, such as high-contrast ratio, ultra-fast response, vivid visual full-color appearance, and mechanically, a thinner and simpler structure compared to the LCD [2]. Recently, new light emitting diodes (LEDs), including quantum-dot light emitting diodes (QDLEDs) [3] and micro-light emitting diodes (μLEDs) [4] have emerged as attractive competitors in next-generation display technologies, ranging from micro-sized virtual/augmented reality to large-sized TVs. LED-based displays are current-driven and bring several attractive benefits such as thinner size, lighter weight, mechanically flexible, and more efficient with better image/video quality.

To realize high information-content, high-resolution displays with LED technologies, active-matrix addressing is required. This can be realized using crystalline silicon field effect transistor (Si-FET) or thin-film transistor (TFT) backplanes to control and drive the light-emitting elements [5, 6]. In contrast to LCDs, which are voltage-driven, LEDs are current-driven. Compared to the simple one-TFT and one-capacitor (1T-1C) pixel circuit used in active-matrix liquid-crystal displays (AMLCDs), the pixel circuit for active-matrix LED displays are more complex, requiring at least one additional TFT to serve as a current source to provide a constant current for the light-emitting elements. In the simplest 2T-1C pixel circuit as shown in the inset of Figure 7.1, T_1 works as the switch and T_2 as the driving TFT. For this, the well-established hydrogenated amorphous silicon (a-Si:H) TFT, which has been widely used in AMLCDs, is not as desirable due to its relatively low mobility and instability [7, 8]. Low temperature polycrystalline silicon (LTPS) and amorphous metal oxide semiconductor (AOS) TFTs have been adopted for the active

Advances in Semiconductor Technologies: Selected Topics Beyond Conventional CMOS, First Edition. Edited by An Chen.
© 2023 The Institute of Electrical and Electronics Engineers, Inc. Published 2023 by John Wiley & Sons, Inc.

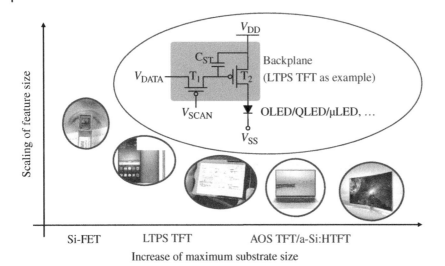

Figure 7.1 Transistor backplane technologies for various display-size applications, including crystalline silicon field effect transistor (Si-FET), thin-film transistors in low temperature polycrystalline silicon (LTPS), amorphous metal oxide semiconductor (AOS), and hydrogenated amorphous silicon (a-Si:H). Inset: Simplest pixel circuit composed of two p-type LTPS TFTs and one capacitor for active-matrix light emitting displays.

matrix, but still issues remain in terms of achieving stable and spatially uniform emissive current. Therefore, either internal or external compensation driving schemes are required to reduce the spatial or temporal luminance variations caused by fabrication processing and/or aging in practical applications [9–12]. Moreover, the light-emitting element could also have instability issues to be compensated. In particular, red-green-blue (RGB) sub-pixels undergo different degradation rates, which cause color mismatch after long-term operation [13–15]. These challenges call for comprehensive solution wherein the backplane provides stable operation of both TFT and LED.

This chapter first introduces the basics of various LED technologies for displays, followed by a summary of the key features of backplane technologies and inherent issues with current driving of LED-based displays. The design of pixel circuits and associated driving methods, developed over the last 20 years to address these issues, will be reviewed. Finally, additional design considerations with displays moving toward higher resolution, faster refresh rate, and lower power will be addressed.

7.2 Light-Emitting Diodes for Displays

This part will introduce the various LED technologies, including thermally evaporated and printed OLEDs, QLEDs, and µLEDs, which are already in commercialize displays and/or under industrial research and development.

7.2.1 Thermally Evaporated OLEDs

The organic layers in OLEDs can be deposited through vacuum thermal evaporation or solution printing processes. The former approach has been widely adopted in manufacturing for commercial applications. A typical OLED structure is illustrated in Figure 7.2a. When a voltage bias is applied, the carriers (hole/electron) inject from the anode/cathode with help of the hole injection

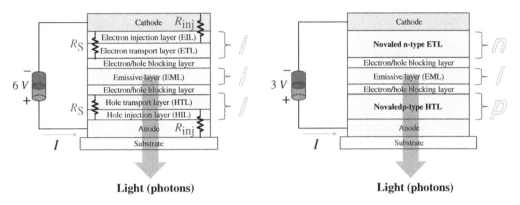

Figure 7.2 (a) Typical OLED structure fabricated by thermal evaporation. (b) The p-i-n structure OLED is designed to avoid injection between ETL or HTL and reduce the operation voltage.

layer/electron injection layer (HIL/EIL), respectively. After crossing through the hole transport layer/electron transport layer (HTL/ETL), the carriers recombine and form excitons. The electron/hole blocking layers confine the injected electrons and holes inside the emissive layer. The excitons diffuse and transit from an excited state to the ground or an intermediate state, resulting in photon generation and light radiation. The luminance efficiency of OLEDs has been continuously improved yielding low-operating power. The external quantum efficiency can be expressed as follows:

$$\eta_{ext} = \eta_c \times \eta_{int} = \gamma \times \eta_r \times \eta_{F-P} \tag{7.1}$$

where η_c is the out-coupling efficiency, η_{int} the internal quantum efficiency, γ the fraction of charge carriers recombining to form the excitons (i.e. the electron/hole charge carrier balance factor), η_r the exciton production efficiency, and η_{F-P} the fraction of total excitons formed resulting in radiative decay (i.e. internal quantum efficiency of luminescence).

By lowering the hole/electron injection barrier using HIL/EIL, and optimizing the charge transport capability of the HTL/ETL, a good balance between the two types of charges can be achieved for maximizing η_r. The injected holes/electrons are confined at the emissive layer (EML) with the electron blocking layer/hole blocking layer (EBL/HBL) to improve exciton production efficiency (η_r). One-fourth of the formed excitons are singlets, and others are triplets. By adopting the phosphorescent EML to replace the fluorescent one, OLEDs can be made to efficiently harvest the energy from both singlet and triplet excitons to produce light, making η_{F-P} close to 100%. Further incorporation of doped HTL and ETL layers can avoid using injection layer between ETL or HTL and electrodes and reduce the operation voltage for higher luminous efficiency (Figure 7.2b) [16]. Although vacuum evaporation has been well commercialized for manufacturing small-size OLED displays, it has drawbacks of large material waste and high capital infrastructure, leading to high production cost, especially for large-size displays.

7.2.2 Realization of Full Color Displays

Several strategies were proposed for fabricating full-color OLED displays [17]. The three main methods are illustrated in Figure 7.3. To realize full-color small-size displays, the method adopted to form R/G/B subpixels is by using vacuum evaporation through fine metal mask (FMM), as illustrated in Figure 7.3a. The R/G/B method can help achieve excellent color quality and

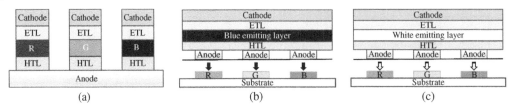

Figure 7.3 Different approaches for realization of full-color displays: (a) red/green/blue (R/G/B) subpixels; (b) white OLED with color; (c) filtersblue OLED with color conversion layers.

high-power efficiency. However, very precise alignment of the FMM is required for selective deposition of RGB materials. This becomes very challenging when manufacturing large-size displays, since both the thin glass substrate and FMM are prone to sagging when the size becomes too large. Another issue with the RGB architecture is the differential aging properties of the R/G/B subpixels which can cause color shift after long-term operation [18]. Since blue OLEDs have lower efficiency and poorer lifetime than the red and green counterparts, a larger area is normally designed for the blue subpixels to be operated at low current density [19].

For large-size active-matrix organic light-emitting diode (AMOLED) displays (e.g. TV), the color filter method with white OLEDs by vacuum evaporation is used to realize full color, as depicted in Figure 7.3b [20]. Compared to the RGB method, this method offers a few attractive features such as simple manufacturing, equal lifetime of all subpixels; and most importantly, the ability to use the developed high-efficiency stable white organic light emitting diode display (WOLED) formulation. However, approximately 70% of the white emission is absorbed by each of the RGB color filter, significantly increasing the power consumption. To reduce the power consumption, a white-emitting sub-pixel is added into each pixel to form the red-green-blue-white (RGBW) structure [20]. With this structure, the power consumption is dependent on the frequency of use of the highly efficient unfiltered white subpixel. For images with high probability of neutral color content, the RGBW display will have lower power consumption than the RGB one. For images with a high probability of saturated color content, the RGBW display will have a power consumption nearly equivalent to the RGB display. To further reduce the power consumption, color down conversion layers are introduced between the color filters and the WOLED to recycle emission from blue to green and from yellow to red [21].

Another approach for full color is using blue OLEDs for all subpixels, while red and green subpixels are realized with color down conversion layers (Figure 7.3c) [22]. The blue sub-pixel can simply pass the blue light with minimal losses. The green and red sub-pixels, each with a layer of green or red quantum dots (QDs) instead of an absorbing color filter, absorbs the blue light and down-converts it to green and red light, respectively. Not only does each green and red subpixel solely emit the desired color (and thus provide a saturated color primary for the display), the light throughput of each sub-pixel can in principle be much higher than in a conventional LCD. Compared to the R/G/B method, using the same blue OLED also makes the process much easier and avoids the issue of different aging properties with the subpixels. The power loss is less than the white OLED architecture since less light absorbs in the color conversion layers. Recently, quantum dots (QDs) have been used in the color conversion layers with high photoluminescence efficiency and excellent saturated color. These are being referred to as "QD-OLED" and used in for large-screen displays [23]. However, the relatively low efficiency and poor lifetime of the blue OLED is still an issue.

7.2.3 Printed Displays

To achieve cost-competitive manufacturing for large-screen OLED displays, solution-based processes would be an alternative for higher yield and economic material utilization. Ink-jet printing is a preferred technique to realize RGB subpixels for large-sized, full color displays. In practical manufacturing, the printed OLEDs use a simpler structure compared to the evaporated counterparts (Figure 7.4a), and inkjet printing for the HIL, HTL, and EML followed by evaporation process for other layers is used, as depicted in Figure 7.4b [24]. With advances in efficiency and lifetime of soluble OLED materials and related processing techniques, there has been great progress in printing AMOLED displays [25–29]. A case in hand is J-OLED and its efforts with commercialization of printed AMOLED panels [28].

With the tremendous progress on QLED efficiency and lifetime achieved with solution printing processes, this is being regarded as a promising choice technology for large-area, energy-saving, ultra-thin, and flexible display screens [3]. Quantum dots are a unique class of emitters with size-tunable emission wavelengths, saturated emission colors, near-unity luminance efficiency, and inherent photo- and thermal-stability. Based on a structure similar to the printed OLEDs, using QDs as the emissive layer (Figure 7.5) can offer distinct advantages of pure saturated colors with narrow full width at half maximum (FWHM) and potentially better air stability [30–32]. Therefore, excellent color performance can be achieved without the need for microcavities or

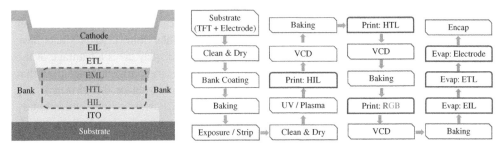

Figure 7.4 The OLED structure and process flow used in practical manufacturing for printed OLED displays [C.-L. Chen, et al. SID 2013 DIGEST 760].

Structure of cadmium-free blue
Quantum Dot ligth-emitting diodes (QLEDs)

Blue QD double emitting layer

Figure 7.5 Cross-sectional image of a typical QD-LED device. Source: Taehyung Kim et al., 2020, Springer Nature.

filters, which are often used in sole-OLED displays. A wide color gamut of 90% BT. 2020 coverage has already been reported for QD-LEDs [33]. High resolution active-matrix addressing QD-LED displays have been demonstrated recently, showing great potential [34–36]. However, several challenges need to be addressed for further improving the power conversion efficiency and lifetime of blue QD-LEDs. First, there is lack of nonblinking blue QDs so far. Second, blue QDs have very deep valence-band energy levels and thus new organic HTLs are required to achieve efficient hole injection in devices. Third, the quenching mechanisms, including energy transfer in blue QD layers, interfacial charge transfer between the blue QDs and the charge transport layers and electrical-field-induced quenching in blue QD-LEDs, have not been clearly understood. Therefore, for full-color QD-LED displays, development of efficient and stable blue QLEDs is the holy grail, especially with cadmium-free requirements [30]. Efforts on crystal engineering and surface engineering for the QDs, material chemistry for high-performance HTLs with high work function, device structure engineering, and in-depth understanding of the device-physics are needed.

7.2.4 Micro-LED

Both OLEDs and QD-LEDs rely on new materials development that meet the challenges of efficiency and reliability improvement, especially for blue light emission devices. By leveraging the well-established III–V semiconductor LED industry, emissive micro-LED displays have recently been attracting growing interest in view of their potential advantages such as high brightness and reliability, and the capability of realizing highly transparent displays [36, 37]. For larger displays (>100″ diagonal) such as TVs and indoor signage, modular mini-LED products have been demonstrated for commercial applications. The micro-LED display takes advantage of the exceptional luminance by spreading the generated photons over a larger area or space than that occupied by the micro-LED itself (generally smaller than $50\,\mu m \times 50\,\mu m$ or $0.0025\,mm^2$) [36]. As shown in Figure 7.6, one approach is hybridizing the micro-LED arrays onto complementary metal-oxide semiconductor (CMOS) backplanes to form highly integrated microdisplays, which disperses light optically into a larger space. Active matrix micro-LED displays with a resolution larger than 5000 pixels per inch (PPI) have been demonstrated for augmented reality and compact projection [38]. The other approach is to distribute individual micro-LED die spatially onto an active-matrix TFT backplane using different mass transfer techniques [39, 40]. Such TFT-based active matrix micro-LED display prototypes ranging from small-size to large-size have been demonstrated on both rigid and flexible backplanes to form emissive RGB pixels [41–43].

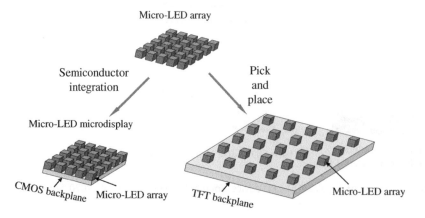

Figure 7.6 Illustration of two ways for making micro-LED displays.

7.3 TFT Backplanes

The TFT technologies currently available on a manufacturing-scale are a-Si, LTPS, and AOS TFT (see comparison in Table 7.1). The development a-Si:H TFT was a major milestone in the FPD history, enabling mass production of high-information content AMLCDs [44]. The a-Si:H TFT as depicted in Figure 7.7a can be manufactured over large-area and at low cost, while having uniform and reproducible device performance to match the driving requirements of LCDs.

With the continuously increasing demand on display resolution, refresh rate, and power consumption, TFTs with higher electron mobility, good electrical stability, and uniform performance are becoming increasingly mandatory. The a-Si:H TFT, having large parasitic capacitance and low

Table 7.1 Comparison of the industry available display backplane technologies: a-Si, LTPS, and AOS TFT.

TFT Type Parameter	a-Si:H TFT	LTPS TFT	AOS TFT
Structure	BCE	Coplanar	BCE, ES, Coplanar
Number of photo masks	4–5	7 (PMOS)–9 (CMOS)	6 (BCE)–8 (coplanar)
Maximum substrate Temperature (°C)	~350	450–500	~350
Substrate size	Gen 10	Gen 6	Gen 8
Mobility (cm²/V/s)	0.5	50–100	>10
S Slope (V/dec)	0.4–0.5	0.2–0.3	0.09–0.2
Normalized I_{off} (A/µm)	10^{-14}	10^{-12}	$<10^{-20}$
Stability	Bad	Very good	Not good
Short-range uniformity	Good	Not good	Very good
Run-to-run variation	Excellent	Good	Issue

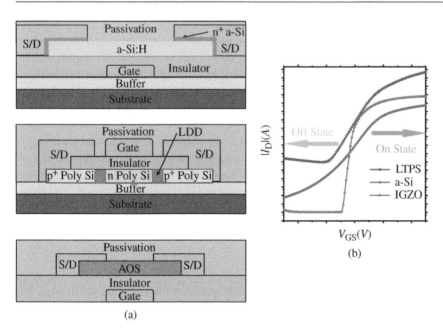

Figure 7.7 (a) Typical device structures of a-Si:H, LTPS, and AOS TFTs. (b) Comparison of the typical current-voltage characteristics of the three types of TFTs.

electron mobility, fails to meet these stringent requirements. Moreover, the bias instability of the a-Si:H TFT makes it difficult for current-driven AMOLED displays [8, 45, 46]. To overcome the poor mobility and bias stress instability of a-Si TFTs, LTPS TFTs with much higher mobility and very stable electrical performance were developed [47, 48]. This is now being successfully applied to commercial high-end AMLCD displays and AMOLED displays of high image quality. The LTPS TFT as shown in Figure 7.7a not only enables a much higher pixel density but also allows integration of the gate drivers and the source multiplexing circuits on the same substrate to achieve a narrow bezel. However, the LTPS TFT requires the additional high-cost of laser recrystallization and doping processes, and suffers from poor spatial uniformity in electrical properties [48]. Manufacturing of the LTPS TFTs is limited to be scaled up to large area (e.g. >Gen 6), which adds to high cost for large-area applications.

Since 2004, the AOS semiconductor-based TFTs have attracted wide attention for their advantages of relatively high mobility, steep subthreshold slope and importantly, ultra-low leakage current [49, 50]. In addition, it can be manufactured at low cost and its large area manufacturability attributes are similar to a-Si:H TFTs. With these attractive features, the quaternary indium–gallium–zinc–oxide (IGZO) AOS has been quickly adopted for display manufacturing for high-resolution, low-power AMLCD panels, and large-size AMOLED panels [51–55]. The AOS TFTs can be fabricated in different structural forms according to the needs of performance and/or cost, as shown in Figure 7.7a [56]. The amorphous nature of the AOS semiconductor leads to better short-range uniformity of the TFT compared to that of the LTPS counterpart. The manufacturing process of AOS TFTs is more compatible with that of the a-Si infrastructure, and thus has the advantage of production scalability and cost over LTPS. Although its stability and mobility is not as high as that of the LTPS TFT, the AOS TFT shows great potential for improvement and functionalization through tuning material composition, processes, and interfaces [57–67]. In addition, the AOS TFT has an extremely low leakage current, enabling low refresh rate displays for low-power consumption [68]. A comparison of the typical current–voltage characteristics of a-Si:H, AOS, and LTPS TFTs is shown in Figure 7.7b.

In all, the most mature technologies in the industry that are suitable for current-driven displays are LTPS and AOS TFTs. The former is currently limited to small-size high resolution displays due to difficulties in scaling beyond G6, while the latter is used in large area AMOLED display manufacturing. However, because the LTPS TFT suffers large leakage current with severe voltage dependence, the AOS TFT has recently been integrated with the LTPS TFT to form a hybrid backplane technology called LTPO for low refresh rate mobile AMOLED displays [69]. For micro-displays, the normally adopted technology is the Si-FET backplane, which is beyond the current scope. The organic thin-film transistor (OTFT) has attracted wide attention as a potential technology for creating truly flexible displays in a highly customizable way, owning to the advantages of low temperature, low-cost manufacturing, and excellent intrinsic mechanical flexibility [70]. Despite of many efforts on developing OTFT-driven AMOLED displays [71–75], many technical challenges remain to be addressed for the required device performance, stability, and manufacturability [76].

7.4 Driving Schemes and Pixel Circuits

Gray scales can be represented for displays in two ways: analog driving and digital driving. To realize active matrix addressing for a current-driven display, the pixel circuits require not only one switching TFT but also another driving TFT to provide constant current for the light-emitting element. Details of the implementation of pixel circuits for these driving techniques, as shown in Figure 7.8a–c will be introduced in the following.

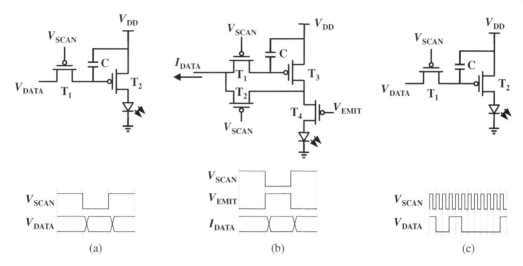

Figure 7.8 Classification of various driving schemes for LED-based displays: (a) analog driving with voltage programming, (b) analog driving with current programming, and (c) digital driving.

7.4.1 Analog Driving

This is also known as pulse-amplitude modulation (PAM). Since the luminance is linearly proportional to the current density for most, if not all, LEDs, the gray scales can be tuned by modulating the current via changing the amplitude of the input data signals. The input data can be either a voltage signal or current signal, for which the pixel circuits are different, as shown in Figure 7.8a,b, respectively. For voltage programming, the pixel circuits require at least two TFTs and one storage capacitor (2T-1C). This includes one switching TFT and one driving TFT to provide a constant current to the light-emitting element. When the pixel is selected through turning on the switching TFT T_1, the data voltage signal from the source driver is programmed into the pixel and stored at the gate of the driving TFT T_2 during the frame to provide the desired OLED current. Similar to that in AMLCDs, the switching TFT is required to have high carrier mobility for a large turn-on current at small device dimensions and low leakage current to ensure sufficient charge retention during the whole frame time.

The driving TFT T_2 is operated in the saturation regime and converts the input data voltage (V_{DATA}) to the pixel emission current (I_{GL}), representing the gray level, simply described as follows:

$$I_{GL} = \frac{1}{2}\mu C_G \frac{W}{L}(V_{DATA} - V_{th})^2 \tag{7.2}$$

where μ is the mobility, C_G the specific gate dielectric capacitance, W and L the channel width and channel length, respectively, and V_{th} the threshold voltage of the driving TFT.

Therefore, the emission current is very sensitive to the drive TFT characteristics, which might vary spatially over the display panel or temporally over time, causing display nonuniformity (e.g. mura) or image sticking issues [9]. Moreover, the light-emitting devices, especially OLEDs and QD-LEDs, could have aging issues after long-term operation. As a result, in practical applications, more complicated pixel circuits and driving schemes are required to compensate these non-idealities. This will be discussed in detail in the next section.

The pixel circuit for the current programming method is depicted in Figure 7.8b. When the pixel is selected (T_4 switched off and T_1/T_2 turned on), the gate voltage of the drive TFT T_3 is programmed by the data current from the source driver to reproduce the current signal for light emission. The reproduced emission current is theoretically equal to the input data current, independent of process variations of the drive TFT. However, this method suffers from long settling time for programming due to small current level and large parasitic capacitances on the data lines, especially in lower gray scales. Although there have been lots of efforts on improving the settling rate based on different ways, it is still impractical for high resolution displays [77–80].

7.4.2 Compensation for Voltage Programming

Voltage programming analog driving methods have been adopted in practical AMOLED display panels. However, for the simplest 2T-1C pixel circuit as depicted in Figure 7.8a, the pixel current is sensitive to performance variation of the driving TFT. Therefore, compensation is required to suppress the influence of the manufacturing process shortcomings or aging induced spatial or temporal performance variations. Process-induced spatial variations result in after-fabrication nonuniformity of the displays, and can affect the yield significantly. The aging induced temporal instability may increase the nonuniformity over time, and shorten the lifetime. The compensation methods developed to address these issues are generally classified as internal compensation and external compensation.

7.4.2.1 Internal Compensation

Various pixel circuit designs have been proposed to implement the internal compensation driving scheme [81–93]. These designs are generally based on three mechanisms, as described with the simple circuit schematics in Figure 7.9. A compensating voltage (V_{comp}) is precharged into a storage capacitor (C_S), and then discharged through the diode-connected drive TFT until it turns off, generating a gate-source voltage equal to V_{th} of the driving TFT. In the programming period, a

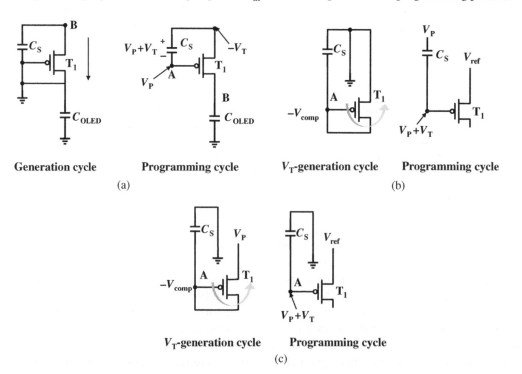

Figure 7.9 Illustration of various circuit techniques for implementing the internal compensation driving scheme for AMOLED displays: (a) stacked-compensation, (b) boot-strapping-compensation, and (c) parallel-compensation.

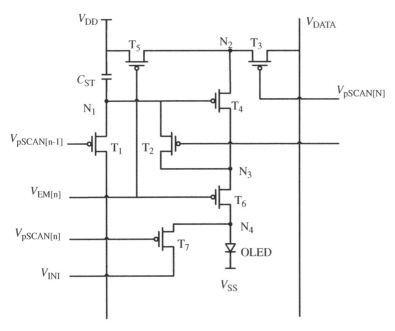

Figure 7.10 The 7T–1C internal compensation pixel circuit being used in commercial mobile AMOLED displays based on p-type LTPS TFTs.

programming voltage (V_P) is added to the generated V_{th}, resulting in a gate-source voltage as $V_P + V_{th}$. Therefore, the pixel emissive current during driving can be independent of the V_T of the driving TFT. As seen in Figure 7.9, different ways can be adopted of generating the V_{th}, and adding the programming voltage (V_P), including stacking, boot-strapping, and parallel-compensation. With the parallel-compensation method, V_{th}-generation and data programming occur simultaneously, which can help simply the timing and achieve fast driving. Currently, for commercialized mobile displays, the LTPS TFT is used for its high mobility and high-resolution integration. A 7T-1C pixel circuit using the parallel-compensation method has been popularly used in mobile AMOLED display products to suppress the influence of process induced spatial variations with LTPS TFTs (Figure 7.10). The presence of mobility variations, and other nonideal characteristics of TFTs, such as leakage, hysteresis, and kickback with the switching TFT, and kink with the driving TFT, might cause imperfect compensation and is thus needed to be considered [94–98]. Moreover, as the OLED luminance efficiency increases and the resolution improves, the required emissive current is significantly decreased. As a result, the driving TFT is operated in the subthreshold regime for most gray-levels and in the deep subthreshold regime for lower ones, resulting in difficulties of accurate gray-level representation [99–101]. There is not only the threshold voltage variation but also the subthreshold swing variation, which might obviously affect display uniformity and in turn need more design considerations for compensation [102].

7.4.2.2 External Compensation
The internal compensation as we saw needs relatively complex pixel circuit design, which requires a large design effort, facing subsequent challenges for high yield and further resolution improvement. Moreover, the internal compensation schemes are mostly only for V_{th} compensation, and are difficult for compensating other nonideal TFT characteristics and aging of the LEDs. External compensation schemes were proposed to simplify the pixel circuits by partially or fully

moving the compensation function to the external driver, and also enhance the compensation functions for not only the TFT but also the LEDs [103–107].

The external compensation methods can be classified into two types: closed-loop and open-loop. With the closed-loop method as shown in Figure 7.11a, the pixel properties are sensed, and then compensated through a feedback system with the external driver at each frame [95, 108–111]. Such external compensation methods can perform real-time compensation of all of the nonideal characteristics of the drive TFT, but face long settling time issues for feedback loop convergence in the limited pixel-selected time.

With open-loop compensation methods (Figure 7.11b), the panel properties are sensed during the nondisplay periods using external source drivers. The input data signals are then updated based on the sensed information to drive the pixels in subsequent frames [112–115]. Since the LTPS TFT is difficult to be scalable for large-area manufacturing and the a-Si:H TFT has issues of low mobility and poor stability, the AOS TFT is regarded as the mainstream backplane technology for large-size AMOLED displays and potentially also for medium-size ones [56, 103]. However, AOS TFTs have bias stress instabilities under long-term operation. Therefore, to compensate the aging effects of both the AOS TFT and the OLED, the open-loop external compensation method has been applied to large-size AMOLED TV panels with a simple 3T-1C pixel circuit, but requires a relatively long nondisplay waiting period to sense all of the pixel information across the panel. To reduce the waiting time, inserting blank frames between two neighboring normal frames for sensing has been proposed [114, 115]. This, however, also suffers from timing issues with the fresh rate limitation. A displaying-synchronous open-loop (DSOL) external compensation method has been proposed, with which the sensing and compensation is performed synchronously during several continuous display frames with normal displaying [116].

The external compensation methods can in theory provide better compensation capabilities than the internal schemes and can also simplify the pixel circuit. However, analog sensing circuitry, additional memories, and external logic blocks are required to sense and store the electrical information as well as modulate the input video signals. This results in increased system costs. Therefore, the external compensation methods are currently only applied for large-sized AMOLED displays. With increase in computing power and memory size in mobile devices, there have been proposals of using external compensation methods with AOS TFTs for medium- and small-sized AMOLED displays [103].

7.4.3 Digital Driving

Digital driving has been proposed as an alternative to mitigate variations in TFT performance using with the simplest 2T-1C pixel structure (Figure 7.8c). It is also useful for true black-levels since the LEDs can be completely turned off. In digital driving, the input data voltage signals from the source driver are in two levels "1" and "0," representing light on and off states, and the gray-levels are realized by modulating the light emission time during the frame.

Pulse-width modulation (PWM) is the most common approach used for digital driving due to its compatibility with the traditional frame-based refreshing technique. Each frame is divided into several subframes to generate gray-scales as illustrated in Figure 7.12. The displayed gray-scales are represented by summation of the pulse width in each subframe. Each subframe is also composed of a programming period and emitting period. During the programming period, the data value, which is 1 or 0 for each pixel in this subframe, is programmed and stored in the storage capacitor (C_s). Then, during the emitting period, the whole display is lighted up by powering on V_{DD}. This method is named as simultaneous emission (SE), since the whole display is turned on simultaneously after

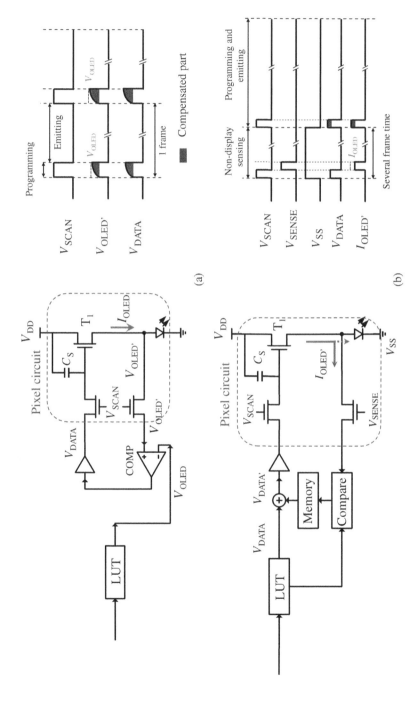

Figure 7.11 Illustration of basic operation mechanisms with the two types of external compensation methods: (a) close-loop and (b) open-loop.

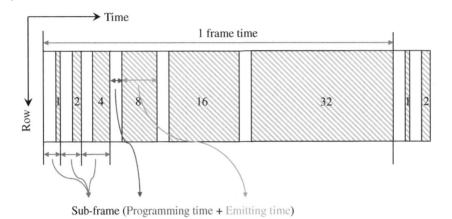

Figure 7.12 Illustration of the PWM-based digital driving method: one frame is divided into several subframes, and each subframe is composed of programming and emitting periods.

data programming. In contrast, most of the analog driving methods are progressive emission (PE), for which light emission occurs row by row in the display array right after data programming.

However, PWM suffers from a false-image-contour problem caused by an incompatibility with the human visual system (Figure 7.13a). A nonframe-refreshing digital-driving technique based on pulse-density modulation (PDM) using a Δ-Σ modulation scheme was proposed to suppress false image contours without increasing system complexity (Figure 7.13b), while providing the same or better resolution and relaxed gate-scan time as compared with PWM [117].

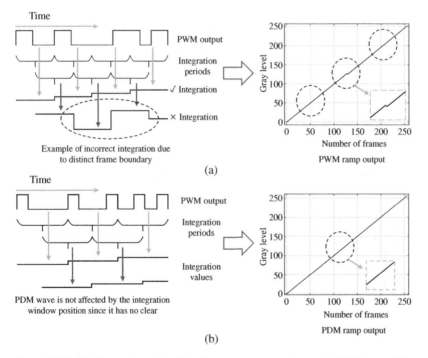

Figure 7.13 (a) Illustration of the false-image-contour problem with PWM caused by incompatibility with the human visual system; (b) Illustration of the pulse density modulation (PDM) method. Source: Modified from Jang et al. [117].

Generally, a digital driver needs to charge and discharge each pixel more frequently than the analog equivalent, and thus consumes higher dynamic power for panel driving. Moreover, it brings challenges in the operation speed of the TFTs to realize high-resolution and high gray-scales.

7.4.4 Hybrid Driving

The AMOLED display commonly uses voltage programming based on the analog driving method, which converts the input data to represented gray-levels in one programming frame utilizing the voltage-to-current transduction characteristics of the TFTs. However, for state-of-the-art mobile AMOLED displays with efficient light emission and high-resolution pixels, the required pixel emissive current is very small and in the level of 10^{-11}–10^{-8} A, causing the LTPS drive TFT to be operated in the deep subthreshold regime at the lower gray-levels. As a result, the lower gray-levels are difficult to be accurately modulated with analog driving. To address this issue, an analog/digital hybrid driving scheme is implemented to have PWM-type modulation for lower gray-levels using the pixel circuit as depicted in Figure 7.10. The low gray-levels are represented by a fixed data voltage signal but with different emission period. However, because it is difficult to realize modulation at a high enough frequency (i.e. 1250 Hz in IEEE standard), the PWM digital driving suffers low-frequency flickering issues.

For micro-LEDs, there is wavelength shift of light emission with variation of the driving current amplitude [118]. The current density ramp-up with increase of the applied voltage is also much sharper than that of the OLEDs, which causes a narrower data voltage range for defining the gray levels. Therefore, it is difficult to use analog driving for micro-LEDs to realize high gray-level and full color displays. In order to solve these issues, a hybrid pixel circuit for grayscale control using LTPS TFTs, composed of PAM and PWM parts, has been proposed for micro-LED displays [119, 120]. As shown in Figure 7.14, the PAM part samples the data voltage Sig(m) as gate-to-source voltage of the drive TFT via the scan pulse, Scan_PAM, and V_{th} external compensation of T_1 is realized via the Sense signal. A switch TFT inside the PWM part can modulate the gate voltage of the driving TFT T_1 to turn-off the driving current according to the required operating duration, so that the gray scale can be controlled well with PWM. The pull-up TFT in the PWM part has V_{th} internal compensation with the Reset, Ref, and Control signals.

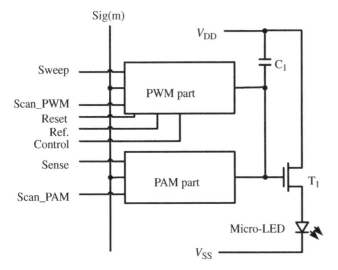

Figure 7.14 Hybrid driving pixel circuits composed of PAM and PWM parts with the PAM part being externally compensated. Source: Modified from Kim et al. [119].

7.5 Conclusion

This chapter presented some of the unique attributes and design considerations pertinent to flat panel active-matrix electroluminescent displays based on the LED. Stability issues related to the TFT backplane and LED were addressed, along with pixel circuits and driving methods placing emphasis on TFT and LED compensation carried out both internal and external means. The novel properties of thin-film organic- and quantum-dot-LED technology such as low weight, mechanical flexibility and durability, simple device integration, along with low-cost and large-area processibility, allow flat panel electroluminescent displays to be utilized in a wide range of applications from smart phones and tablets to notebook computers and TVs. Future developments, particularly in the area of flexible displays, are likely to enhance the performance of the devices discussed here, leading to more widespread applications. In particular, the challenges that remain include the ability to scale up to large areas, high-density integration in terms of the PPI, and increased LED efficiency and stability to reduce operating voltage and power consumption.

References

1 C. -H. Oh, H. -J. Shin, W. -J. Nam, B. -C. Ahn, S. -Y. Cha, and S. -D. Yeo, "Technological Progress and Commercialization of OLED TV," SID Symposium Digest of Technical Papers, vol. 44, no. 1, pp. 239–242, 2013.

2 H. -K. Chung, H. -D. Kim, and B. Kristal, "AMOLED Technology for Mobile Displays," SID Symposium Digest of Technical Papers, vol. 37, no. 1, pp 1447–1450, 2006.

3 X. Dai, Y. Deng, X. Peng, and Y. Jin, "Quantum-dot light-emitting diodes for large-area displays: towards the dawn of commercialization," *Advanced Materials*, vol. 29, no. 14, 1607022, 2017.

4 Y. Huang, G. Tan, F. Gou, M. -C. Li, S. -L. Lee, and S. -T. Wu, "Prospects and challenges of mini-LED and micro-LED displays," *Journal of the Society for Information Display*, vol. 27, no. 7, pp. 387–401, 2019.

5 D. Armitage, I. Underwood, and S. -T. Wu, *Introduction to Microdisplays*, Wiley, 2006.

6 R. Chaji and A. Nathan, *Thin Film Transistor Circuits and Systems*, Cambridge University Press, 2013, pp. 180.

7 G. R. Chaji and A. Nathan, "Low-power low-cost voltage-programmed a-Si:H AMOLED display for portable devices," *Journal of Display Technology*, vol. 4, no. 2, pp. 233–237, 2008.

8 A. Nathan, A. Kumar, K. Sakariya, P. Servati, S. Sambandan, and D. Striakhilev, "Amorphous silicon thin film transistor circuit integration for organic LED displays on glass and plastic," *IEEE Journal of Solid-State Circuits*, vol. 39, no. 9, pp. 1477–1486, 2004.

9 A. Nathan, G. R. Chaji, and S. J. Ashtiani, "Driving schemes for a-Si and LTPS AMOLED displays," *IEEE/OSA Journal of Display Technology*, vol. 1, no. 2, pp. 267–277, 2005.

10 R. Chaji and A. Nathan, "LTPS vs Oxide Backplanes for AMOLED Displays: System Design Considerations and Compensation Techniques", SID Symposium Digest of Technical Papers, vol. 45, no. 1, pp. 153–156, 2014.

11 H. -J. Shin, S. Takasugi, and K. -M. Park, "Technological Progress of Panel Design and Compensation Methods for Large-Size UHD OLED TVs," SID Symposium Digest of Technical Papers, vol. 45, no. 1, pp. 720–723, 2014.

12 Y. H. Jang, D. H. Kim, W. Choi, M. -G. Kang, K. I. Chun, J. Jeon, Y. Ko, U. Choi, S. M. Lee, J. U. Bae, K. Park, S. Y. Yoon, and I. B. Kang, "Internal Compensation Type OLED Display

Using High Mobility Oxide TFT," SID Symposium Digest of Technical Papers, vol. 48, no. 1, pp. 76–79, 2017.

13 A. Laaperi, "OLED lifetime issues from a mobile-phone-industry point of view," *Journal of the Society for Information Display*, vol. 16, no. 11, pp. 1125–1130, 2008.

14 F. Chesterman, B. Piepers, T. Kimpe, P. De Visschere, and K. Neyts, "Impact of long-term stress on the light output of a WRGB AMOLED display," *Journal of Display Technology*, vol. 12, no. 12, pp. 1672–1680, 2016.

15 S. Chen, W. Cao, T. Liu, S. -W. Tsang, Y. Yang, X. Yan, and L. Qian, "On the degradation mechanisms of quantum-dot light-emitting diodes", *Nature Communications*, vol. 10, no. 1, p. 765, 2019.

16 G. He, O. Schneider, D. Qin, X. Zhou, M. Pfeiffer, and K. Leo, "Very high-efficiency and low voltage phosphorescent organic light-emitting diodes based on a p-i-n junction," *Journal of Applied Physics* vol. 95, no. 10, 5773, 2004.

17 P. E. Burrows, G. Gu, V. Bulovic, Z. Shen, S. R. Forrest, and M. E. Thompson, "Achieving full-color organic light-emitting devices for lightweight, flat-panel displays," *IEEE Transactions on Electron Devices*, vol. 44, no. 8, pp. 1188–1203, 1997.

18 C. -W. Han, K. -M. Kim, S.-J. Bae, H. -S. Choi, J. -M. Lee, T. -S. Kim, Y. -H. Tak, S. -Y. Cha, and B. -C. Ahn, "55-Inch FHD OLED TV employing New Tandem WOLEDs," SID Symposium Digest of Technical Papers, vol. 43, no. 1, pp. 279–281, 2012.

19 J. Wang and J. Chung, "Light-emitting display device provided with light-emitting portions having varying areas," US Patent 7443093, 2009.

20 A. D. Arnold, P. E. Castro, T. K. Hatwar, M. V. Hettel, P. J. Kane, J. E. Ludwicki, M. E. Miller, M. J. Murdoch, J. P. Spindler, S. A. Van Slyke, K. Mameno, R. Nishikawa, T. Omura, and S. Matsumoto, "Full-color AMOLED with RGBW pixel pattern," *Journal of the Society for Information Display*, vol. 13, no. 6, pp. 525–535, 2005.

21 W. -Y. So, M. S. Weaver, and J. J. Brown, "Power Efficient RGBW AMOLED Displays Incorporating Color-Down-Conversion Layers," SID Symposium Digest of Technical Papers, vol. 43, no. 1, pp. 282–285, 2012.

22 C. W. Tang, D. J. Williams, and J. C. Chang, "Organic electroluminescent multicolor image display device," US Patent 5294870, 1994.

23 J. H. Lee, "QD display: a game-changing technology for the display industry," *Information Display*, vol. 36, no. 6, pp. 9–13, 2020.

24 Y. Li, "RGB Printed AMOLED: Challenges and Opportunities," in: *International Conference on Display Technology (ICDT 2018)*, 9–12 April 2018, vol. 49, no. S1, pp. 95–98.

25 D. Zhao, W. Huang, L. Dong, Q. Jin, Y. Tian, G. Yuan, J. Ryu, and L. Wang, "5.5 Inch Full Screen Flexible High-Resolution OLED Display Fabricated by Ink Jet Printing Method," SID Symposium Digest of Technical Papers, vol. 50, no. 1, pp. 945–948, 2019.

26 L. -Q. Shao, T. Dong, J. -S. Liang, Z. -T. Bi, Z. Li, J. Song, and S. -T. Huo, "The Development of 403ppi Real RGB Printing AMOLED," SID Symposium Digest of Technical Papers, vol. 50, no. 1, pp. 1943–1945, 2019.

27 J. Kang, Y. Koo, J. Ha, and C. Lee, "Recent Developments in Inkjet-printed OLEDs for High Resolution, Large Area Applications," SID Symposium Digest of Technical Papers, vol. 51, no. 1, pp. 591–594, 2020.

28 K. Noda, T. Komatsu, T. Fukuda, and M. Goto, "Recent Technology of Printed OLED Display and Its World's First Commercialization," SID Symposium Digest of Technical Papers, vol. 51, no. 1, pp. 587–590, 2020.

29 Z. Gao, L. Yu, Z. Li, W. Shi, C. Li, L. G. Yuan, X. Sun, and D. Fu, 31 Inch Rollable OLED Display Fabricated by Inkjet Printing Technology, in *International Conference on Display Technology (ICDT 2020)*, 11–20 Oct. 2020, vol. 52, no. S1, pp. 312–314.

30 T. Kim, K. -H. Kim, S. Kim, S. -M. Choi, H. Jang, H. -K. Seo, H. Lee, D. -Y. Chung, and E. Jang, "Efficient and stable blue quantum dot light-emitting diode," *Nature*, vol. 586, no. 7829, pp. 385–389, 2020.

31 Y. -H. Won, O. Cho, T. Kim, D. -Y. Chung, T. Kim, H. Chung, H. Jang, J. Lee, D. Kim, and E. Jang, "Highly efficient and stable InP/ZnSe/ZnS quantum dot light-emitting diodes," *Nature*, vol. 575, no. 7784, pp. 634–638, 2019.

32 X. Dai, Z. Zhang, Y. Jin, Y. Niu, H. Cao, X. Liang, L. Chen, J. Wang, and X. Peng, "Solution-processed, high-performance light-emitting diodes based on quantum dots," *Nature*, vol. 515, no. 7525, pp. 96–99, 2014.

33 J. R. Manders, J. Hyvonen, A. Titov, K. P. Acharya, J. Tokarz-Scott, Y. Yang, W. Cao, Y. Zheng, L. Qian, J. Xue, and P. H. Holloway, "48-1: Invited Paper: High Efficiency and Ultra-Wide Color Gamut Quantum Dot LEDs for Next Generation Displays," SID Symposium Digest of Technical Papers, vol. 47, no. 1, pp. 644–647, 2020.

34 Y. Li, Z. Chen, B. Kristal, Y. Zhang, D. Li, G. Yu, X. Wang, L. Wang, Y. Shi, Z. Wang, Y. Chen, J. Yu, and Y. He, "Developing AMQLED Technology for Display Applications," SID Symposium Digest of Technical Papers, vol. 49, no. 1, pp. 1076–1079, 2018.

35 Y. Nakanishi, T. Takeshita, Y. Qu, H. Imabayashi, S. Okamoto, H. Utsumi, M. Kanehiro, E. Angioni, E. A. Boardman, I. Hamilton, A. Zampetti, V. Berryman-Bousquet, T. M. Smeeton, and T. Ishida, Active Matrix QD-LED with Top Emission Structure by UV Lithography for RGB Patterning, SID Symposium Digest of Technical Papers, vol. 51, no. 1, pp. 862–865, 2020.

36 V. W. Lee, N. Twu, and I. Kymissis, "Micro-LED technologies and applications," *Information Display*, vol. 32, no. 6, pp. 16–23, 2016.

37 Z. Chen, S. Yan, and C. Danesh, "MicroLED technologies and applications: characteristics, fabrication, progress, and challenges," *Journal of Physics D: Applied Physics*, vol. 54, no. 12, 123001, 2021, doi: 10.1088/1361-6463/abcfe4.

38 L. Zhang, F. Ou, W. C. Chong, Y. Chen, and Q. Li, "Wafer-Scale Monolithic Hybrid Integration of Si-Based IC and III–V Epi-Layers – A Mass Manufacturable Approach for Active Matrix Micro-LED Micro-Displays," SID Symposium Digest of Technical Papers, vol. 26, no. 3, pp. 137–145, 2018.

39 X. Zhou, P. Tian, C. -W. Sher, J. Wu, H. Liu, R. Liu, and H. -C. Kuo, "Growth, transfer printing and colour conversion techniques towards full-colour micro-LED display," *Progress in Quantum Electronics*, 100263, 2020, doi: 10.1016/j.pquantelec.2020.100263.

40 K. Kajiyama, Y. Suzuki, T. Hirano, Y. Yanagawa, K. Fukaya, N. Okura, T. Okuno, and A. Shimoura, "Manufacturing Process for Mass Production of Micro-LED Displays and High-Speed and High-Yield Laser Lift-Off Systems," SID Symposium Digest of Technical Papers, vol. 51, no. 1, pp. 100–103, 2020.

41 S. Nakamitsu, H. Ito, T. Suzuki, M. Nishide, K. Imaizumi, K. Yamanoguchi, F. Rahadian, K. Aoki, S. Matsuda, and R. Yokoyama, "High PPI Micro LED Display for Small and Medium Size," SID Symposium Digest of Technical Papers, vol. 50, no. 1, pp. 137–140, 2019.

42 H. -M. Kim, J. Lee, S. Park, W. Choi, S. Kim, K. -S. Jeon, Y. -H. Choi, J. -S. Lee, and J. Jang, "Full Color, Active-Matrix Micro-LED Display with Dual Gate a-IGZO TFT Backplane," SID Symposium Digest of Technical Papers, vol. 51, no. 1, pp. 335–338, 2020.

43 S. -L. Lee, C. -C. Cheng, C. -J. Liu, C. -N. Yeh, Y. -C. Lin, "9.4-inch 228-ppi flexible micro-LED display," *Journal of Society for Information Display*, vol. 29, no. 5, pp. 360–369, 2021.

44 A. J. Snell, K. D. Mackenzie, W. E. Spear, P. G. LeComber, and A. J. Hughes, "Application of amorphous silicon field effect transistors in addressable liquid crystal display panels," *Applied Physics*, vol. 24, no. 4, pp. 357–362, 1981.

45 A. Nathan, A. Kumar, K. Sakariya, P. Servati, K. S. Karim, D. Striakhilev, and A. Sazonov, "Amorphous silicon back-plane electronics for OLED displays," *IEEE Journal of Selected Topics in Quantum Electronics*, vol. 10, no. 1, , pp. 58–69, 2004.

46 K. Sakariya, C. Ng, P. Servati, and A. Nathan, "Accelerated stress testing of a-Si:H pixel circuits for AMOLED displays," *IEEE Transactions on Electron Devices*, vol. 52, no. 12, , pp. 2577–2583, 2005.

47 S. D. Brotherton, D. J. Mcculloch, J. B. Clegg and J. P. Gowers, "Excimer-laser-annealed poly-Si thin-film transistors," *IEEE Transactions on Electron Devices*, vol. 40, no. 2, pp. 407–413, 1993.

48 S. D. Brotherton, *Introduction to Thin Film Transistors: Physics and Technology of TFTs*. UK: Springer, 2013.

49 K. Nomura, H. Ohta, A. Takagi, T. Kamiya, M. Hirano, and H. Hosono, "Room-temperature fabrication of transparent flexible thin-film transistors using amorphous oxide semiconductors," *Nature*, vol. 432, no. 7016, pp. 488–492, 2004.

50 T. Kamiya, K. Nomura, and H. Hosono. "Origins of high mobility and low operation voltage of amorphous oxide TFTs: electronic structure, electron transport, defects and doping," *Journal of Display Technology*, vol. 5, no. 7, pp. 273–288, 2009.

51 J. Lee, D. Kim, D. Yang, S. Y. Hong, K. S. Yoon, P. S. Hong, C. O. Jeong, H. S. Park, S. Y. Kim, S. K. Lim, and S. S. Kim, "World's Largest (15-inch) XGA AMLCD Panel Using IGZO Oxide TFT," SID Symposium Digest of Technical Papers, vol. 39, no. 1, pp. 625–628, 2008.

52 C. Ha, H. J. Lee, J. W. Kwon, S. -Y. Seok, C. -II. Ryoo, K. -Y. Yun, B. -C. Kim, W. -S. Shin, and S. -Y. Cha, "High Reliable a-IGZO TFTs with Self-aligned Coplanar Structure for Large-Sized Ultrahigh-Definition OLED TV," SID Symposium Digest of Technical Papers, vol. 46, no. 1, pp. 1020–1022, 2015.

53 N. Saito, T. Ueda, K. Miura, S. Nakano, T. Sakano, Y. Maeda, H. Yamaguchi, and I. Amemiya, "10.2-Inch WUXGA Flexible AMOLED Display Driven by Amorphous Oxide TFTs on Plastic Substrate," SID Symposium Digest of Technical Papers, vol. 44, no. 1, pp. 443–446, 2013.

54 A. Chida, K. Hatano, T. Inoue, N. Senda, T. Sakuishi, H. Ikeda, S. Seo, Y. Hirakata, S. Y. S. Yasumoto, M. Sato, and Y. Yasuda, "A 3.4-Inch Flexible High-Resolution Full-Color Top-Emitting AMOLED Display," SID Symposium Digest of Technical Papers, vol. 44, no. 1, pp. 196–199, 2013.

55 C. I. Park, M. Seong, M. A. Kim, D. Kim, H. Jung, M. Cho, S. H. Lee, H. Lee, S. Min, J. Kim, and M. Kim, "World's first large size 77-inch transparent flexible OLED display," *Journal of the Society for Information Display*, vol. 26, no. 5, pp. 287–295, 2018.

56 T. Arai, "Oxide-TFT technologies for next-generation AMOLED displays," *Journal of the Society for Information Display*, vol. 20, no. 3, pp. 156–161, 2012.

57 K. Ghaffarzadeh, A. Nathan, J. Robertson, S. Kim, S. Jeon, C. Kim, U. -I. Chung, and J. -H. Lee, "Persistent photoconductivity in Hf–In–Zn–O thin film transistors," *Applied Physics Letters*, vol. 97, no. 14, 143510, 2010.

58 A. Nathan, A. Ahnood, M. T. Cole, S. Lee, Y. Suzuki, P. Hiralal, F. Bonaccorso, T. Hasan, L. Garcia-Gancedo, A. Dyadyusha, and S. Haque, "Flexible electronics: the next uniquitous platform," *Proceedings of the IEEE*, vol. 100, no. Special Centennial Issue, pp. 1479–1510, 2012.

59 R. Martins, A. Nathan, R. Barros, L. Pereira, P. Barquinha, N Correia, R. Costa, A. Ahnood, I. Ferreira, and E. Fortunato, "Complementary metal oxide semiconductor technology with and on paper," *Advanced Materials*, vol. 23, no. 39, pp. 4491–4496, 219, 2011.

60 S. -E. Ahn, I. Song, S. Jeon, Y.W. Jeon, Y. Kim, C. Kim, B. Ryu, J. -H. Lee, A. Nathan, S. Lee, G.T. Kim, and U. -I. Chung, "Metal oxide thin film phototransistor for remote touch interactive displays," *Advanced Materials*, vol. 24, no. 19, pp. 2631–2636, 2012.

61 S. Jeon, S. -E. Ahn, I. Song, C. J. Kim, U. -I. Chung, E. Lee, I. Yoo, A. Nathan, S. Lee, J. Robertson, and K. Kim, "Gated three-terminal device architecture to eliminate persistent photoconductivity in oxide semiconductor photosensor arrays," *Nature Materials*, pp. 301–305, 2012, doi: 10.1038/NMAT3256.

62 Y. Cong, D. Han, X. Zhou, D. Han, X. Zhou, L. Huang, P. Shi, W. Yu, Y. Zhang, S. Zhang, X. Zhang, and Y. Wang, "High-performance Al–Sn–Zn–O thin-film transistor with a quasi-double-channel structure," *IEEE Electron Device Letters*, vol. 37, no. 1, pp. 53–56, 2015.

63 W. Song, L. Lan, P. Xiao, Z. Lin, S. Sun, Y. Li, E. Song, P. Gao, P. Zhang, W. Wu, and J. Peng, "High-mobility and good-stability thin-film transistors with scandium-substituted indium oxide semiconductors," *IEEE Transactions on Electron Devices*, vol. 63, no. 11, pp. 4315–4319, 2016.

64 S. Lee and A. Nathan, "Subthreshold Schottky-barrier thin film transistors with ultralow power and high intrinsic gain," *Science*, vol. 354, pp. 302–304, 2016.

65 T. Kizu, S. Aikawa, T. Nabatame, A. Fujiwara, K. Ito, M. Takahashi, and K. Tsukagoshi, "Homogeneous double-layer amorphous Si-doped indium oxide thin-film transistors for control of turn-on voltage," *Journal of Applied Physics*, vol. 120, no. 4, 045702, 2016.

66 J. H. Yang, J. H. Choi, S. H. Cho, J. E. Pi, H. O. Kim, C. S. Hwang, K. Park, and S. Yoo, "Highly stable AlInZnSnO and InZnO double-layer oxide thin-film transistors with mobility over $50\,cm^2 \cdot V^{-1} \cdot s^{-1}$ for high-speed operation," *IEEE Electron Device Letters*, vol. 39, no. 4, pp. 508–511, 2018.

67 D. B. Ruan, P. T. Liu, M. C. Yu, T. C. Chien, Y. C. Chiu, K. J. Gan, and S. M. Sze, "Performance enhancement for tungsten-doped indium oxide thin film transistor by hydrogen peroxide as cosolvent in room-temperature supercritical fluid systems," *ACS Applied Materials & Interfaces*, vol. 11, 22521, 2019.

68 S. Steudel, J. L. P. J. van der Steen, M. Nag, T. H. Ke, S. Smout, T. Bel, K. Van Diesen, G. de Haas, J. Maas, J. de Riet, and M. Rovers, "Power saving through state retention in IGZO-TFT AMOLED displays for wearable applications," *Journal of the Society for Information Display*, vol. 25, no. 4, pp. 222–228, 2017.

69 T. K. Chang, C. W. Lin, and S. Chang, "LTPO TFT Technology for AMOLEDs," SID Symposium Digest of Technical Papers, vol. 50, no. 1, pp. 545–548, 2019.

70 X. Guo, Y. Xu, S. Ogier, T. N. Ng, M. Caironi, A. Perinot, L. Li, J. Zhao, W. Tang, R. A. Sporea, and A. Nejim, "Current status and opportunities of organic thin-film transistor technologies," *IEEE Transactions on Electron Devices*, vol. 64, no. 5, pp. 1906–1921, 2017.

71 I. Yagi, N. Hirai, Y. Miyamoto Y, M. Noda, A. Imaoka, N. Yoneya, K. Nomoto, J. Kasahara, A. Yumoto, and T. Urabe, "A flexible full-color AMOLED display driven by OTFTs," *Journal of the Society for Information Display*, vol. 16, no. 1, pp. 15–20, 2008.

72 G. Gelinck, P. Heremans, K. Nomoto, and T. D. Anthopoulos, "Organic transistors in optical displays and microelectronic applications," *Advanced Materials*, vol. 22, no. 34, pp. 3778–3798, 2010.

73 K. Nomoto, M. Noda, N. Kobayashi, M. Katsuhara, A. Yumoto, S. I. Ushikura, R. I. Yasuda, N. Hirai, G. Yukawa, and I. Yagi, "Rollable OLED Display Driven by Organic TFTs," SID Symposium Digest of Technical Papers, vol. 42, no. 1, pp. 488–491, 2011.

74 M. F. Chiang, C. Y. Liu, C. H. Tu, W. H. Chen, C.H. Tsai, A. J. Wu, S. H. Hsu, K. H. Liu, and Y. C. Lin, "6-Inch AMOLEDs Driven by High Stability Organic Thin-Film Transistor Backplanes," SID Symposium Digest of Technical Papers, vol. 47, no. 1, pp. 1786–1788, 2016.

75 L. Feng, Y. Huang, and J. Fan, J. Zhao, S. Pandya, S. Chen, W. Tang, S. Ogier, and X. Guo, "Solution processed high performance short channel organic thin-film transistors with excellent uniformity and ultra-low contact resistance for logic and display," in *2018 IEEE International Electron Devices Meeting (IEDM)*, 2018, pp. 38.3.1–38.3.4.

76 X. Guo, L. Han, Y. Huang, and W. Tang, "Development of Organic Thin-Film Transistor Technology for Active-Matrix Display Backplane," SID Symposium Digest of Technical Papers, vol. 52, no. 1, pp. 9–12, 2021.

77 Y. C. Lin, H. P. Shieh, and J. Kanicki, "A novel current-scaling a-Si:H TFTs pixel electrode circuit for AM-OLEDs," *IEEE Transactions on Electron Devices*, vol. 52, pp. 1123–1132, 2005.

78 S. Ono and Y. Kobayashi, "An accelerative current-programming method for AMOLED," *IEICE Transactions on Elections*, vol. E88-C, no. 2, pp. 264–269, 2005.

79 S. J. Ashtiani, P. Servati, D. Striakhilev, A. Nathan, "A 3-TFT current-programmed pixel circuit for active-matrix organic light-emitting diode displays," *IEEE Transactions on Electron Devices*, vol. 52, no. 7, pp. 1514–1518, 2005.

80 J. H. Baek, M. Lee, J. H. Lee, H. S. Pae, C. J. Lee, J. Kim, J. Kim, C. S. Choi, H. K. Kim, T. J. Kim, and H. K. Chung, "A current-mode display driver IC using sample-and-hold scheme for QVGA full-color active-matrix organic LED mobile displays," *IEEE Journal of Solid-State Circuits*, vol. 41, no. 12, pp. 2974–2982, 2006.

81 R. M. A. Dawson, Z. Shen, D. A. Furst, S. Connor, J. Hsu, M. G. Kane, R.G. Stewart, A. Ipri, C. N. King, P. J. Green, and R. T. Flegal, "A Polysilicon Active Matrix Organic Light Emitting Diode Display with Integrated Drivers," SID Symposium Digest of Technical Papers, 1999.

82 S. W. Tam, Y. Matsueda, M. Kimura, H. Maeda, T. Shimoda, and P. Migliorato, "Poly-Si driving circuits for organic EL displays," *Proceedings of SPIE*, vol. 4295, pp. 125–133, 2001.

83 J. L. Sanford and F. R. Libsch, "TFT AMOLED Pixel Circuits and Driving Methods," SID International Symposium-Digest of Technical Papers Papers, Baltimore, pp. 10–13, 2003.

84 J. C. Goh, J. Jang, K. S. Cho, and C. K. Kim, "A new a-Si:H thin-film transistor pixel circuit for active-matrix organic light-emitting diodes," *IEEE Electron Device Letters*, vol. 24, no. 9, pp. 583-585, 2003.

85 J. C. Goh, C. K. Kim, and J. Jang, "A novel pixel circuit for active-matrix organic light-emitting diodes," SID Symposium Digest of Technical Papers, Baltimore, pp. 494–497, 2003.

86 J. C. Goh, J. Jang, K. S. Cho, and C. K. Kim, "A new pixel circuit for active matrix organic light emitting diodes," *IEEE Electron Device Letters*, vol. 23, pp. 583–585, 2002.

87 S. H. Jung, W. J. Nam, and M. K Han, "A new voltage-modulated AMOLED pixel design compensating for threshold voltage variation in poly-Si TFTs," *IEEE Electron Device Letters*, vol. 25, pp. 690–692, 2004.

88 G. R. Chaji and A. Nathan, "A stable voltage-programmed pixel circuit for a-Si:H AMOLED displays," *Journal of Display Technology*, vol. 2, no. 4, pp. 347–358, 2006.

89 G. R. Chaji and A. Nathan, "Parallel addressing scheme for voltage-programmed active matrix OLED displays," *IEEE Transactions on Electron Devices*, vol. 54, no. 5, pp. 1095–1100, 2007.

90 Y. W. Kim, W. K. Kwak, J. Y. Lee, W. S. Choi, K. Y. Lee, S. C. Kim, and E. J. Yoo, "40 Inch FHD AM-OLED Display with IR Drop Compensation Pixel Circuit," SID Symposium Digest of Technical Papers, vol. 40, no. 1, pp. 85–87, 2009.

91 C. -L. Lin, P. -C. Lai; P. -C. Lai, P. -S. Chen, W. -L. Wu, Pixel circuit with parallel driving scheme for compensating luminance variation based on a-IGZO TFT for AMOLED displays, *Journal of Display Technology*, vol. 12, no. 12, pp. 1681–1687, 2016.

92 Y. H. Jang, D. H. Kim, W. Choi, M. G. Kang, K. I. Chun, J. Jeon, Y. Ko, U. Choi, S. M. Lee, J. U. Bae, and K. S. Park, "Internal Compensation Type OLED Display Using High Mobility Oxide TFT," SID Symposium Digest of Technical Papers, vol. 48, no. 1, pp. 76–79, 2017.

93 C. -L. Lin, P. -C. Lai, L. -W. Shih, C. -C. Hung, P. -C. Lai, T. -Y. Lin, K. -H. Liu, and T. -H. Wang, "Compensation pixel circuit to improve image quality for mobile AMOLED displays," *IEEE Journal of Solid-State Circuits*, vol. 54, no. 2, pp. 489–500, 2019.

94 K. Sakariya and A. Nathan, "Leakage and charge injection optimization in a-Si AMOLED displays," *Journal of Display Technology*, vol. 2, no. 3, pp. 254–257, 2006.

95 J. -H. Lee, S. -G. Park, S. -M. Han, M. -K. Han, and K. -C. Park, "New PMOS LTPS–TFT pixel for AMOLED to suppress the hysteresis effect on OLED current by employing a reset voltage driving," *Solid-State Electronics*, vol. 52, no. 3, pp. 462–466, 2008.

96 W. Liu, G. Yao, C. Jiang, Q. Cui, and X. Guo, A new voltage driving scheme to suppress non-idealities of polycrystalline thin-film transistors for AMOLED displays, *Journal of Display Technology*, vol. 10, no. 12, pp. 991–994, 2014.

97 S. -J. Song and H. Nam, In-pixel mobility compensation scheme for AMOLED pixel circuits, *Journal of Display Technology*, vol. 11, no. 2, pp. 209–213, 2015.

98 C. -L. Lin, P. -C. Lai, J. -H. Chang, S. -C. Chen, C. -L. Tsai, J. -L. Koa, M. -H. Cheng, L. -W. Shih, and W. -C. Hsu, Leakage-prevention mechanism to maintain driving capability of compensation pixel circuit for low frame rate AMOLED displays, *IEEE Transactions on Electron Devices*, vol. 68, no. 5, pp. 2313–2319, 2021.

99 X. Xu, B. Huang, J. Fan, J. Zhao, and X. Guo, "Employing drain-bias dependent electrical characteristics of poly-Si TFTs to improve gray level control in low power AMOLED displays," *IEEE Journal of the Electron Devices Society*, vol. 7, no. 1, pp. 489–494, 2019.

100 X. Y. Lin, Y. H. Tai, C. Y. Chang, and C. W. Lu, "AMOLED Driving Circuit with Subthreshold Current Compensating Capability for High PPI Display Panel," SID Symposium Digest of Technical Papers, vol. 47, no. 1, pp. 1189–1192, 2016.

101 W. Qin, K. J. Peng, D. Liu, Z. Z. Yang, Z. Q. Xu, W. P. Ten, S. N. Li, X. L. Li, T. S. Wang, J. Yang, and L. L. Zhang, "Study of AMOLED Short-Term Image Sticking Mechanism and Improvement," SID Symposium Digest of Technical Papers, vol. 50, no. 1, pp. 1899–1902, 2019.

102 N. -H. Keum, C. -C. Chai, S. -K. Hong, and O. -K. Kwon, "A compensation method for variations in subthreshold slope and threshold voltage of thin-film transistors for AMOLED displays," *IEEE Journal of the Electron Devices Society*, vol. 7, pp. 462–469, 2019.

103 Y. Takeda, S. Kobayashi, S. Murashige, K. Ito, I. Ishida, S. Nakajima, H. Matsukizono, and N. Makita, "Development of High Mobility Top Gate IGZO-TFT for OLED Display," SID Symposium Digest of Technical Papers, vol. 50, no. 1, pp. 516–519, 2019.

104 G. R. Chaji, C. Ng, A. Nathan, A. Werner, J. Birnstock, O. Schneider, and J. Blochwitz-Nimoth, "Electrical compensation of OLED luminance degradation," *IEEE Electron Device Letters*, vol. 28, no. 12, pp. 1108–1110, 2007.

105 S. J. Ashtiani, G. R. Chaji, and A. Nathan, "AMOLED pixel circuit with electronic compensation of luminance degradation," *IEEE/OSA Journal of Display Technology*, vol. 3, no. 1, pp. 36–39, 2007.

106 A. Nathan, G. R. Chaji, and P. Servati, "Method and system for programming, calibrating and driving a light emitting device display," US Patent 7619597, 2009

107 A. Nathan, G. R. Chaji, and P. Servati, "Method and system for programming and driving active matrix light emitting device pixel," US Patent 7800565, 2010.

108 S. J. Ashtiani and A. Nathan, "A driving scheme for active-matrix organic light-emitting diode displays based on feedback," *Journal of Display Technology*, vol. 2, no. 3, pp. 258–264, 2006.

109 J. Ashtiani and A. Nathan, "A driving scheme for active-matrix organic light-emitting diode displays based on current feedback," *Journal of Display Technology*, vol. 5, no. 7, pp. 257–264, 2009.

110 J. Jeon, Y. Jeon, Y. Son, and G. Cho, "A direct fast feedback current driver using an inverting amplifier for high-quality AMOLED displays," *IEEE Transactions on Circuits and Systems II: Express Briefs*, vol. 59, no. 7, pp. 414–418, 2012.

111 L. Lu, L. Deng, J. Ke, C. Liao, and S. Huang, "A fast ramp-voltage based current programming driver for AMOLED display," *IEEE Transactions on Circuits and Systems II: Express Briefs*, vol. 66, no. 7, pp. 1129–1133, 2019.

112 H. -J. In and O. -K. Kwon, "External compensation of nonuniform electrical characteristics of thin-film transistors and degradation of OLED devices in AMOLED displays," *IEEE Electron Device Letters*, vol. 30, no. 4, pp. 377–379, 2009.

113 G. R. Chaji, S. Alexander, J. M. Dionne, Y. Azizi, C. Church, J. Hamer, J. Spindler, and A. Nathan, "Stable RGBW AMOLED display with OLED degradation compensation using electrical feedback," in *IEEE International Solid-State Circuits Conference, ISSCC 2010, Digest of Technical Papers*, Feb. 2010, pp. 118–119.

114 J. -S. Bang, H. -S. Kim, K. -D. Kim, O. -J. Kwon, C. -S. Shin, J. Lee, and G. -H. Cho, "A hybrid AMOLED driver IC for real-time TFT nonuniformity compensation," *IEEE Journal of Solid-State Circuits*, vol. 51, no. 4, pp. 966–978, 2016.

115 T. Charisoulis, C. Reiman, D. Frey, and M. Hatalis, "Current feedback compensation circuit for 2T1C LED displays: analysis and evaluation," *IEEE Transactions On Circuits and Systems I: Regular Paper*, vol. 66, no. 1, pp. 175–188, 2019.

116 J. Huang, X. X. Fan, and X. Guo, "Displaying-synchronous open-loop external compensation for active-matrix light emitting diode displays," *IEEE Transactions on Circuits and Systems II: Express Briefs*, vol. 67, no. 10, pp. 1790–1794, 2020.

117 J. H. Jang, M. Kwon, E. Tjandranegara, K. Lee, and B. Jung, "A digital driving technique for an 8b QVGA AMOLED display using $\Delta\Sigma$ modulation," in *2009 IEEE International Solid-State Circuits Conference - Digest of Technical Papers*, 8–12 Feb. 2009, pp. 269–271.

118 Z. Gong, S. Jin, Y. Chen, J. McKendry, D. Massoubre, I. M. Watson, E. Gu, M. D. Dawson, "Size-dependent light output, spectral shift, and self-heating of 400 nm InGaN light-emitting diodes," *Journal of Applied Physics*, vol. 107, no. 1, 013103, 2010.

119 J. -H. Kim, S. Shin, K. Kang, C. Jung, Y. Jung, T. Shigeta, S. -Y. Park, H. S. Lee, J. Min, J. Oh, and Y. -S. Kim, "PWM Pixel Circuit with LTPS TFTs for Micro-LED Displays," *SID Symposium Digest of Technical Papers*, vol. 50, no.1, pp. 192–195, 2019.

120 M. Tamaki, T. Suzuki, K. Aoki, R. Yokoyama, H. Ito, S. Nakamitsu, K. Imaizumi, K. Yamanoguchi, M. Nishide, F. Rahadian, S. Matsuda, E. Lang, and L. Hoeppel, A 3.9-inch LTPS TFT Full Color MicroLED Display with Novel Driving and Reflector Cavity Process, vol. 51, no. 1, pp. 111–114, 2020.

8

Organic and Macromolecular Memory – Nanocomposite Bistable Memory Devices

Shashi Paul

Emerging Technologies Research Centre, De Montfort University, Leicester, UK

8.1   Introduction

It goes without saying that without the development of electronic memory devices, information technology (IT) would not have been at its current peak, and the future of computer science (including artificial intelligence [AI] and Internet of things [IoT]) will not be able to exploit its full benefits without further development of electronic memory. Electronic memory storage platforms have also become an integral part of consumer electronics, for instance, personal computers, tablets, or smartphones. There is a demand for a memory that can cope with the requirements of these new applications and researchers are constantly exploring new materials and device structures to meet this demand. Organic electronic memory is one of these memories undergoing an exploratory phase. Before, we get into organic memory and in particular nanocomposite memory, we would like to go through a few basic concepts and the definitions used to describe electronic memory.

8.1.1   What Is an Electronic Memory Device?

A device is called a memory device that can store/hold information for a given period of time, for example, the hard disk of computers or memory card in cameras, or USB flash drive. Furthermore, electronic information storage devices can be classified as volatile and nonvolatile. The pen drive (Flash Memory) is an example of nonvolatile memory, which does not require electrical potential or some other form of stimulation to preserve/retain the information. An external electrical potential or stimulus is needed to retain the information of volatile memory devices, and the information will be lost once the stimulus is switched off. An example of volatile memories is Random Access Memory (RAM) of desktops or laptops.

Before we embark on the understanding of organic memory materials and devices [1], we would like to invest some time to understand ethos of *bistability* [2, 3] on which most current nonvolatile memory devices are based on. Bistability is usually defined as a dynamic system which can rely on two stable states [4]. To pass from equilibrium **state 1** to equilibrium **state 2**, one must pass through a certain type of energy barrier. From the point of view of data storage, we can say that state-1 is "**0**" and state-2 is "**1**." This will allow us to build a binary system. A representation of this system is shown in Figure 8.1. A bistable system can be symmetrical (Figure 8.1a) or asymmetrical (Figure 8.1b) about the energy barrier.

Advances in Semiconductor Technologies: Selected Topics Beyond Conventional CMOS, First Edition. Edited by An Chen.
© 2023 The Institute of Electrical and Electronics Engineers, Inc. Published 2023 by John Wiley & Sons, Inc.

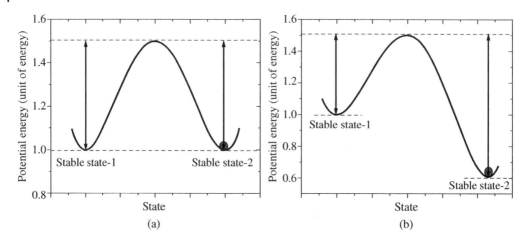

Figure 8.1 The schematic of (a) symmetric and (b) asymmetric bistable systems. In the symmetric system, the energy required to move between stable states is equal, whereas in the asymmetric system, it is unequal.

In order to better understand the bistable system, we would like to invoke the following differential equation in relation to a dynamic system:

$$\frac{dx}{dt} = x(r - x^2)$$

(8.1)

where x is some physical quantity (it can be electrical resistivity of stable states, resultant direction of ferromagnetic or ferroelectric domain or spin state (up or down) of a electrons in a system) and r is a number. Usually, r is referred to as a bifurcation parameter. This can have an enormous impact on changing the state "x" for electronic memory devices. Ideally, states (0 or 1) should remain unchanged over a long period of time for nonvolatile memory devices. To understand stable and unstable points (where $\frac{dx}{dt} = 0$, for $r < 0$, $r = 0$, and $r > 0$), the graphical presentation is given in Figure 8.2. It is evident from the graphical representation that there are two stable points for $r > 0$.

The general solution for Eq. (8.1) can be determined using the partial fractional method:

$$x = \pm\sqrt{r\,\frac{e^{2r(t+c)}}{e^{2r(t+c)} - 1}}$$

(8.2)

where c is an arbitrary constant.

The plot of quantity (x) vs. time, for $r > 0$, is shown in Figure 8.3a, and it is quite clear that binary states (0 and 1) are separable and stable for $t \in [0,\infty)$. However, for $r < 0$, the separation between binary states is increased over time and states become unstable or reaching into chaos (Figure 8.3b). Therefore, if our ultimate aim is to look for an electronic memory device that can preserve information for an infinite time, i.e. immortal memory, it is necessary to understand the dynamic behavior of such devices or systems in the light of physics and to understand their "r" values. This will indicate whether a system/device is capable of providing immortal memory or not. Existing solid-state memory devices (for example, flash memory) can keep the information for up to few years. There are a number of reasons behind losing the stored information.

So far we have discussed the bistability and its importance in realizing electronic memory devices. This bistable behavior is present in ferroelectric memory devices [6, 7], ferromagnetic memory devices [8], flash memory [9], spin-electronic memory devices [10], etc.

Polymers are a category of natural or synthetic materials made up of very large molecules, often called macromolecules, through the repetition of subunits called monomers. As an example, polystyrene is a polymer, and the monomer is styrene. Moreover, most monomers generally consist

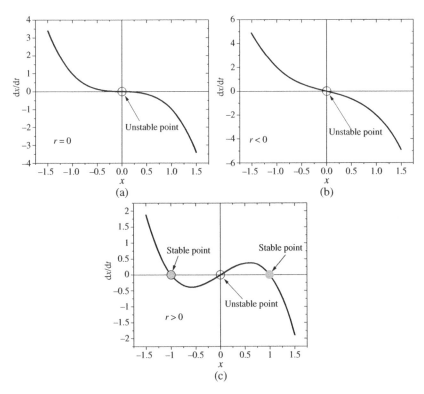

Figure 8.2 Schematic illustrations of equilibria of a one-dimensional, continuous-time dynamical system $\frac{dx}{dt} = x(r - x^2)$ a, where $\frac{dx}{dt} = 0$ continuous- for different values of r. The solid dots represent a stable equilibrium, whereas the open dot represents an unstable equilibrium. The value r is a determining parameter to look for bistable memory systems. More details on how to decide stable and unstable points can be found somewhere else [5].

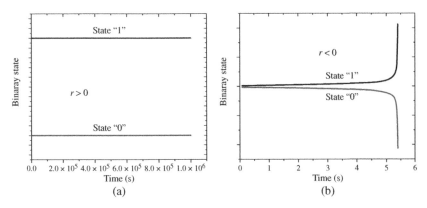

Figure 8.3 The dynamical behavior of bistable system for (a) $r > 0$; both states are stable for a long period time and (b) $r < 0$; unstable system enters into chaos after a short period of time and such systems will not be used in memory devices.

of carbon, nitrogen, oxygen, and hydrogen atoms. Figure 8.4 shows some examples of polymers and arrangement of atoms in their long chain structure.

Small organic molecules are typically composed of carbon, nitrogen, oxygen, and hydrogen atoms. Figure 8.5 shows some examples of small organic molecules. The molecules can be donor or acceptor meaning either gain electrons or lose electrons [14, 15].

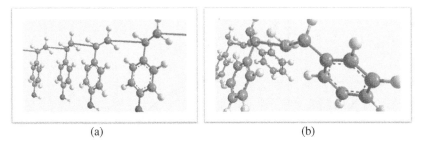

Figure 8.4 Long chain polymers are often used in nanocomposite/macromolecules memory devices (a) polystyrene and (b) polyvinyl alcohol.

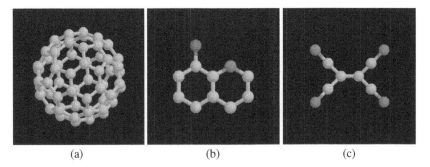

Figure 8.5 The molecules structures of small organic molecules (a) C_{60}; it can accept electrons, (b) 8-hydroxyquinoline; electron donor molecule (c) tetracyanoethylene (TCNE); electron acceptor molecule. Donor–acceptor combination is often used in polymer matrix to fabricate exploratory polymer memory devices [11–13].

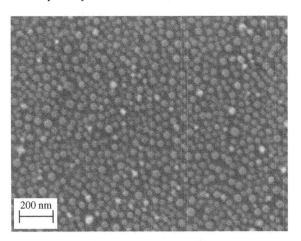

Figure 8.6 The scanning electron microscope image of tin nanoparticle and metal nanoparticles often used in making nanocomposite memory devices; e.g. gold nanoparticles used by a number of researchers in this field [16].

Nanoparticles (NPs) are particles in a nanorange size (1–100 nm) and give rise to some interesting physical and chemical properties. Figure 8.6 shows the scanning electron microscope image of tin nanoparticles.

Since 1960s, there has been strong interest to understand electrical behavior of polymers and it was observed that a sandwich structure made of Pb/polydivinylbenzene/Pb [18] shows electrical

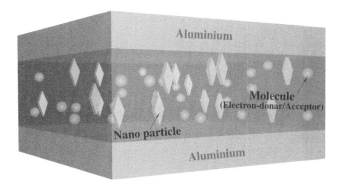

Figure 8.7 Schematic depiction of the nanocomposite (mixture of polymer and small molecules and/or nanoparticles) sandwiched between two metal electrodes to form a memory device.

bistable behavior and such behavior was proposed to be utilized for information storage. Since then, various combinations of polymers, molecules, and nanoparticles have been proposed, with varying degrees of merits and demerits, for information storage [19–22]. It is to be noted that the class of memory devices that are discussed and presented in this chapter is based on the admixture of small molecules and/or nanoparticles, and polymers, and such devices is called *organic memory devices or devices based on organic materials containing small molecules and/or nanoparticles* [15, 23–25]. The admixture is also called nanocomposite. The typical structures of such devices are shown in Figure 8.7.

8.2 Organic Memory and Its Evolution

There was considerable interest in organic electronics because of the ease of production and the applicability of inexpensive substrates [26, 27] over the last three decades. In this period, there have been substantial advances in materials and devices such as organic light-emitting diodes [28, 29], organic field effect transistors [27], and solar cells [30–32], with some devices being developed to the point of commercialization. The factors driving this growth can be attributed to the overall low cost of organic devices, material engineering for specific properties, simple device structures, low temperature manufacturing processes (such as printing), almost zero wastage of materials, and compatibility of organic materials with cheap flexible substrates.

The organic memory devices can be broadly split into two categories, namely molecular memory devices and polymer memory devices.

8.2.1 Molecular Memory

In the case of molecular memory, a monolayer of molecules is deposited between metal electrodes. Several types of single molecule devices have been demonstrated; molecular memories have indeed shown promise of technological feasibility [33–35]. The main approach employed in molecular memory technologies is the crossbar or gap cell architecture which involves deposition of a molecular monolayer sandwiched between two metal electrodes shown in Figure 8.8. The molecules exhibit "switching" characteristics, which can be exploited in emerging memory devices. Examples of molecules that exhibit appropriate switching behavior are catenanes,

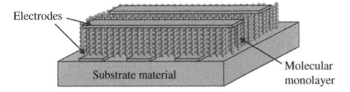

Figure 8.8 Schematic diagram of molecular memory devices; a monolayer of molecules sandwiched between metal electrodes, forming a cross-point memory array. Source: The diagram is reused with permission from Prime and Paul [1], Royal Society.

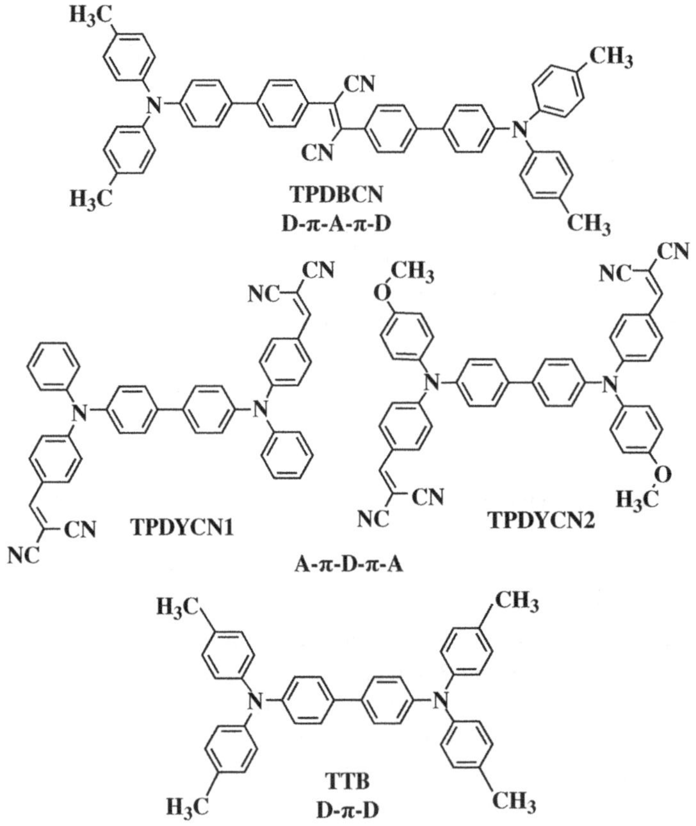

Figure 8.9 The Donor–Acceptor (D–A) molecules of structures of bis{4-[4-[di(p-tolyl)amino]phenyl]phenyl} fumaronitrile (TPDBCN), N,N′-bis[4-(1,1-dicyanovinyl)phenyl]-N,N′-diphenylbenzidine (TPDYCN1), N,N′-bis[4-(1,1-dicyanovinyl)phenyl]-N,N′-bis(4-methoxyphenyl)benzidine (TPDYCN2) and D–Π–D structural molecule (N,N,N′,N′-tetra(4-methylphenyl)-(1,1′-biphenyl)-4,4′-diamine (TTB)) used in molecular memory. Source: Ma et al. [36], from Wiley.

rotaxanes, porphyrins, etc. When voltage is applied on these molecules, they "switch" between two electrical conductance states and behave as a bistable system.

Figure 8.9 shows the molecular structure of "donor–acceptor molecules" used in molecular memory devices and their electrical switching behavior is shown in Figure 8.10. The molecular electronic memory (predominately organic molecules) has the potential to miniaturize up to a few atoms. However, molecular devices are extremely difficult to assemble and show variability in

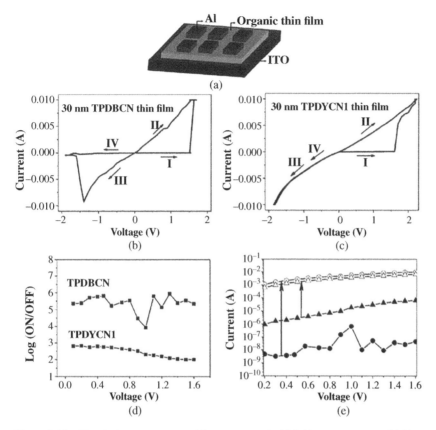

Figure 8.10 The device structure used for macroscopic (a) *I–V* measurements; (b) Macroscopic *I–V* characteristics of ITO/TPDBCN (30 nm thick)/Al; (c) macroscopic *I–V* characteristics of ITO/TPDYCN1 (30 nm thick)/Al; (d) the comparison of the ON/OFF current ratio of b and c; (e) the comparison of ON (open symbols) and OFF (filled symbols) state current in b (circles) and c (triangles); the arrows indicate transitions from OFF to ON state in each of the devices. Source: Reused with permission from Ma et al. [36], from Wiley.

electrical behavior from devices to devices, thus rendering the mass production of these structures difficult with existing technologies.

8.2.2 Polymer Memory Devices

There had been a few reports of polymer memory devices in the first few years at the beginning of the twenty-first century [37, 38]. In a polymer memory device an admixture (a blend) of small organic molecules and/or metal nanoparticles in a polymer matrix is deposited between metal electrodes to form an array of memory elements as shown in Figure 8.11. Such memory devices can be switched between two conductivity states upon the application of write and erase voltages, with the state being sensed by an intermediate read voltage. This bistable behavior of the devices renders them suitable for nonvolatile organic memory.

The first fully organic class of materials for memory devices, based on nanocomposite (a blend of poly-vinyl-phenol [PVP] and Bucky-ball [C_{60}]), in this field was presented by Kanwal et al. in Materials Research Society, Fall Meeting in 2004, subsequently published in Materials Research Society Symposium Proceedings [37]. Around the same time, memory devices using gold nanoparticles and 8-hydroxyquinoline, dispersed in a polystyrene matrix, were demonstrated [38].

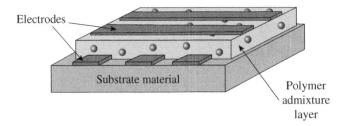

Electrodes

Substrate material

Polymer
admixture
layer

Figure 8.11 Schematic diagram of a nanocomposite, a mixture of polymer and small molecule and/or nanoparticles, inserted between two metallic electrodes in a cross-point structure. It is the most common structure used in exploratory memory devices. Source: Reused with permission from Prime and Paul [1], Royal Society.

An in-depth analysis of all-organic memory devices was published by Paul et al. [39]. Since then, the interest to use an admixture of nanoparticles, small molecules, and polymers in the manufacture of electronic memory devices is on the rise both for exploring new materials and for understanding the basic science governing their electrical properties.

Macromolecular (or polymer) memory devices are fabricated by depositing a blend (an admixture of organic polymer, organic molecules, and nanoparticles) between two metal electrodes mostly by the spin-coating technique. These devices show two electrical conductance states, as in molecular memory devices, when voltage is applied. These two states can then be used for nonvolatile memory devices. Table 8.1 shows blend of various polymers and nanoparticles.

All single-crystal semiconductor technologies are ultimately constrained by the fact that they are restricted to two dimensions, because single crystals cannot currently be grown on top of an amorphous substrate (e.g. glass or a flexible plastic substrate); hence, multiple active memory layers cannot be built on a single wafer. Thus, one is led to surmise that the only way to enhance memory density in these structures is to reduce the feature size in the two dimensional plane, i.e. miniaturization, which solicits increased production expenditure and some additional technological and scientific challenges.

Organic memory technology harbors the potential to radically improve on this situation, because it is based on the deposition of polymer films of an active organic material by a spin-on mechanism and/or other techniques (e.g. printing, dip-coating). A simple crossbar architecture for organic memories may eliminate the need for transistors as selectors in a bit-cell array. This in turn can greatly reduce the cost of manufacturing these devices because in this technique, the bit-cell array can be created by using a single (or a few) lithographic step, as compared to multiple lithographic

Table 8.1 Some previous studies of nanocomposite bistable memory devices

Article	Nanoparticle/small molecule	Polymer
[19, 37]	Bucky ball (C_{60})	Polyvinyl phenol (PVP)
[17, 38, 40, 41]	Gold	Polystyrene, Poly (4- vinyl phenol) (PVP)
[42]	Zinc oxide (ZnO)	Polystyrene or poly(3-hexylthiophene)
[43]	$BaTiO_3$ (Barium titanite) (annealed)	Polystyrene (PS) and polyvinyl acetate (PVAc)
[16]	Gold	Conjugate-polymers
[44]	Gold	Al
[45]	ZnO	Polymethylmethacrylate polymer

steps. In addition, the requirement for a diffusion step in the manufacturing process of transistors is eliminated. Furthermore, the polymer memory devices are better from the point of view of ease of fabrication and the crossbar architecture may potentially have lower costs than complementary metal oxide semiconductor (CMOS)-based memory devices because the substrate material for the bit-cell array does not need a transistor to sense electrical current. Thus, this approach of memory fabrication allows for the creation of inexpensive memory devices. In light of the lower production costs of the spin-coating or printing technique, organic memories also have the option of moving along the Z-direction, which opens up the possibility of creating inexpensive three-dimensional memories.

It merits mention that a prerequisite for using this type of architecture is that the storage medium must have enough intrinsic electrical hysteresis in its switching mechanism (at least a few order of magnitude), in order for the memories to work without transistors. Figure 8.12 shows one of the typical electrical behaviors of simulated devices. Figure 8.12a shows a control device without any incorporation of nanoparticles and/or small organic molecules and no hysteresis is observed and Figure 8.12b with nanoparticles and small organic molecules in polymer showing clearly a hysteresis. The hysteresis is an indication of electrical bistability of the system. Now, if we apply Write, Read, and Erase pulses, as shown in Figure 8.13, we can clearly observe two distinctive values of current at the same voltage of $+2\,V$ (in the form of State "0" and State "1") after writing at $+5\,V$ and erasing at $-5\,V$.

There is a wide variety of electrical behavior observed in these memory devices. The current–voltage behavior in polymer memory devices is summarized [1] into one of three general shapes, namely, N-shaped, S-shaped, and O-shaped as shown in Figure 8.14. The electrical conduction mechanism in polymer memory devices is fundamental to the understanding of the operation of such device. In insulators and high bandgap semiconductors, high electric field conduction can occur by a number of mechanisms which are dependent upon the trap density and the trap depth below the conduction band.

These mechanisms include Simmon–Verderber-like switching, thermionic field emission, space charge limited conduction, Fowler–Nordheim tunneling, Poole–Frenkel, Thermionic Emission, Ohmic conduction, charge transfer mechanism, internal electric field formation based mechanism, and conducting filament formation. However, sometimes there are contradictions in the reported mechanisms in identical device structures made of the same materials. Whatever the mechanism, if the device shows electrical bistability, as discussed before, without an iota of doubt it may be

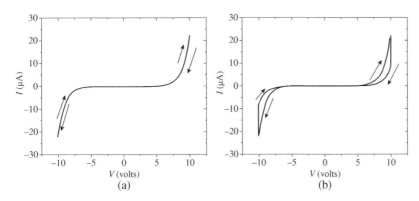

Figure 8.12 (a) Current–voltage (*I*–*V*) characteristics of a layer of pure polymer (control device) and (b) *I*–*V* characteristics of nanocomposite.

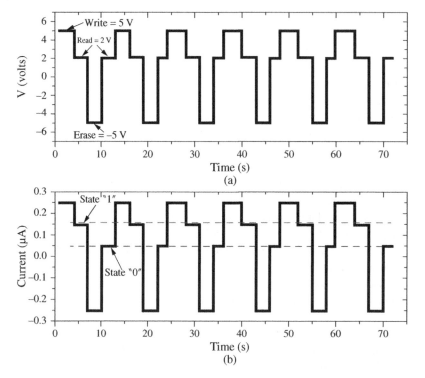

Figure 8.13 The simulated behavior (a) Write-Read-Erase (WRE) voltage pulse; (b) electrical current response of nanocomposite memory device. The similar electrical current response, of a real memory device, can be found in Paul et al. [39]. Source: Based on Paul et al. [39].

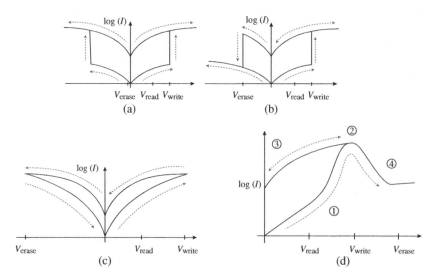

Figure 8.14 Different types of electrical behavior reported in polymer devices (a) Symmetric S-shaped and (b) asymmetric S-shaped current–voltage characteristics (c) O-shaped current–voltage characteristics (d) N-shaped current–voltage characteristic. Source: Reused with permission from Prime and Paul [1], Royal Society.

Table 8.2 Various electrical conduction processes in insulators (both organic and inorganic insulators).

Process	Full expression	Temperature dependence	Voltage dependence
Schottky emission	$J = A^* T^2 \exp\left[\dfrac{-e(\phi_B - \sqrt{e(V/d)/4\pi\varepsilon_i})}{kT}\right]$	$\ln\left(\frac{J}{T^2}\right) \sim \frac{\alpha}{T}$	$\ln(J) \sim \alpha V^{1/2}$
Poole–Frenkel emission	$J \sim \dfrac{V}{d} \exp\left[\dfrac{-e(\phi_B - \sqrt{e(V/d)/\pi\varepsilon_i})}{kT}\right]$	$\ln(J) \sim \frac{2\alpha}{T}$	$\ln\left(\frac{J}{V}\right) \sim 2\alpha V^{1/2}$
Direct tunneling	$J \sim V \exp\left[\dfrac{-2d\sqrt{2m^*\phi_B}}{\hbar}\right]$	None	$J \sim V$
Fowler–Nordheim tunneling	$J \sim \left(\dfrac{V}{d}\right)^2 \exp\left[\dfrac{-4d\phi_B^{3/2}\sqrt{2m^*}}{3e\hbar V/d}\right]$	None	$\ln\left(\frac{J}{V^2}\right) \sim \frac{\beta}{V}$
Space charge limited	$J = \dfrac{9}{8}\dfrac{\varepsilon_i \mu V^2}{d^3}$	None	$J \sim V^2$
Ohmic	$J \sim \dfrac{V}{d} \exp\left[\dfrac{-\Delta E_{ae}}{kT}\right]$	$\ln(J) \sim \frac{\gamma}{T}$	$J \sim V$
Ionic	$J \sim \dfrac{V}{dT} \exp\left[\dfrac{-\Delta E_{ai}}{kT}\right]$	$\ln(JT) \sim \frac{\delta}{T}$	$J \sim V$

These are often used to explain the observed electrical behavior of polymer memory devices. With the help of these equations and the measured *I–V* characteristics, the dominant conduction mechanism responsible for the observed electrical behavior can be determined. This table is compiled (with permission from Wiley and Sons) and the original table is available in Sze and Kwok [46].

used for electronic memory devices. The mathematical relationships between current, electrical field, and temperature of these mechanisms are listed in Table 8.2.

Where *J* is the current density, A^* the Richardson constant, ϕ_B the barrier height, ε_i the insulator permittivity, *d* the insulator thickness, m^* the effective mass, μ the charge carrier mobility and E_{ae}, and E_{ai} the activation energy of electrons and ions, respectively. Device dependent constants are α, β, γ, and δ.

The memory devices can exhibit symmetrical or asymmetrical *I–V* characteristics for negative and positive applied voltages. The symmetrical *I–V* characteristics are typical of a bulk-limited mechanism. The metal/nanocomposite interface does not play any significant role in determining the electrical behavior of these devices.

If the Schottky conduction mechanism is predominant, then current is limited by thermionic emission over the barrier. Alternatively, if the trap density dominates the bulk conduction mechanism, the current density is likely to be bulk-limited by the Poole–Frenkel effect.

For thin insulating/composite layers, the tunneling mechanisms can become dominant, with direct tunneling through a potential barrier and Fowler–Nordheim tunneling through a triangular barrier, often observed at higher applied electric fields.

Another mechanism often reported in electronic memories is space charge limited conduction. It is a result of charge carriers being injected into the insulating polymer from metal electrodes when there is no neutralizing charge present in the insulating matrix. Ohmic conduction results from thermally excited electrons at higher temperatures. The ionic conduction is a result of ionic impurities moving through the insulating matrix under high electric fields. A number of research papers also explained the electrical behavior due to electroforming in metal-polymer(insulating polymer)-metal structure that resulted in filamentary formation in the polymer material creating a low electrical resistance pathways between the metal electrodes [47–50]. The following possible explanations are given by various researchers for the formation of conducting filaments within the polymer matrix:

- Formation of filaments of carbonaceous material from the polymer
- Formation of defect states within energy bandgap of polymer due to migration of metal atoms/ions from the electrodes upon the application of electric field
- Formation of Conducting filaments (CF) due to some electrolytic processes
- Metal islands/clusters formation due to electroforming process within the polymer matrix and the tunneling between adjacent metal islands/clusters.

As previously mentioned, memory devices using gold nanoparticles and 8-hydroxyquinoline dispersed in a polystyrene matrix, have also been demonstrated [38]. These devices show a large hysteresis in their current–voltage characteristics. There has also been a report on the use of the Al/Pentacene/Al structure for memory devices [51]. In this report, memory behavior is observed without the use of gold nanoparticles and small organic molecules in the pentacene matrix. Surprisingly, the electrical behavior of these devices is quite similar to what is reported in [16, 38]. The authors claimed that the physical mechanism that determines the electrical behavior is the diffusion of aluminum particles from the top and/or bottom contacts into the pentacene matrix. However, further evidence and investigation are needed. A recent paper [52] highlights some of the problems and disagreements that continue to plague the field despite the decades of works. Albeit, the newer materials and device structures have been proposed by various researchers, the review article draws an attention to several older problems associated with polymer memory devices. In this article, it is mentioned that devices made from Tris(8-hydroxyquinoline)aluminum (III) (commonly known as Alq_3) by different research groups with similar device structure have different electrical switching mechanisms been proposed. Thus, a lack of agreement on the switching mechanisms needs further scrutiny (Table 8.3).

The on/off current ratio from 10 to 10^9 is reported [1, 52]. It is often considered that the higher the on/off ratio, the better a memory device. If the difference between on- and off-state is large, the states may take longer time to overlap, i.e. longer retention. One way to describe the on/off ratio is the nonzero Euclidean distance between on and off states. There are a number of methods to find the distance between two sets of data (on and off states), such as Hellinger distance, Bhattacharyya distance, Mahalanobis distance.

Figure 8.15 shows the Euclidean distance (calculated by the difference between average values of each state) between on- and off-state of simulated polymer memory devices. It is clear that there is less variation within each state in Figure 8.15a as compared with Figure 8.15b. Although the Euclidean distances are very close to each other in these two figures, we can "feel" that the device behavior in Figure 8.15a is much "better" than device behavior in Figure 8.15b.

In order to quantify this "better" device quality, we can invoke the idea of Bhattacharyya distance [63], given by

$$D_{\mathrm{B}(p,q)} = \frac{1}{4} \ln \left[\frac{1}{4} \left(\frac{\sigma_p^2}{\sigma_q^2} + \frac{\sigma_q^2}{\sigma_p^2} + 2 \right) \right] + \frac{1}{4} \left[\frac{(\mu_p - \mu_q)^2}{\sigma_p^2 + \sigma_q^2} \right]$$

where $D_{\mathrm{B}(p,q)}$ is the Bhattacharyya distance between state-1 (say p) and state-0 (say q). We also assume both states have normal distribution (statistically speaking). The μ_p, μ_q are the mean of state-1 and state-0, respectively. The σ_p and σ_q are the standard deviations of state-1 and state-0,

Table 8.3 Memory devices made using Tris(8-hydroxyquinoline)aluminum (III) (known as Alq$_3$) and suggested switching type and mechanism reported by various researchers.

Switching type	Device structure	Switching mechanism	ON/OFF ratio	Retention	References
Switching by either polarity, negative differential resistance (NDR)	Al/Alq$_3$/Al/Alq$_3$/Al, Cr, Cu, ITO, Au, Ni	Simmon–Verderber-like switching	—	—	[53]
	Al/Alq$_3$/Ag, Cr, Mg, CuPC/Alq$_3$/Al	Simmon–Verderber-like switching	—	—	[53]
	Ag, Al, Au/Alq$_3$/ITO, Au	—	—	—	[54]
	Al/Alq$_3$/Ni/Alq$_3$/Al	SCLC, thermionic-field-emission	10^3	—	[55]
	Al/α-NPD/Alq$_3$/Ni/Alq$_3$/α-NPD /Al	SCLC	~10^2	~10 yr	[56]
	ITO/Alq$_3$/Ag	CF formation	~10^5	2 yr	[47]
	Al/Alq$_3$/Al/Alq$_3$/ITO	Fowler–Nordheim tunneling (ON state) Poole–Frenkel (OFF state)	>10^5	>3500 s	[57, 58]
	Al/Alq$_3$/MoO$_3$ NPs/Alq$_3$/ITO	Thermionic emission (OFF state) Ohmic conduction (ON state)	~10^3	~4500 s	[59]
WORM	Al/Alq$_3$/Al, Au, ITO	—	—	—	[60]
S-curve	Au/Alq$_3$/Au, Al/Alq$_3$/Al	SCLC	10^4	4 h	[44]
	Al/Alq$_3$/Al/Alq3/Al Al/ZnSe/Al/Alq3/Al	Thermionic emission (ON state) Ohmic conduction (OFF state)	~10^4	8000 s	[61]
	Al/Alq$_3$/Al	CF formation	10^5	~20 min	[62]
	ITO/Alq$_3$/Ag	CF formation	—	—	[50]

The ON/OFF ratio and retention time are also quite different.
Source: Modified from Paul [52], Wiley.

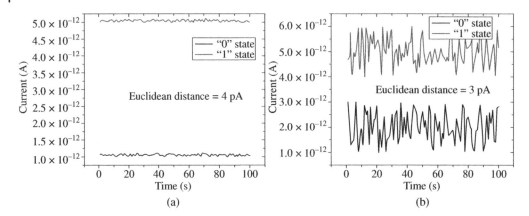

Figure 8.15 The retention data (simulated) of polymer devices showing the Euclidean distance between two states (a) showing less variation within each memory state while (b) showing larger variation within each memory state. Although the Euclidean distance is very close to each other, we may qualitatively say that the retention data present in (a) are better than (b). To quantify "goodness," we need to use some statistical concepts of distance.

Table 8.4 Bhattacharyya distance calculated between states "0" and "1" of data presented in Figure 8.15.

Device	State	Mean (μ)	Standard deviation (σ)	$D_{B(p,q)}$
Device (a)	State-1	5.05×10^{-12}	3.13×10^{-14}	2246
	State-0	1.05×10^{-12}	2.83×10^{-14}	
Device (b)	State-1	5.02×10^{-12}	5.39×10^{-13}	3.6
	State-0	2.01×10^{-12}	5.83×10^{-13}	

respectively. The values of mean (μ), standard deviation, and the calculated values of Bhattacharyya distance ($D_{B(p,q)}$) are listed in Table 8.4.

It is quite clear from the values of $D_{B(p,q)}$ that device (a) is "better" than device (b) since the Bhattacharyya distance of device (a) is significantly higher.

8.3 Summary

In organic memory devices, nanoparticles (such as Au, BaTiO$_3$, ZnO) are incorporated into the polymer matrix. The control of the nanoparticle size uniformity is an issue. In fact, the very use of nanoparticles may be disadvantageous, primarily because the size and solubility of the nanoparticles cannot be ideally controlled. Therefore, efforts are needed to develop a manufacturing method to produce nanoparticles with tight size distribution.

Furthermore, although devices in macroscopic scales (few mm^2) have been reported to demonstrate the memory effect (high and low conduction states due to charge storage in nanoparticles), there is only one report of memory effects at the nanoscale [37, 64], to the best of our knowledge.

One additional flaw in organic memory devices reported in the literature is their inability to perform memory operations at high speeds. Switching speed of seconds/milliseconds is often reported, which is too slow for many memory applications.

Organic memories may provide an opportunity to combine the advantages of molecular electronics (high-packing densities) and organic electronics (fabrication that promises to be inexpensive and on large area substrates) to enable low-cost, high-density, and flexible memory technologies.

To improve memory retention, materials/device scientists should explore materials with a strong internal electric field, e.g. ferro-electric polymers, nanocomposite with a higher value of electrical polarization, or a polymer blend with nanoparticles (exploiting quantized states), to enable charge storage capability for an extended period of time.

It will not be possible to replace silicon-based memory devices in the foreseeable future. However, there are a number of applications where cheap electronic memory devices can play a vital role. For example, nanocomposite-based memory devices can be directly printed on medicine packaging, and the information about the patient and schedule of taking medicine can be stored in the printed devices.

Acknowledgment

The author gratefully acknowledges a number of research students and postdoctoral fellows who have worked with him on this topic over the years. The author would like to say his heartfelt thank you to the late Mr. Hareesh Mareedu for drawing the schematic diagram presented in Figure 8.7.

References

1 D. Prime and S. Paul, "Overview of organic memory devices," *Philosophical Transactions of the Royal Society A: Mathematical, Physical and Engineering Sciences*, vol. 367, no. 1905, 2009, doi: https://doi.org/10.1098/rsta.2009.0165.

2 S. L. Lim, Q. Ling, E. Y. H. Teo, C. X. Zhu, D. S. H. Chan, E. T. Kang, and K. G. Neoh, "Conformation-induced electrical bistability in non-conjugated polymers with pendant carbazole moieties," *Chemistry of Materials*, vol. 19, no. 21, 2007, doi: 10.1021/cm071520x.

3 B. Pradhan, S. K. Batabyal, and A. J. Pal, "Electrical bistability and memory phenomenon in carbon nanotube-conjugated polymer matrixes," *Journal of Physical Chemistry B*, vol. 110, no. 16, 2006, doi: 10.1021/jp060122z.

4 C. Downing, "Advanced quantum mechanics: a Practical Guide, by Yuli V. Nazarov and Jeroen Danon," *Contemporary Physics*, vol. 54, no. 3, 2013, doi: 10.1080/00107514.2013.818065.

5 S. H. Strogatz, *Nonlinear Dynamics and Chaos*, 2018, doi: 10.1201/9780429492563.

6 T. Mikolajick, U. Schroeder, and S. Slesazeck, "The past, the present, and the future of ferro-electric memories," *IEEE Transactions on Electron Devices*, vol. 67, no. 4, 2020, doi: 10.1109/TED.2020.2976148.

7 A. K. Tripathi, A. J. Van Breemen, J. Shen, Q. Gao, M. G. Ivan, K. Reimann, E. R. Meinders, and G. H. Gelinck, "Multilevel information storage in ferroelectric polymer memories," *Advanced Materials*, vol. 23, no. 36, 2011, doi: 10.1002/adma.201101511.

8 J. De Boeck, W. Van Roy, J. Das, V. Motsnyi, Z. Liu, L. Lagae, H. Boeve, K. Dessein, and G. Borghs, "Technology and materials issues in semiconductor-based magnetoelectronics," *Semiconductor Science and Technology*, vol. 17, no. 4, 2002, doi: 10.1088/0268‐1242/17/4/307.

9 K. Saranti and S. Paul, Charge-trap-non-volatile memory and focus on flexible flash memory devices. In *Charge-Trapping Non-volatile Memories*, 2017, doi: 10.1007/978-3-319-48705-2_2.

10 E. Y. Tsymbal, O. N. Mryasov, and P. R. LeClair, "Spin-dependent tunnelling in magnetic tunnel junctions," *Journal of Physics Condensed Matter*, 2003, doi: 10.1088/0953-8984/15/4/201.

11 C. L. Liu and W. C. Chen, "Donor–acceptor polymers for advanced memory device applications," *Polymer Chemistry*, 2011, doi: 10.1039/c1py00189b.

12 G. Liu, X. Zhuang, Y. Chen, B. Zhang, J. Zhu, C. X. Zhu, K. G. Neoh, and E. T. Kang, "Bistable electrical switching and electronic memory effect in a solution-processable graphene oxide-donor polymer complex," *Applied Physics Letters*, vol. 95, no. 25, 2009, doi: 10.1063/1.3276556.

13 A. D. Yu, T. Kurosawa, Y. C. Lai, T. Higashihara, M. Ueda, C. L. Liu, and W. C. Chen, "Flexible polymer memory devices derived from triphenylamine-pyrene containing donor-acceptor polyimides," *Journal of Materials Chemistry*, vol. 22, no. 38, 2012, doi: 10.1039/c2jm33852a.

14 I. Salaoru and S. Paul, "Memory devices based on small organic molecules donor-acceptor system," *Thin Solid Films*, vol. 519, no. 2, 2010, doi: 10.1016/j.tsf.2010.07.009.

15 I. Salaoru and S. Paul, "Small organic molecules for electrically re-writable non-volatile polymer memory devices," in *MRS Online Proceedings Library*, 2010, 1250, 411, doi: 10.1557/PROC-1250-G04-11.

16 A. Prakash, J. Ouyang, J. L. Lin, and Y. Yang, "Polymer memory device based on conjugated polymer and gold nanoparticles," *Journal of Applied Physics*, vol. 100, no. 5, 2006, doi: 10.1063/1.2337252.

17 D. Prime, S. Paul, and P. W. Josephs-Franks, "Gold nanoparticle charge trapping and relation to organic polymer memory devices," *Philosophical Transactions of the Royal Society A: Mathematical, Physical and Engineering Sciences*, vol. 367, no. 1905, 2009, doi: 10.1098/rsta.2009.0141.

18 L. v. Gregor, "Polymer dielectric films," *IBM Journal of Research and Development*, vol. 12, no. 2, 1968, doi: 10.1147/rd.122.0140.

19 A. Kanwal and M. Chhowalla, "Stable, three layered organic memory devices from C_{60} molecules and insulating polymers," *Applied Physics Letters*, vol. 89, no. 20, 2006, doi: 10.1063/1.2388131.

20 G. Khurana, P. Misra, and R. S. Katiyar, "Multilevel resistive memory switching in graphene sandwiched organic polymer heterostructure," *Carbon*, vol. 76, 2014, doi: 10.1016/j.carbon.2014.04.085.

21 Q. Lai, Z. Zhu, Y. Chen, S. Patil, and F. Wudl, "Organic nonvolatile memory by dopant-configurable polymer," *Applied Physics Letters*, vol. 88, no. 13, 2006, doi: 10.1063/1.2191874.

22 L. Li, Z. Zhu, Y. Chen, S. Patil, and F. Wudl, "A flexible polymer memory device," *Organic Electronics*, vol. 8, no. 4, 2007, doi: 10.1016/j.orgel.2007.02.002.

23 S. Paul, "Realization of nonvolatile memory devices using small organic molecules and polymer," *IEEE Transactions on Nanotechnology*, vol. 6, no. 2, 2007, doi: 10.1109/TNANO.2007.891824.

24 D. Prime and S. Paul, "Electrical and morphological properties of polystyrene thin films for organic electronic applications," *Vacuum*, vol. 84, no. 10, 2010, doi:10.1016/j.vacuum.2009.10.033.

25 I. Salaoru and S. Paul, "Non-volatile memory device- using a blend of polymer and ferroelectric nanoparticles," *Journal of Optoelectronics and Advanced Materials*, vol. 10, no. 12, 2008.

26 S. R. Forrest, "The path to ubiquitous and low-cost organic electronic appliances on plastic," *Nature*, 2004, doi: 10.1038/nature02498.

27 J. A. Rogers, T. Someya, and Y. Huang, "Materials and mechanics for stretchable electronics," *Science*, 2010, doi:10.1126/science.1182383.

28 J. H. Burroughes, D. D. Bradley, A. R. Brown, R. N. Marks, K. Mackay, R. H. Friend, P. L. Burns, and A. B. Holmes, "Light-emitting diodes based on conjugated polymers," *Nature*, vol. 347, no. 6293, 1990, doi: 10.1038/347539a0.

29 J. Saghaei, M. Koodalingam, P. L. Burn, A. Pivrikas, and P. E. Shaw, "Light-emitting dendrimer: exciplex host-based solution-processed white organic light-emitting diodes," *Organic Electronics*, vol. 100, 2022, doi: 10.1016/j.orgel.2021.106389.

30 D. Black, I. Salaoru, and S. Paul, "Route to enhance the efficiency of organic photovoltaic solar cells – by adding ferroelectric nanoparticles to P3HT/PCBM admixture," *EPJ Photovoltaics*, vol. 5, 2014, doi: 10.1051/epjpv/2014010.

31 H. Hoppe and N. S. Sariciftci, "Organic solar cells: an overview," *Journal of Materials Research*, vol. 19, no. 7, 2004, doi: 10.1557/JMR.2004.0252.

32 B. C. Thompson and J. M. J. Fréchet, "Polymer-fullerene composite solar cells," *Angewandte Chemie – International Edition*, 2008, doi: 10.1002/anie.200702506.

33 Y. Chen, G. Y. Jung, D. A. Ohlberg, X. Li, D. R. Stewart, J. O. Jeppesen, K. A. Nielsen, J. F. Stoddart, and R. S. Williams, "Nanoscale molecular-switch crossbar circuits," *Nanotechnology*, vol. 14, no. 4, 2003, doi: 10.1088/0957-4484/14/4/311.

34 J. S. Lindsey and D. F. Bocian, "Molecules for charge-based information storage," *Accounts of Chemical Research*, vol. 44, no. 8, 2011, doi: 10.1021/ar200107x.

35 Y. Luo, C. P. Collier, J. O. Jeppesen, K. A. Nielsen, E. DeIonno, G. Ho, J. Perkins, H. R. Tseng, T. Yamamoto, J. F. Stoddart, and J. R. Heath, "Two-dimensional molecular electronics circuits," *ChemPhysChem*, vol. 3, no. 6, 2002, doi: 10.1002/1439-7641(20020617)3:6%519::AID-CPHC519%3.0.CO;2-2.

36 Y. Ma, X. Cao, G. Li, Y. Wen, Y. Yang, J. Wang, S. Du, L. Yang, H. Gao, and Y. Song, "Improving the ON/OFF ratio and reversibility of recording by rational structural arrangement of donor-acceptor molecules," *Advanced Functional Materials*, vol. 20, no. 5, 2010, doi: 10.1002/adfm.200901692.

37 A. Kanwal, S. Paul, and M. Chhowalla, "Organic memory devices using C_{60} and insulating polymer," in *Materials Research Society Symposium Proceedings*, 2005.

38 J. Ouyang, C. W. Chu, C. R. Szmanda, L. Ma, and Y. Yang, "Programmable polymer thin film and non-volatile memory device," *Nature Materials*, vol. 3, no. 12, 2004, doi: 10.1038/nmat1269.

39 S. Paul, A. Kanwal, and M. Chhowalla, "Memory effect in thin films of insulating polymer and C_{60} nanocomposites," *Nanotechnology*, vol. 17, no. 1, 2006, doi: 10.1088/0957-4484/17/1/023.

40 D. C. Prime, Z. Al Halafi, M. A. Green, I. Salaoru, and S. Paul, "Electrical conductivity bistability in nano-composite," in *ECS Transactions*, 2013, doi: 10.1149/05304.0141ecst.

41 D. Prime and S. Paul, "Gold nanoparticle based electrically rewritable polymer memory devices," in *CIMTEC 2008 – Proceedings of the 3rd International Conference on Smart Materials, Structures and Systems – Smart Materials and Micro/Nanosystems*, 2008, doi: 10.4028/www.scientific.net/AST.54.480.

42 F. Verbakel, S. C. J. Meskers, and R. A. J. Janssen, "Electronic memory effects in diodes of zinc oxide nanoparticles in a matrix of polystyrene or poly(3-hexylthiophene)," *Journal of Applied Physics*, vol. 102, no. 8, 2007, doi: 10.1063/1.2794475.

43 I. Salaoru and S. Paul, "Electrical bistability in a composite of polymer and barium titanate nanoparticles," *Philosophical Transactions of the Royal Society A: Mathematical, Physical and Engineering Sciences*, vol. 367, no. 1905, 2009, doi:10.1098/rsta.2009.0167.

44 X. Liu, Z. Ji, L. Shang, H. Wang, Y. Chen, M. Han, C. Lu, M. Liu, and J. Chen, "Organic programmable resistance memory device based on Au/Alq_3 gold-nanoparticle/Alq_3Al structure," *IEEE Electron Device Letters*, vol. 32, no. 8, 2011, doi: 10.1109/LED.2011.2158055.

45 D. I. Son, C. H. You, W. T. Kim, J. H. Jung, and T. W. Kim, "Electrical bistabilities and memory mechanisms of organic bistable devices based on colloidal ZnO quantum dot-polymethylmethacrylate polymer nanocomposites," *Applied Physics Letters*, vol. 94, no. 13, 2009, doi:10.1063/1.3111445.

46 S. M. Sze and K. Ng Kwok, *Physics of Semiconductor Devices*, 3rd Edition. Wiley, 2007.

47 Y. Busby, S. Nau, S. Sax, E. J. W. List-Kratochvil, J. Novák, R. Banerjee, F. Schreiber, and J. J. Pireaux, "Direct observation of conductive filament formation in Alq$_3$ based organic resistive memories," *Journal of Applied Physics*, vol. 118, no. 7, 2015, doi: 10.1063/1.4928622.

48 M. Cölle, M. Büchel, and D. M. de Leeuw, "Switching and filamentary conduction in non-volatile organic memories," *Organic Electronics*, vol. 7, no. 5, 2006, doi: 10.1016/j.orgel.2006.03.014.

49 W. J. Joo, T. L. Choi, J. Lee, S. K. Lee, M. S. Jung, N. Kim, and J. M. Kim, "Metal filament growth in electrically conductive polymers for nonvolatile memory application," *Journal of Physical Chemistry B*, vol. 110, no. 47, 2006, doi: 10.1021/jp0649899.

50 F. Santoni, A. Gagliardi, M. A. der Maur, A. Pecchia, S. Nau, S. Sax, E. J. List-Kratochvil, and A. Di Carlo, "Modeling of filamentary conduction in organic thin film memories and comparison with experimental data," *IEEE Transactions on Nanotechnology*, vol. 15, no. 1, 2016, doi: 10.1109/TNANO.2015.2496726.

51 D. Tondelier, K. Lmimouni, D. Vuillaume, C. Fery, and G. Haas, "Metal/organic/metal bistable memory devices," *Applied Physics Letters*, vol. 85, no. 23, 2004, doi: 10.1063/1.1829166.

52 S. Paul, "To be or not to be – review of electrical bistability mechanism(s) in polymer memory devices," *Small*, 2022, doi: 10.1002/smll.202106442.

53 L. D. Bozano, B. W. Kean, V. R. Deline, J. R. Salem, and J. C. Scott, "Mechanism for bistability in organic memory elements," *Applied Physics Letters*, vol. 84, no. 4, 2004, doi: 10.1063/1.1643547.

54 W. Tang, H. Z. Shi, G. Xu, B. S. Ong, Z. D. Popovic, J. C. Deng, J. Zhao, and G. H. Rao, "Memory effect and negative differential resistance by electrode-induced two-dimensional single-electron tunneling in molecular and organic electronic devices," *Advanced Materials*, vol. 17, no. 19, 2005, doi: 10.1002/adma.200500232.

55 W. S. Nam, S. H. Seo, and J. G. Park, "Effect of small-molecule layer thickness on nonvolatile memory characteristics for small-molecule memory-cells," *Electrochemical and Solid-State Letters*, vol. 14, no. 7, 2011, doi: 10.1149/1.3582801.

56 W. S. Nam, S. H. Seo, K. H. Park, S. H. Hong, G. S. Lee, and J. G. Park, "Nonvolatile memory characteristics of small-molecule memory cells with electron-transport and hole-transport bilayers," *Current Applied Physics*, vol. 10, no. 1 Suppl. 1, 2010, doi: 10.1016/j.cap.2009.12.009.

57 V. S. Reddy, S. Karak, S. K. Ray, and A. Dhar, "Carrier transport mechanism in aluminum nanoparticle embedded Alq$_3$ structures for organic bistable memory devices," *Organic Electronics*, vol. 10, no. 1, 2009, doi: 10.1016/j.orgel.2008.10.014.

58 V. S. Reddy, S. Karak, and A. Dhar, "Multilevel conductance switching in organic memory devices based on Alq$_3$ and Al/Al$_2$O$_3$ core-shell nanoparticles," *Applied Physics Letters*, vol. 94, no. 17, 2009, doi: 10.1063/1.3123810.

59 T. Abhijith, T. V. A. Kumar, and V. S. Reddy, "Organic bistable memory devices based on MoO$_3$ nanoparticle embedded Alq$_3$ structures," *Nanotechnology*, vol. 28, no. 9, 2017, doi: 10.1088/1361-6528/28/9/095203.

60 A. K. Mahapatro, R. Agrawal, and S. Ghosh, "Electric-field-Induced conductance transition in 8-hydroxyquinoline aluminum (Alq$_3$)," *Journal of Applied Physics*, vol. 96, no. 6, 2004, doi: 10.1063/1.1778211.

61 K. Onlaor, B. Tunhoo, T. Thiwawong, and J. Nukeaw, "Electrical bistability of tris-(8-hydroxyquinoline) aluminum (Alq_3)/ZnSe organic-inorganic bistable device," *Current Applied Physics*, vol. 12, no. 1, 2012, doi: 10.1016/j.cap.2011.07.004.

62 C. H. Tu, Y. S. Lai, and D. L. Kwong, "Memory effect in the current–voltage characteristic of 8-hydroquinoline aluminum salt films," *IEEE Electron Device Letters*, vol. 27, no. 5, 2006, doi: 10.1109/LED.2006.872915.

63 G. B. Coleman and H. C. Andrews, "Image segmentation by clustering," *Proceedings of the IEEE*, vol. 67, no. 5, 1979, doi: 10.1109/PROC.1979.11327.

64 D. Prime and S. Paul, "First contact-charging of gold nanoparticles by electrostatic force microscopy," *Applied Physics Letters*, vol. 96, no. 4, 2010, doi:10.1063/1.3300731.

Further Reading/Resources

S. M. Sze and K. Ng Kwok, *Physics of Semiconductor Devices*. Wiley, 2007, ISBN: 9780471143239, Online ISBN:9780470068328, doi: 10.1002/0470068329, All rights reserved.

W. -C. Chen, (ed.), "*Electrical Memory Materials and Devices*," 2015, Print publication Print ISBN 978-1-78262-116-4, PDF eISBN 978-1-78262-250-5, ePub eISBN 978-1-78262-739-5.

S. Paul. Theme Issue 'Making nano-bits remember: a recent development in organic electronic memory devices' compiled, *Philosophical Transactions of Royal Society A (Mathematical, Physicial and Engineering Science)*, vol. 267, no. 1905, 2009.

Related Articles (See Also)

1 F. Paul and S. Paul, To be or not to be – review of electrical bistability mechanism(s) in polymer memory devices, *Small*, 2022, doi: 10.1002/smll.202106442.

2 D. Prime and S. Paul, "Overview of organic memory devices," *Philosophical Transactions of the Royal Society A: Mathematical, Physical and Engineering Sciences*, vol. 367, no. 1905, pp. 4141–4157, 2009, doi: 10.1098/rsta.2009.0165.

3 Q. D. Ling, D. J. Liaw, C. Zhu, D. S. H. Chan, E. T. Kang, and K. G. Neoh, "Polymer electronic memories: materials, devices and mechanisms," *Progress in Polymer Science*, vol. 33, no. 10, pp. 917–978, 2008, doi: 10.1016/j.progpolymsci.2008.08.001.

4 B. Hwang and J. S. Lee, "recent advances in memory devices with hybrid materials," *Advanced Electronics Materials*, vol. 5, no. 1, pp. 1–22, 2019, doi: 10.1002/aelm.201800519.

5 C. -U. Pinnow and T. Mikolajick, "Material aspects in emerging nonvolatile memories," *Journal of the Electrochemical Society*, vol. 151, no. 6, p. K13, 2004, doi: 10.1149/1.1740785.

6 A. Chen, "A review of emerging non-volatile memory (NVM) technologies and applications," *Solid State Electronics*, vol. 125, pp. 25–38, 2016, doi: 10.1016/j.sse.2016.07.006.

7 T. C. Chang, K. C. Chang, T. M. Tsai, T. J. Chu, and S. M. Sze, "Resistance random access memory," *Materials Today*, vol. 19, no. 5, pp. 254–264, 2016, doi: 10.1016/j.mattod.2015.11.009.

8 Y. Fujisaki, "Review of emerging new solid-state non-volatile memories," *Japanese Journal of Applied Physics*, vol. 52, no. 4R, p. 40001, 2013.

9 V. S. Makarov and S. Selberherr, "Emerging memory technologies: trends, challenges, and modeling methods," *Microelectronics Reliability*, vol. 52, no. 4, pp. 628–634, 2012, doi: 10.1016/j.microrel.2011.10.020.

10 K. Saranti and S. Paul, *Charge-Trap-Non-volatile Memory and Focus on Flexible Flash Memory Devices,* Springer Science, Charge-Trap Memory, 2017, doi: 10.1007/978-3-319-48705-2_2.

9

Next Generation of High-Performance Printed Flexible Electronics

Abhishek S. Dahiya, Yogeenth Kumaresan, and Ravinder Dahiya

University of Glasgow, James Watt School of Engineering, Bendable Electronics and Sensing Technologies (BEST) Group, Glasgow, UK

9.1 Introduction

Advances in flexible and printed large-area electronics have enabled novel applications across numerous areas including wearable systems, robotics, displays, Internet of Things (IoT), and healthcare [1–17]. The devices fabricated on flexible substrates can conform to different curvy surfaces, which is required for many of the above applications [18–23]. Along with the flexible form factor, applications such as IoT, smart healthcare demand high-device performance (fast data processing) leading to myriad machine-to-machine and/or human-to-machine connectivity at 5G communications. Accordingly, significant research efforts are on-going to manufacture electronic devices/systems with flexible form factor and high-performance [11, 24–39]. Different manufacturing strategies have been implemented to fabricate devices with such features including heterointegration using off-the-shelf electronic devices on flexible substrates, printing functional inks and materials to realize active devices and circuits, etc. Among these technologies, the printed electronics, defined as the printing of circuits on diverse planar and nonplaner substrates such as paper, polymers, and textiles, has seen rapid development motivated by the promise of low-cost, high-volume, high-throughput production of electronic devices [34, 40].

Presently, the printed electronics is not considered as the substitute for conventional high-performance Si-based electronics because state-of-the-art all-printed devices/circuits are encountering low integration density, long switching time, and low-device performance. Limited device performance offered by printed electronics is due to (i) use of low-mobility materials such as organics, amorphous, and poly-crystalline semiconductors, and (ii) large channel lengths ($\sim$10 μm) of thin-film transistors (TFTs) (due to limitations in the printing resolution) [41]. This limits the use of current printed electronics, despite inherent advantages in terms of lesser material waste, etc. Currently, the niche market for printed electronics is considered to lie in the low-cost printed circuits based on conducting, organic semiconducting, and dielectric materials based inks aiming at high-volume market segments, where the high performance of conventional electronics is not required (in general, the cost of printed electronics is expected to be 2 or 3 orders of magnitude cheaper than Si per unit area) [42–46]. The chemical instability of organic materials also limits their use to the low-end applications [47, 48].

On the other hand, as we have discussed above, several emerging applications (e.g. IoT, mobile healthcare, smart cities, robotics) require fast computation and communication, large-scale integration, for example to achieve a processing engine [49, 50]. As a result, alternative approaches

Advances in Semiconductor Technologies: Selected Topics Beyond Conventional CMOS, First Edition. Edited by An Chen.
© 2023 The Institute of Electrical and Electronics Engineers, Inc. Published 2023 by John Wiley & Sons, Inc.

such as metal-oxide-based, TFT technology are being explored with complicated layouts [49]. This is because in the absence of viable p-type materials, only n-type field-effect transistors (FETs) can be fabricated. Likewise, metal-oxide based printed sensors have been reported [51, 52]. The single-crystalline inorganic semiconducting materials-based printed electronics is the other route that has been explored recently with innovative printing of nano (e.g. nanowires [NWs], nanoribbons [NRs]) to macro (e.g. ultrathin chips [UTCs]) scale structures [11, 18, 29, 31, 36–39, 53–64]. The transfer printing and contact printing methods developed for this purpose provide an alternative to the traditional thermal budget issue associated with the inorganic semiconducting materials, i.e. due to high-temperature processing requirements, it is difficult to fabricate devices directly over flexible polymeric substrates [29, 30, 53, 56, 65]. Combining the two properties of cheap unit cost and high-performance, inorganic semiconducting materials-based printed electronics could provide a means to implement innovative solutions for faster data computing in flexible devices such as electronic-skin (eSkin) [8–10, 35, 66].

This chapter presents a comprehensive discussion about latest advances in printed electronics technologies and the opportunities they offer through high-performance devices. Following a brief introduction of various printing technologies, a detailed discussion is presented about the

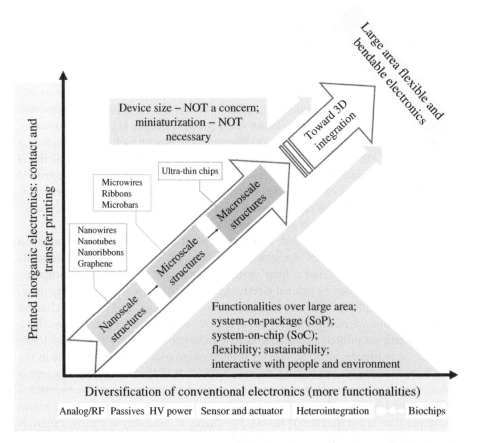

Figure 9.1 Printed electronics enabling higher diversification and functionalities than conventional electronics including heterointegration of biochips, microelectromechanical sensors, power electronic, analog/RF devices in large-area electronics. Printing technologies have been developed to provide versatile routes for assembly of nano-to-chip scale structure and 3D integration of inorganic functional materials/inks into well-organized arrangements onto various substrates for high-performance flexible electronics. Source: Dahiya et al. [40]/Springer Nature/CC BY 4.0.

high-performance printed devices using nano-to-chip scale inorganic semiconducting structures. Finally, some of the challenges in the way of developing next generation of high-performance printed electronics are discussed along with potential solutions (Figure 9.1).

9.2 Printing Technologies

The printing methods can be divided into contact and noncontact methods depending on whether the printing materials are in direct contact with the substrate, as shown in Figure 9.2 [34]. In contact printing process, the patterned structures with inked surfaces are brought in physical contact with the substrate. In a noncontact process, the solution is dispensed through via openings or nozzles and structures are defined by moving the stage (substrate holder) in a preprogrammed pattern. The prominent noncontact printing techniques include slot-die coating, screen printing, and inkjet printing. The contact-based printing technologies comprise of gravure printing, gravure-offset printing, flexographic, microcontact or simply contact, transfer, and nanoimprinting. For the commercialization of printed devices, it is important to develop a process that enables mass production with a roll-to-roll (R2R) type process. Both contact and noncontact printing techniques will promote the large-scale production process, R2R printing, which is one of the manufacturing methods to obtain large-area electronics. Each of these techniques has its own advantages and limitations. Characteristics of each printing technique are summarized in Table 9.1.

The noncontact printing techniques have received attractions due to their distinct capabilities such as simplicity, affordability, speed, adaptability to the fabrication process, reduced material wastage, high resolution of patterns, and easy control by adjusting few process parameters. These are generally used to manufacture printed devices mainly using organic materials for applications toward low-end. Nevertheless, noncontact printing approaches such as screen and inject printing have been exploited to realize high-performing devices over large areas [67–69]. Screen printing technique typically consists of a screen and squeegee in which the substrate and screen are not in direct contact. During printing steps, the ink placed on the top of screen passes through the screen mesh and transferred to the other side when squeegee blade is moved under optimal force and pressure. The printed structures using screen printing technique are useful in fabrication of various devices such as RFID antennas [70], touch sensors [71], TFTs [72], electrochemical sensors, biofuel cells [73], and array of microelectrodes [74]. For example, a capacitive touch sensor array was fabricated on flexible polyethylene terephthalate (PET) and paper substrate by screen printing silver (Ag) flake ink-based electrodes and polyvinylidene fluoride (PVDF) ink-based piezoelectric touch-sensitive layer [71]. The touch sensor array on PET and paper substrate demonstrated a sensitivity of 1.2 and 0.3 V/N, respectively. Similarly, highly stretchable electro chemical sensor array was achieved by replacing the nonstretchable commercial screen-printing ink with intrinsically stretchable inks [73]. Figure 9.3a1 shows the

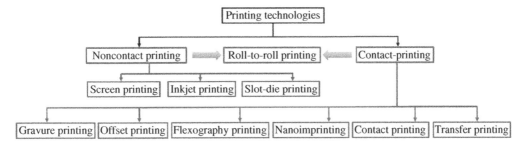

Figure 9.2 The classification of common printing techniques.

Table 9.1 Comparison of various printing techniques.

Parameter	Gravure	Offset	Flexographic	Screen	Contact and nanoimprint	Transfer	Inkjet	Slot-die
Print resolution (µm)	50–200	20–50	30–80	30–100	1–20	4–50	15–100	200
Print thickness (µm)	0.02–12	0.6–2	0.17–8	3–30	0.18–0.7	0.23–2.5	0.01–0.5	0.15–60
Printing speed (m/min)	8–100	0.6–15	5–180	0.6–100	0.006–0.6	NA	0.02–5	0.6–5
Req. solution viscosity (Pa S)	0.01–1.1	2–5	0.01–0.5	0.5–5	0.1	NA	0.001–0.1	0.002–5
Solution surface tension (mN/m)	41–44		13.9–23	38–47	22–80	NA	15–25	65–70
Material wastage	Yes	Yes	Yes	Yes	Yes	No	No	Yes
Controlled environment	Yes	Yes	Yes	No	No	Yes	No	No
Experimental approach	Contact	Contact	Contact	Contact	Contact	Contact	Contact-less	Contact-less
Process mode (sample pattern line)	Multi-steps	Multi-steps	Multi-steps	Multi-steps	Multi-steps	Multi-steps	Single step	Single step
R2R compatibility	Yes	Yes	Yes	Yes	Yes	No	No	Yes
Hard mask requirements	No	No	No	Yes	No	Yes	No	No
Printing area	Large	Large	Large	Medium	Medium	Medium	Large	Large

Source: Adapted from Khan et al. [34].

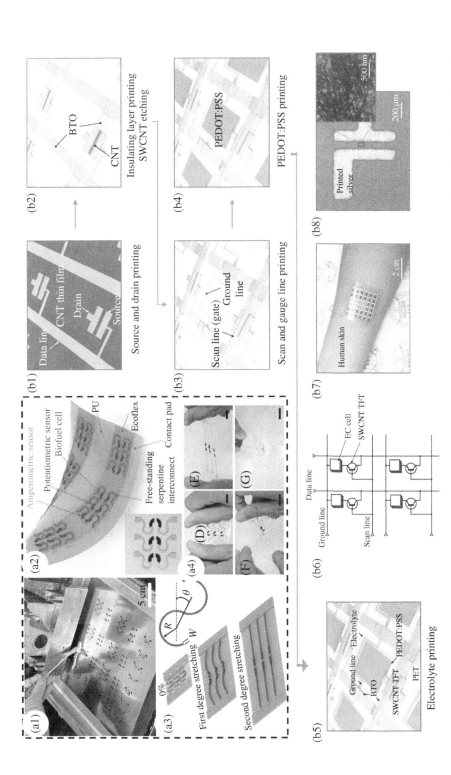

Figure 9.3 (a) Screen printed devices: (a1) image of the stencil employed for printing the stress-enduring stretchable devices; schematics showing (a2) large-scale printed stretchable device arrays along with their various applications (inset shows an image of a printed CNT-based array device); (a3) the two degrees of stretching–design-induced (first stretching) and intrinsic (second stretching) stretchability, enabling the printed arrays to accommodate high levels of strains along with parameters defining the curvature of a free-standing serpentine interconnect (top right); and (a4) photographs of stretchable array under (D) 0% and (E) 175% linear, (F) 180° torsional, and (G) 5 mm indention strains. Scale bar for images D–G = 1 cm. Source: From Bandodkar et al. [73]/with permission of American Chemical Society. (b1–b5) schematic diagram showing the fabrication process and structure of a fully printed AMECD; (b6) circuit diagram showing the configuration of as-printed AMECD; (b7) a photograph of 6 × 6 pixel flexible AMECD laminated on human skin displaying English letter "U"; and (b8) scanning electron microscope (SEM) image showing the printed silver electrodes. The inset shows an SEM image of the SWCNT network in the channel region. Source: From [72]/with permission of American Chemical Society.

large-scale fabrication of carbon nanotube (CNT)-based electrochemical sensors and biofuel cells array using screen printing technique. The development of intrinsically stretchable CNT-based ink, by combining the electrical and mechanic property of CNT with elastomeric polyurethane binder, along with the structural engineering of serpentine shaped interconnects (Figure 9.3a2) enabled the printed stretchable sensors to sustain two degrees of stretchability as shown in Figure 9.3a3,a4. When sensor is subjected to external strain, the serpentine pattern unwinds its structural geometry to accommodate the first-degree strain, which is followed by intrinsic property of stretchable CNT-based ink to handle the second-degree strain. Accordingly, the fabricated devices demonstrated 500% stretchability [73]. Likewise, a fully printed active-matrix electrochromic display (AMECD) was realized using screen printing of CNT-based TFTs [72]. The process flow of fully screen printed AMECD is given in Figure 9.3b, in which single-walled carbon nanotube (SWCNT) TFTs consist of SWCNT as active channel, barium titanate (BTO) as dielectric layer, and silver electrode with poly(3,4-ethylenedioxythiophene):polystyrene sulfonate (PEDOT:PSS) as an active material in electrochromic pixels to demonstrate a color change. First, the silver contact electrode for source, drain, and data line region were printed on SWCNT incubated PET substrate. Sequentially, printing of BTO layer as hard mask to etch CNT and passive the data lines with BTO printing, followed by scan line, ground line, PEDOT:PSS layer, and electrolyte screen printing were performed to obtain AMECD. Figure 9.3b7 shows the flexile fully screen-printed, AMECD wrapped on human skin displaying the letter "U." The scanning electron microscope image of SWCNT channel region and printed source/drain Ag electrode is given in Figure 9.3b8.

Recently, the newly emerging contact-based printing methods such as contact/micro-contact printing and transfer printing have attracted significant interest, and especially for inorganic monocrystalline semiconductors based high-performance flexible electronics [75]. Wide range of inorganic structures, from nano-to-chip scale, have been printed using these techniques to obtain high-performance printed electronics. Generally, contact printing is suitable for printing *vertically aligned* nanostructures (NSs) (grown via top-down or bottom-up approaches) [31, 62]. On the other hand, transfer printing technique enables the transfer of *laterally aligned* structures from a donor substrate to a receiver substrate using soft elastomeric stamp [18]. The transfer printing provides a promising solution to transfer nano-to-chip scale structures. The mechanisms for contact and transfer printing techniques have been discussed and presented in detail elsewhere [30, 31, 40]. In Section 9.3, advances in high-performance printed devices using nano-to-chip scale printed structures using contact and transfer printing techniques are presented.

9.3 High-Performance Printed Devices and Circuits Using Nano-to-Chip Scale Structures

9.3.1 Nanoscale Structures

Various top-down and bottom-up approaches have allowed large-area synthesis of nanoscale materials at wafer scale [31, 33, 76–81]. Both contact and transfer printing has been used to print nanoscale inorganic materials to fabricate high-performance devices such as FETs, photodetectors (PDs), energy generators. Using contact printing, mainly NWs are printed from the donor substrates over flexible receiver substrate. The printing process has three main steps: (i) NW bending; (ii) alignment of NWs by the applied shear force; and (iii) breakage and transfer of NWs upon anchoring to the receiver substrate through surface chemical and physical interactions. The advantage of contact printing is that the as-printed NWs are transferred onto the receiver substrate in an aligned manner in a single step, dictated by the direction of sliding. Such alignment is enabled by the shearing force generated during sliding and is favorable for the fabrication of FETs capable of delivering high ON currents. For instance, high-performance (field-effect mobility, $\mu_e > 2000 \, cm^2/V \, s$) FET based on InAs printed NWs as the channel material has been developed [82]. The FET device consists of 400 printed NWs in a 3 μm long channel and able to deliver an ON current of ~6 mA at V_{DS} of 3 V (Figure 9.4a–c). The contact printing could also offer 3D assembly of NW for vertical electronics on both planar and flexible substrates through monolithic printing steps [58, 59, 62, 84]. Regarding this, the 10 functional device layers of Ge/Si NWs, stacked to form a 3D electronic structure are noteworthy (Figure 9.4d,e) [62]. Notably, the fabricated transistors show minimal variation in the threshold voltage and exhibit a large average on-current of 4 mA. Using contact printing, multifunctional circuitry that utilizes both the sensory and electronic functionalities of NWs has also been demonstrated. For example, using 3D stacking methodology with contact printed NWs leads to ultrahigh-performance electronics not accessible by scaled complementary metal-oxide-semiconductor (CMOS) [62]. By repeating the printing process, up to 10 layers of active NW-FET devices were assembled, and a bilayer structure consisting of logic in layer 1 and nonvolatile memory in layer 2 was demonstrated. In an another example, using the sensory and electronic functionalities of nanoscale structures, multifunctional circuitry was realized using contact printed optically active CdSe NWs and high-mobility Ge/Si NWs [59]. The NW-based photo sensors and electronic devices are then interfaced to enable an all-NW circuitry with on-chip integration, capable of detecting and amplifying an optical signal with high sensitivity and precision. Approximately 80% of the circuits demonstrated successful photo response operation. The capabilities of contact printing technique were further demonstrated by large area (49 cm²) printing of aligned NW arrays as the active-matrix backplane of a flexible pressure-sensor array (18 × 19 pixels) [85]. The integrated sensor array effectively functions as an eSkin to monitor the applied pressure profiles with high spatial resolution. The eSkin system can provide fast mapping of normal pressure distributions in the range from 0 to 15 kPa.

Contact printing of inorganic semiconducting NWs can also be used to form optoelectronic devices such as PDs, light-emitting diodes (LEDs) in a mechanically flexible format [31, 86]. For example, by printing ZnO and Si NWs (Figure 9.4f), ultraviolet (UV) PDs with Wheatstone bridge (WB) configuration on rigid and flexible substrates were fabricated (Figure 9.4g–i) [31]. The fabricated PDs demonstrate high efficiency, a high photocurrent to dark current ratio ($>10^4$) and reduced thermal variations because of inherent self-compensation of WB arrangement. Due to statistically lesser dimensional variations in the ensemble of NWs, the UV PDs made from them have exhibited uniform response. Similarly, nanoscale structures have been exploited to fabricate visible [59] and near infrared PDs [87] on flexible/stretchable substrates. High-performance energy harvesting devices were also realized using printed inorganic nanoscale materials [83, 88]. For example, parallel ZnO NWs arrays enable fabrication of piezoelectric nanogenerators (PENGs) (Figure 9.4j,k) [83]. To enhance the output power in PENGs, all NWs must be oriented in same direction. This requirement is difficult to be satisfied with most of the nanostructure assembly/integration approaches such as Langmuir–Blodgett (LB) deposition [89], solution shearing methods including blown bubble approach [90], and capillary force assembly [91]. In this regard, contact printing method ensures that the crystallographic orientations of the horizontal NWs are aligned along the sweeping direction. Consequently, the polarity of the induced piezopotential is also aligned, leading to a macroscopic potential contributed constructively by all the NWs.

Transfer printing has also shown enormous potential for realizing high-performance flexible electronic devices and circuits using nanoscale materials. The gained popularity of transfer

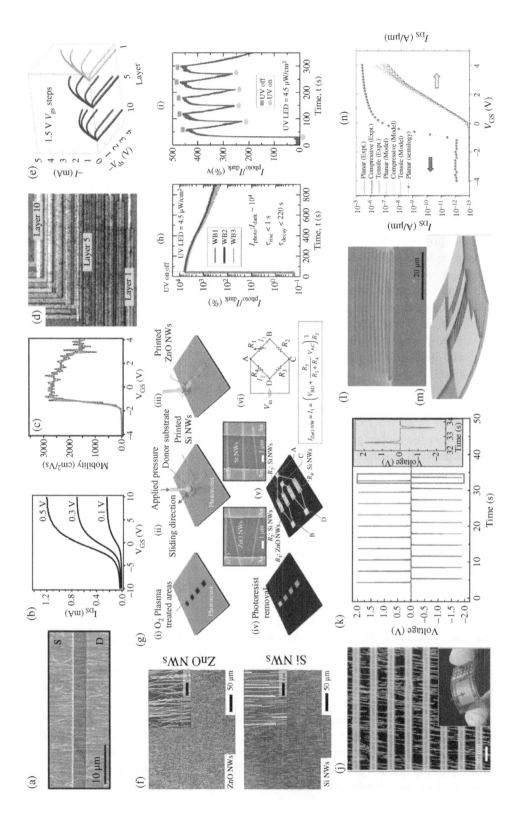

(a) S D — 10 μm

(b) I_{DS} (mA) vs V_{GS} (V); 0.5 V, 0.3 V, 0.1 V

(c) Mobility (cm²/Vs) vs V_{GS} (V)

(d) Layer 1, Layer 5, Layer 10

(e) $-I$ (mA) vs $-V_{ds}$ (V); 1.5 V V_{gs} steps; Layer

(f) ZnO NWs — 50 μm; Si NWs — 50 μm

(g)
(i) O₂ Plasma treated areas; Photoresist
(ii) Sliding direction; Applied pressure; Donor substrate; Printed Si NWs; Photoresist
(iii) Printed ZnO NWs; Printed Si NWs
(iv) Photoresist removal
(v) R_1: ZnO NWs; R_2: Si NWs — 1 μm; Au; R_3: Si NWs; R_4: Si NWs — 1 μm; Au; A, B, C, D
(vi) V_{in}; $I_{ZnO NW} = I_1 = \left(V_{BD} + \dfrac{R_3}{R_3 + R_4} V_{AC}\right)\dfrac{1}{R_2}$

(h) I_{photo}/I_{dark} (%) vs Time, t (s); WB1, WB2, WB3; UV on-off; UV LED = 4.5 μW/cm²; $I_{photo}:I_{dark} \sim 10^4$; $\tau_{rise} < 1$ s; $\tau_{decay} < 220$ s

(i) I_{photo}/I_{dark} (%) vs Time, t (s); UV off, UV on; UV LED = 4.5 μW/cm²

(j) — inset

(k) Voltage (V) vs Time (s); inset Voltage (V) vs Time (s)

(l) — 20 μm

(m)

(n) I_{DS} (A/μm) vs V_{GS} (V); Planar (Expt.), Compressive (Expt.), Tensile (Expt.), Planar (Model), Compressive (Model), Tensile (Model), Planar (semilogy)

printing could be dedicated to its distinguishing capabilities in integration and assembly of laterally aligned silicon and compound semiconducting nanostructures (NSs) such as nanomembranes (NMs), NRs, NWs over large areas. Using transfer printed silicon NRs, high-performance FETs (NR-FETs) were successfully developed over fully flexible polyimide (PI) substrates (Figure 9.4l–n) [18]. The unique feature of these devices is that the high-quality dielectric (silicon nitride [SiN$_x$]) was deposited at room temperature (RT). The electrical characterizations of NR-FETs have shown high performance (mobility $\approx 656\,\text{cm}^2/\text{V}\,\text{s}$ and on/off ratio $>10^6$) which is on par with the highest performance of similar devices reported with high-temperature processes, and significantly higher than devices reported with low-temperature processes. It is known that the ohmicity of metal–semiconductor (MS) contacts deteriorates while depositing dielectrics at high-temperature which affects the transistor performance and its reliability. High-performance achieved from NR-FET devices was attributed to the RT dielectric deposition process with negligible degradation of MS contacts. The measured breakdown field strength ($>2.2\,\text{MV/cm}$) further confirms the excellent quality of the RT deposited dielectric. The reported NR-FETs are mechanically robust, with the ability to withstand mechanical bending cycles (100 cycles tested) without performance degradation. High-performance transistors based on printed nano/microscale structures (NWs and NRs) of compound semiconductor such as GaAs, GaN, and InAs on plastic substrates have also been demonstrated, making them potentially useful platforms for ultra-high frequency electronics [63, 92].

Transfer printing of graphene sheets can also produce high-mobility transistors on flexible substrates (hole mobility $\mu_p \sim 3700\,\text{cm}^2/\text{V}\,\text{s}$) [93]. However, one of the major limitations of graphene is attributed to its zero-band gap which restricts it from being used in digital applications [94]. Nevertheless, transfer printed graphene sheets could be used as an excellent material to develop flexible sensors. For example, large-area transfer printing of graphene layer on a photovoltaic (PV) cell resulting in energy autonomous tactile sensitive system for soft robotics (Figure 9.5) [32].

Figure 9.4 Nanoscale printed structures for the fabrication of high-performance flexible electronics: (a–c) SEM image and electrical characterization data of a back-gated FET fabricated on a contact printed, parallel array of InAs nanowires. Source: [82] A. C. Ford et al., 2008, Springer Nature. (d) 3D electronics with 10 layers of Ge/Si NW FETs. Each device is offset in *x* and *y* to facilitate imaging. Source: Reproduced with permission from [62], American Chemical Society. (e) Current vs. drain-source voltage characteristics (with 1.5 V gate step) for NW FETs from layers 1, 5, and 10. Source: Javey et al. [62]/with permission of American Chemical Society. (f–i) 3D schema of UV photodetector fabrication using contact printed ZnO and Si NWs. (f) Contact printing of Si and ZnO NWs. Source: (f) Reproduced with permission from Ref. [31], (2018) Springer Nature. (g) Fabrication steps of UV PDs based on ZnO and Si NWs, comprising: (i) definition of 20 mm^2 areas on a S1818 photoresist layer by photolithography, followed by an O$_2$ plasma treatment (100 W and 0.3 mbar for one minute); (ii) contact printing of ZnO and Si NWs; (iii) schematic showing printed NWs; (iv) removal of the photoresist in warm acetone (50 °C for two minutes); (v) definition of Ti (4 nm)/Au (200 nm) interdigitated electrodes by photolithography and lift-off, where (e1) and (e2) show SEM images of printed ZnO and Si NWs, respectively, bridging a pair of Ti/Au electrodes with a 5 μm gap; (vi) WB equivalent circuit and the expression determining the electric current flowing through ZnO NWs. (h) Single cycle and (i) multi-cycles measured over time and using a UV LED power density of 4.5 μW/cm^2 and a V_{in} of 0.05 V, keeping a distance between UV LED and the PD surface of 5 cm. Source: (g–i) [31]/with permission of Springer Nature. (j, k) ZnO NW based piezoelectric nanogenerator (PENG). (j) SEM image of ZnO NW arrays bonded by Au electrodes. Inset: demonstration of an as-fabricated PENG. The arrowhead indicates the effective working area. Source: Reproduced with permission from Ref. [83] (2010) American Chemical Society. (k) Open-circuit voltage. Source: Zhu et al. [83]/with permission of American Chemical Society. (l–n) NR-FET device using transfer printed nanoribbons (NRs): (l) cross-section SEM image of suspended NRs array, (m) schematic 3D illustration of the array of NR-FETs on flexible substrate, and (n) transfer characteristics (experimental (line) vs. model (dashed) simulations). Source: Reproduced with permission from Ref. [18] (2020) John Wiley & Sons.

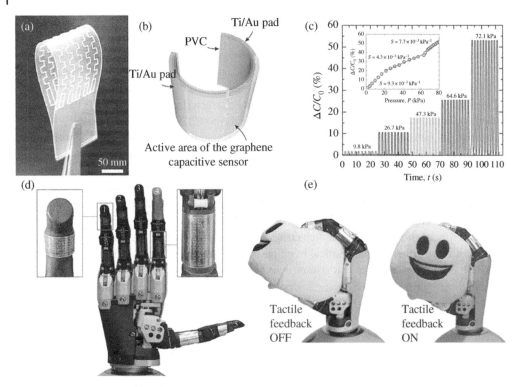

Figure 9.5 Large area graphene-based capacitive tactile sensors: (a) photograph and (b) 3D schema of a flexible graphene capacitive touch sensor. (c) Touch sensor response vs. applied pressure. (d) eSkin with capacitive sensors integrated onto a robotic hand. (e) Self-powered eSkin used as tactile feedback for a robotic hand. Source: (a, b, d, e) From [32]/with permission of John Wiley & Sons.

Transfer of single graphene layer was demonstrated with the transfer of 4-in. Chemical vapor deposition (CVD) grown monolayer of graphene from Cu foil to 125-μm-thick polyvinyl chloride (PVC) substrates by using a hot-lamination method at 125 °C. The transfer printing led to the fabrication of large-area flexible graphene-based capacitive touch sensors. The fabricated sensors showed high sensitivity (4.3 kPa^{-1}) to a wide range of pressures (0.11–80 kPa). One of the key features of the fabricated eSkin relied on its great transparency, i.e. a sunlight absorption below 5%, which allowed the effective energy harvesting of light energy by a PV cell underneath the eSkin. The viability of graphene-based skin sensors is also analyzed by means of a dynamic characterization consisting in the grabbing of a soft object. The response obtained from the capacitive sensors was successfully used as tactile feedback in an artificial hand, allowing the manipulation of rigid and soft objects with different shapes.

9.3.2 Microscale Structures

The microscale structures of inorganic materials on various substrates offer broad range of devices, including solar cells [95], LEDs [96], biosensors [97], mechanical energy harvesting device [98], piezoelectric touch sensors [71], and photonic crystals [67]. In particular, an array of highly ordered microscale structure ($\leq$999 μm) over large areas have numerous potential applications, but it is challenging to achieve versatile, scalable, and precise assembly of such structures over flexible substrates using conventional microfabrication methods [67]. Among different contact and noncontact-based methods, transfer printing technique has emerged as a potential solution to achieve integration of high-performance inorganic functional micro components or devices (i.e. inks of various inorganic semiconductors and metals) on soft and flexible receiver substrate from its donor using an intermediate elastomeric stamp [99]. In general, the transfer printing has two process steps, namely retrieval process and printing process, with three-layer system (substrate/ink/stamp) which contains two interfaces between substrate/ink and ink/stamp [30]. The adhesion between the interfaces, for example a strong adhesion between stamp/ink at the retrieval stage and a weak interface between stamp/ink than the ink/substrate during the printing process, plays a significant role in achieving successful transfer with high yield. Based on the adhesion modulation at the interfaces, the transfer printing process are classified into various techniques; the detailed description regarding the transfer mechanism of various transfer printing techniques has been reported elsewhere [30]. Here, we present the work related to high-performance microscale devices, such as LEDs, displays, PDs, solar cells, and mechanical energy harvesting device, achieved through transfer printing technique. A thin silicon (Si) micro plate array obtained from silicon on insulator (SOI) through series of lithography and etching steps have numerous applications in high-performance device. To obtain Si-micro structure on target substrate, an adhesive-less transfer printing process was demonstrated by controlling the bending radius of the polydimethylsiloxane (PDMS) stamp as shown in Figure 9.6a(1–5) [100]. This transfer printing technique revealed a 100% yield with successful transfer of the micro-Si plates array to various substrates such as glass slide, wristwatch, eye glass, Bluetooth earphone curved surface, and concave/convex surfaces. A photomicrograph image of 7×7 micro-Si plates array (size of each micro plate is $760\,\mu m \times 760\,\mu m$) on various substrates are shown in Figure 9.6a(6–10) [100]. The transistor device fabricated using a transfer printed n-type Si through adhesion-less transfer process demonstrated a high-performance with the mobilities $>325\,cm^2/V\,s$ and on/off ratios $>10^5$ [104]. Similarly, three successive R2R transfer printing process was performed to achieve overlay-aligned transfer of inorganic LEDs and single-crystal Si TFTs in a heterogeneous integration with stretchable interconnects to realize stretchable active matrix (AM) display on elastomeric substrate [101]. Figure 9.6b(1) depicts the R2R process to achieve active-matrix light-emitting diode (AMLED) display. The device performance of the Si-TFT, micro-light emitting diode (μ-LED), and integrated pixel device is given in Figure 9.6b(2–4). The transfer and output characteristics of the printed Si-TFT reveal high-device performance with mobility over $750 \pm 20\,cm^2/V\,s$, on/off ratio of $>10^6$ and subthreshold swing of $0.14\,V\,dec^{-1}$, along with stable performance up to 40% strain. The I–V characteristics of μ-LED demonstrated stable device performance up to 30% strain with slight degradation at 40% strain. However, the resulting integrated pixel enables stable emission at 40% strain demonstrating the significant importance of transfer printing in achieving the stretchable AMLED display using inorganic materials.

Microscale structures are also widely used in various optoelectronic applications, for example, heterogeneous integration of microscale patterns of Al_2O_3-capped PbS quantum dot (QD) films over a large area will help to realize PD array [102]. Figure 9.6c depicts the heterogeneous integration of PD array over a large area through micro transfer printing approach. The device performance was characterized by illuminating the normalized surface with $10\,nW$ at $2.1\,\mu m$ wavelength and the time-dependent photo response and cyclic response was measured as shown in Figure 9.6c(2–3). The photoconductor demonstrated maximum of $85\,A/W$ for $140\,nm$ thick PbS QD film with fast response and recovery time. Transfer printing has also been used to transfer lead zirconate titanate (PZT) ribbons to fabricate flexible mechanical energy harvester [98]. Electromechanical characterization of the PZT ribbons by piezo-force microscopy indicates

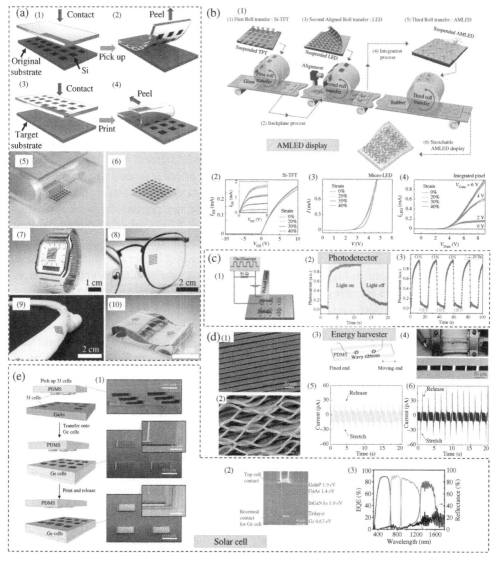

Figure 9.6 Transfer printing of microscale structure for high-performance devices. (a) Schematic illustrations of adhesive-less transfer printing procedure and the optical images after transferring to various substrates: (1–4) schematic illustration demonstrating the transfer of micro Si plates from donor to receiver substrate through elastomeric stamp; (5) optical image while transferring to glass substrate; and (6–10) photograph of Si-micro plates transferred to various substrates such as (6) glass slide, (7) wrist watch, (8) eye glass, (9) curved surface of ear phone, and (10) concave/convex surface. Source: From [100]/with permission of American Chemical Society. (b) Roll-transfer technique for heterogeneous integration of single-crystal Si-TFT and μ-LED arrays for stretchable AMLED display: (1) schematic illustration of roll-transfer technique; and (2–4) device characteristics of (2) Si-TFT, (3) μ-LED, and (4) integrated pixels. Source: Choi et al. [101]/with permission of John Wiley & Sons. (c) Al$_2$O$_3$-capped PbS quantum dot photodetector through micro-transfer printing technique: (1) schematic representation of measurement setup; (2) time-dependent photodetector response; and (3) the cyclic response at 20 Hz frequency. Source: From Mahmoud et al. [102]/with permission of American Chemical Society. (d) Energy harvesting device using transfer printing: (1–2) SEM image of PZT ribbon transferred to (1) PDMS without prestrain and (2) with prestrain to achieve wavy/buckled piezoelectric PZT ribbon; and (3–6) energy conversion from stretching the wavy/buckled PZT ribbons, their (3) experiment setup, (4) photograph demonstrating the device mounted on stretching stage, and (5, 6) short-circuit current measurement for device with 5 and 10 ribbons. Source: From [98]/with permission of American Chemical Society. (e) Multi-stacked solar cell through transfer printing: (1) schematic representation of process flow for assembled 3J/Ge cells and their SEM images (right); (2) cross-sectional SEM of a 3J/trilayer/Ge cell; and (3) EQE and reflectance spectra covering the wide range from 300 to 1800 nm. Source: From [103]/with permission of Elsevier.

that their energy conversion metrics are among the highest reported on a flexible medium. The transfer printing method was also exploited to develop flexible micro thermoelectric generators (μ-TEGs) on PET substrate. A TEG module, consisting of an array of 34 alternately doped p-type and n-type Si microwires, is developed on a SOI wafer using standard photolithography and etching techniques. The TEG modules are transferred from SOI wafer to PET substrate by using transfer printing method. Figure 9.6d(1, 2) shows the PZT ribbon on elastomer substrate transferred under normal and prestretched state, respectively. A maximum of 9.3 mV open-circuit voltage was recorded from the flexible μ-TEG prototype with a temperature difference of 54 °C (Figure 9.6d). In case of solar cells, monolithic stack of multiple junctions tends to enhance the efficiency, but the challenges related epitaxial and current matching in monolithic stacks needs to be addressed. In this regard, transfer printing was adopted to mechanically stack the layers, by transferring triple-junction (3J) solar cells onto Ge cell, to bypass the above existing challenges [103]. Figure 9.6e shows the transfer of 3J cells onto Ge cell and their external quantum efficiency (EQE) value covering the broad spectral range of 300–1800 nm. The performance comparison before and after transfer printing of various high-performance electronic devices, such as PD and LED, have been reported [105]. Wang et al. demonstrated the heat induced adhesion modulation by using thermally expandable micro spears inside the elastomer to facilitate easy transfer [105]. The LED and PD devices revealed similar performance before and after printing, demonstrating the capability of transfer printing technique to transfer an array of microstructures to achieve high-performance devices on flexible/elastomeric substrates.

9.3.3 Chip-Scale (or Macroscale) Structures

Chip-/macroscale structures with the dimension greater than 1 mm is also printable using transfer printing approach. In general, inorganic or silicon-based electronic devices are entirely fabricated using conventional approach and then transferred over to flexible or elastomeric substrates for potential applications without compensating its high-performance [106, 107]. In this regard, a physical method namely layer transfer process was broadly utilized to realize UTCs by removing the top thin layer of the processed silicon wafer using SlimCut process [108, 109]. Figure 9.7a(1–2) demonstrates the fabrication, using SlimCut process, and transfer of UTCs to the carrier substrate. The SlimCut process is also referred as the controlled spalling technique (CST), in which the active device fabricated on the rigid silicon or SOI substrate was separated along with ultrathin body silicon by placing a tensile stressor layer (tensile Ni stressor layer) on top of the active device (Figure 9.7a(1)) [108]. Under the applied upward shear force, stress-induced cracks were created at the edges near the top surface of the rigid silicon wafer, simultaneously cracks propagate parallelly along the top surface to remove the top surface of silicon along with the active device. The thickness of the stressor layer and the applied shear force determines the fracture depth and the thickness of the thinned silicon. After SlimCut process, the transfer of UTCs to the carrier substrate was carried out, followed by the removal of the stressor layer to reveal the active device and circuits as shown in Figure 9.7a(4, 5). The photographic images of flexible circuits before and after transfer process are given in Figure 9.7a(3, 6), respectively [108]. The transistor on the circuit revealed a stable device performance without any performance degradation under 15.8 and 6.3 mm bending radii (Figure 9.7a(7, 8)), which demonstrates CST capability to successfully transfer the wafer scale circuits and devices to target substrate for potential applications. Likewise, inorganic thin-film electronic devices, such as TFTs and gas sensors, fabricated on rigid glass slide were transferred to various surfaces such as cylindrical pen, wrinkled hand glove and cylindrical gas line as shown in Figure 9.7b [110]. The polymethyl methacrylate (PMMA) layer spin-coated

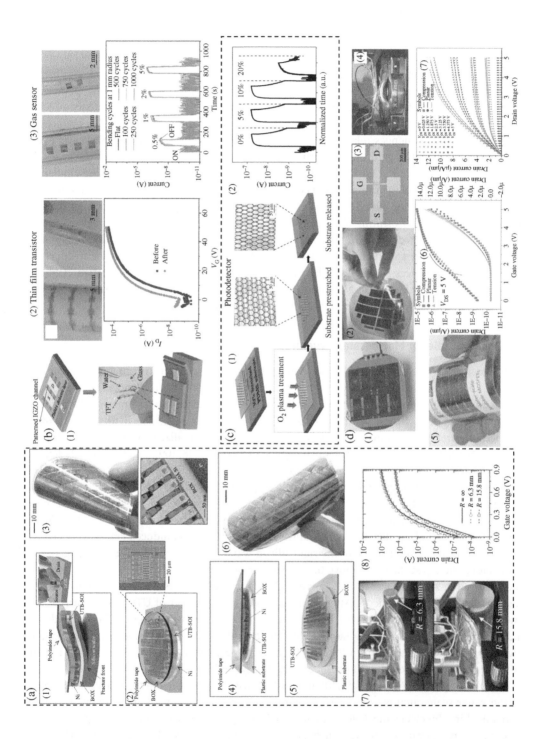

(3) Gas sensor

Bending cycles at 1 mm radius
— Flat
— 100 cycles — 500 cycles
— 250 cycles — 750 cycles
— 1000 cycles

Current (A) vs Time (s)

(2) Thin film transistor

I_D (A) vs V_G (V)

■ Before
● After

(b)
(1) Patterned IGZO channel — TFT — Water — Glass

(c) Photodetector
(1) O$_2$ plasma treatment — Substrate prestretched — Substrate released

(2) Current (A) vs Normalized time (a.u.)
0% 5% 10% 20%

(3) G S D 200 μm

(4)

(6) Drain current (A/μm) vs Gate voltage (V)
Symbols
Compression
Planar
Tension
V_{DS} = 5 V

(7) Drain current (μA/μm) vs Drain voltage (V)
Symbols
Compression
Planar
Tension
V_G = 0 V
V_G = 1.25 V
V_G = 1.875 V
V_G = 3.125 V
V_G = 3.75 V
V_G = 4.375 V
V_G = 5 V

(a)
(1) Polyimide tape — Ni — BOX — UTB-SOI — Fracture front — Silicon wafer — Source — Drain — UTB-SOI

(2) Polyimide tape — BOX — UTB-SOI — Ni — 20 μm

(3) 10 mm — 50 μm — BOX (GLASS) — UTB-SOI

(4) Polyimide tape — Ni — BOX — UTB-SOI — Plastic substrate

(5) UTB-SOI — BOX — Plastic substrate

(6) 10 mm

(7) R = 6.3 mm — R = 15.8 mm

(8) Drain current (A) vs Gate voltage (V)
— R = ∞
–○– R = 6.3 mm
–◇– R = 15.8 mm

on glass slide has poor adhesion, therefore an inorganic high-performance electronic device fabricated on top of PMMA surface could be readily separated from the underlying glass substrate by placing a water-droplet at the interface. Accordingly, the bottom-gated indium gallium zinc oxide (IGZO) TFTs and pd-decorated IGZO gas sensors fabricated on glass slides have been transferred to cylindrical pen and gas tube as shown in Figure 9.7b(2, 3), respectively. The transferred TFTs demonstrated stable performance with a mobility of $10.7\,cm^2/V\,s$, a threshold voltage of 8.4 V, and an on/off current ratio of 5×10^5 even after 50 bending cycles at a 1 mm bending radius. In addition, the pd-decorated IGZO gas sensors wrapped around cylindrical gas tube revealed excellent sensitivity of $1.6 \times 10^6\%$ at a 5% H_2 concentration at room temperature, demonstrating its potential for accurate detection of dangerous H_2 gas [110]. Similar approach has been utilized to transfer a honeycomb patterned gold electrode and IGZO PD array to prestrained PDMS substrate to enhance the stretchability (Figure 9.7c) [53]. Honeycomb gold on PDMS substrate handled 70% stretching strain and the honeycomb-patterned IGZO demonstrated higher responsivity of $58.5\,mA/W$ at $0.2\,mW/cm^2$.

The electronic devices and circuits, such as metal oxide semiconductor (MOS) capacitors and metal oxide semiconductor field-effect transistors (MOSFETs), on 4-in. Si wafer fabricated using conventional microfabrication approach were transferred to the flexible PI substrate through PDMS-assisted wafer scale transfer process (Figure 9.7d(2)) [29]. The fabricated MOSFET device using conventional approach and their respective optical image are shown in Figure 9.7d(1, 3). The n-MOSFET transferred to PI substrate was subjected to tensile and compressive bending strain and the transistor performance was characterized (Figure 9.7d(4, 5)). For Si MOSFET, the effective mass of the carrier will be affected under bending conditions. Accordingly, the current level increases under tensile bending and decreases under compressive bending, which is reflected in the transfer and output characteristics with the effective mobility of $384\,cm^2/V\,s$ under tensile strain and $333\,cm^2/V\,s$ under compressive strain.

Figure 9.7 Macroscale or wafer-scale transfer of processed devices on to flexible substrates. (a) Extremely flexible n-type FET from ultrathin silicon integrated circuits using SlimCut and transfer process on flexible carrier substrate: (1) schematic illustration of the SkimCut process used for removing the prefabricated devices and circuits from the rigid silicon wafer; (2) the postspalling process involving the selective removal of the residual Si; (3) photograph of a 100 mm ultrathin body-silicon on insulator (UTB-SOI) flexible circuit; (4) schematic illustration of transfer of the thin Si film onto another flexible substrate; (5) removal of the polyimide tape and the relatively thick Ni layer; (6) photograph of a 100 mm flexible UTB-SOI circuit prepared using this strategy, illustrating the high level of the mechanical flexibility of the circuits; (7) bending test of n-FET and (8) their transfer characteristics. Source: From [108]/with permission of American Chemical Society. (b) Extremely flexible IGZO-based bottom gated TFT and Pd-decorated IGZO-based gas sensor through water-assisted transfer: (1) schematic illustration of separation of electronic device from ridged glass substrate by placing water at the interface between PMMA and glass slide; (2) optical image of bottom-gated IGZO TFT wrapped around cylindrical surface and their transfer characteristics under different bending radii; and (3) optical image of Pd-decorated IGZO gas sensor wrapped around gas tube and the time-dependent sensor response. Source: From [110]/with permission of John Wiley & Sons. (c) Omnidirectional stretchable IGZO photodetector: (1) a schematic drawing representing the transfer process of honeycomb-patterned stretchable Au electrode on prestretched PDMS and their respective optical images; and (2) the time-resolved photoresponse of honeycomb-patterned IGZO photodetector under different stretching. (d) Transfer of n-MOSFET and MOS capacitor by PDMS assisted-wafer scale transfer process: (1–4) photomicrograph of (1) flexible UTCs transferred on polyimide, (2) n-MOSFET on wafer, (3) magnified image of n-MOSFET, (4) measurement setup under bending condition, and (5) the n-MOSFET device encapsulated by polyimide; and (6) their transfer and (7) output characteristics under various bending condition. Source: From [29]/with permission of John Wiley & Sons.

9.4 Challenges and Future Directions

9.4.1 Integration of Nano-to-Chip Scale Structures

The printing of nanoscale materials is of significant interest for next generation of printed flexible high-performance electronics. To achieve large-area, uniform printing of these structures, many existing hurdles need to be overcome. First, uniform NW growth over large area is required. The exact NW synthesis route, i.e. top-down or bottom-up for the fabrication of future semiconductor nanostructure-based flexible electronic devices is unclear; however, it is possible that the route will exploit both top-down and bottom-up techniques in tandem to allow a scalable process to achieve nanostructure at wafer level with good uniformity [78]. The second barrier for large-scale integration of nanoscale structures is to develop printing techniques that allow transfer of these structures with good uniformity [75]. To exploit the potential of contact printing for large area and uniform printing of NWs; they can be directly grown on cylindrical rolls using bottom-up process which can be used as a stamp [8]. The bottom-up synthesis of NWs on tubes of glass, quartz, and stainless-steel and even on polymers like PDMS, has been demonstrated in the past. One could see new commercial opportunities, for example, commercializing NW rolls just as the Si wafers today. By using such rolls in differential roll printing and roll transfer-printing settings, the contact printing approach can be extended to an R2R-type printing.

As we have seen in this chapter, transfer printing of nanoscale structures has also shown huge potential for realizing high-performance flexible electronic devices and circuits [18]. However, for large area electronics, controllable and reproducible transfer of nanoscale structures from the donor to the receiver substrate is needed, and hence, a precise control over the interface properties (stamp/donor and stamp/receiver) is required during transfer printing. To this end, complete control over printing parameters (e.g. retrieval/pick up velocity, adhesion switchability, stamp surface recovery) and interface properties are required. However, because of the viscoelastic properties of soft stamps, which usually cause unexpected tilt, rotation, and even buckling and drop of nanoscale structures due to the stamp deformation under applied force during printing process, the transferred yield and registration is poor. Further, printing of sub-100 nm thick structures is challenging using conventional transfer printing. This is because decreasing the thickness of a nanostructure layer decreases the strain energy release rate at the stamp/ink interface relative to the ink/substrate interface and therefore, decreases the printing yield [111]. These challenges could be addressed with innovative advancements in transfer printing technique which allows high printing yield with greater registration control over large areas. Although modified transfer printing methods have shown improvements in the yield and reliability of the process, but the use of additional equipment inevitably makes the printing system intricate and thus, not compatible with the R2R electronic manufacturing.

The *3D integration* of printed devices could offer major advantages in the future for miniaturized high-performance flexible devices, just like the 3D integration of conventional CMOS electronics. In this regard, contact printing has shown potential to be used for vertical 3D stacking of printed NWs. As an example, functional device in 10 vertically stacked of Ge/Si NWs have been shown. The advances in multimaterial additive manufacturing could offer new avenues for introducing eSkin like features in prosthesis and robotics. For instance, such 3D manufacturing processes could be employed to develop prosthesis with directly integrated or embedded touch sensors, thereby enabling robust limbs that are also free from wear and tear issues. The ability to simultaneously print multiple materials in 3D will also address the traditional robotic eSkin issue of routing of wiring. Bringing multifunctionality into eSkin like devices or other wearables is important for efficient miniaturization and monitoring of different input parameters using single device. The *heterogeneously integrated* NWs with distinct functionality will represent the future technology, where cost-competitive, scalable strategies allow integration of diverse materials with complementary performance. The need of the hour is to develop printing techniques overcoming the critical issue of multifunctionality and permitting the highly precise integration of individually selected semiconductor NWs from different materials (e.g. InP, GaAs, ZnO, Si) onto a variety of substrates (e.g. polymer, silicon, silica, metals). This will open avenues toward the manufacture of heterogeneous devices, consisting of integrated systems made from pure and/or hybrid inorganic/organic materials.

9.4.2 Technological Challenges

The technological device parameters include channel lengths, MS contacts, etc., which are important factors influencing the device performance. The transfer printing of micro/nano structures results in well-defined structures over target device substrates. However, issues such as poor registration and organic contamination because of the use of soft PDMS stamps results in poor device performance and device-to-device performance variations. For example, in presence of PDMS residues, it is difficult to realize ohmic contacts at the source and drain junctions of a transistor. In this regards, chemical methods have been reported for achieving a compete removal of PDMS residues from the surface of transferred micro/nanostructure on flexible substrate [112]. Moreover, as mentioned above, printing of nanoscale structures is challenging using conventional transfer printing which resulted into poor registration of inks. Due to this one can expect huge variations in electrical performance from device to device. Another challenge is the printing of high-resolution, high-aspect-ratio metal lines for the miniaturization of transistor channel length. At present, printed transistors have channel length in few microns which is far larger than the advanced conventional Si electronics (few nanometers). However, during the initial development stages, i.e. the time when CMOS technologies were at the point where printed electronics presently is, the channel length was longer than 1 μm. Going with the growth trend for CMOS technologies, directly printing submicron channel on printed inorganic semiconductors could be possible in future with advances in printed technologies.

9.4.3 Robustness

Along with the high performance, highly robust printed electronics is needed in future. Printed sensors that have been developed in the past suffer from wear and tear [113–115]. This is because, in the early days, printed sensors were simply wrapped around the rigid structures. This approach has some inherent problems, such as slow displacement of the sensors from the place that was secured, misleading the system on the exact place of contact, wires often are exposed on the surface of the object which can result into a hazard [116–119]. To mitigate these problems researchers have started to implement enclosers for the printed sensors. Some of these attempts are made by using simple polymers that package the printed sensors as an additional layer for protection [120, 121]. Combining such an enclosure approach with recent advancements in 3D printing and new materials have resulted in new tactile sensors made via multimaterial 3D printing [122–124]. These advancements have resulted in 3D printed intelligent fingertips for robots that not only are able to sense but also digitalize the information for maximizing the use of space [125]. Further, fingertip patterns have been developed, using 3D printing, to enhance a Neural Network for classification of different surfaces [126].

9.4.4 Disposability

The "tsunami of electronic waste," reaching 50 million tons alone in 2018, requires a significant number of resources for disposal, reuse, and recycling [127]. Some of the materials (e.g. rare earth) used in electronics today, and wasted away after end-of-use, are not available in abundance. Further, during the development of electronics itself, significant quantity of chemical wastes is generated during processes, such as chemical etching. As a material efficient technology, printed electronics leads to less materials wastage, helps to reduce electronic waste (e-waste), and could potentially allow reuse of some electronic materials (e.g. conductive inks) to open new avenues for circular electronics. In this regard, novel conductive, dielectric, and functional sustainable printing inks are required. Toward this direction, biodegradable materials have been developed to enable sensors that degrade naturally after use [128–132]. Further, a nontoxic, sweat-based flexible supercapacitor for self-powered smart textiles and wearable systems has been reported [47]. Using transfer printing approach, doped silicon NMs or fully formed ultrathin silicon microdevices have been printed over a biodegradable substrate which allowed the fabrication of high-performance biodegradable/transient electronics [133].

9.4.5 Modeling for Flexible Electronics

The flexible integrated circuits (ICs) made from ultrathin and bendable silicon chips or other micro/nanostructures experience bending induced variations in their output, owing to piezoresistive effects. More specifically, bending-induced mechanical stresses affect the crystal structure of silicon and thus the energy band structure. This effect can be macroscopically observed as variations in the electrical parameters of the devices, such as the charge carrier mobility and threshold voltage [134]. Since the bending-induced effects can significantly deviate the response of devices and circuits (up to 10% in some cases) and jeopardize the design specifications, it is essential to understand the behavior of transistors under different bending conditions, model their response, and include an extra step in the verification of circuits. There has been some progress in this direction recently [135, 136]. Despite the progress of flexible and printed electronics, the modeling of bending-induced effects on the performance of devices on mechanically flexible substrates and their inclusion on compact models used in simulation program with integrated circuit emphasis (SPICE) simulations is relatively a new development and will require greater attention in the future. By capturing the bending induced effects, it is also possible to exploit them to solve some of the longstanding issues. For example, the bending-related variation has been captured to obtain drift free response of ion-sensitive FETs on UTCs [137]. Likewise, the response of devices on a bendable substrate could be used to reconstruct the shape of substrate during bending.

9.4.6 Power Consumption

Irrespective of the conventional or printed large area electronics, both have the drawback of high-power consumption. For instance, the modern-day supercomputers consume power of 20 MW to perform the complex operations [138]. In comparison to that human brain is a highly power-efficient computing system. For instance, human brain can perform parallel processing of complex tasks such as pattern recognition, real-time speech/visual computing, and data processing with an ultralow energy consumption ($\approx$20 W) [139–141]. Accordingly, to mimic the ultralow power capabilities of human brain, Si-based CMOS analogue circuits have been proposed in the past. However, they have faced a significant challenge in scaling-up because more than 10 transistors are required to mimic one synapse or neuron. In addition, the energy consumption of a simple synaptic event, e.g. production of the excitatory postsynaptic current (EPSC), in the CMOS-based analog circuit may be a millionfold higher than that of a biological synapse. Recently, the use of one electronic device to mimic synaptic functions has received great attention because it can significantly overcome the limitations of CMOS-based analog circuits. Such a hardware implementation, by using two- or three-terminal electronic devices, aims to construct robust artificial neurons networks. Accordingly, research community has witnessed huge increase in interest in developing memristor as well as transistor-based artificial synapses. Printing of high-performance, and flexible low-dimensional crystalline nanomaterials including one-dimensional (1D) (NWs, nanotubes, etc.) and two-dimensional (2D) (MoS_2, HfS_2, WSe_2, etc.) have huge potential to realize the next-generation of brain-like computing.

9.5 Summary

The chapter summarized a burgeoning printed electronics manufacturing route for large-area high-performance electronics. Focus of the chapter was primarily on printing technologies to realize superior-grade electronic layers from nano-to-chip scale inorganic functional elements. The state-of-the-art devices/circuits realized using such functional elements have shown huge potential to achieve performance at par with the silicon-based electronics. The distinct advantages and disadvantages of the presented printing technique namely, contact and transfer printing, were discussed as well. Finally, we have presented various challenges and potential solution to develop inorganic-material based large area high-performance electronics. Advancement in inorganic-materials based printed electronics open avenues for the fabrication of intricate devices/circuits with performance comparable with the conventional planar ICs. The merging of novel form factors, high-performance, diversification, and functionality by printing technology is an appealing new aspect for electronics to be more interacting with their environment.

References

1 P. Escobedo, M. Bhattacharjee, F. Nikbakhtnasrabadi, and R. Dahiya, "Smart bandage with wireless strain and temperature sensors and battery-less NFC tag," *IEEE Internet of Things Journal*, vol. 8, no. 6, pp. 5093–5100, 2021, doi: https://doi.org/10.1109/JIOT.2020.3048282.

2 M. Bhattacharjee, F. Nikbakhtnasrabadi, and R. Dahiya, "Printed chipless antenna as flexible temperature sensor," *IEEE Internet of Things Journal*, vol. 8, no. 6, pp. 5101–5110, 2021, doi: https://doi.org/10.1109/JIOT.2021.3051467.

3 E. Song, J. Li, S. M. Won, W. Bai, and J. A. Rogers, "Materials for flexible bioelectronic systems as chronic neural interfaces," *Nature Materials*, vol. 19, no. 6, pp. 590–603, 2020, doi: https://doi.org/10.1038/s41563-020-0679-7.

4 Y. Ling, T. An, L. W. Yap, B. Zhu, S. Gong, and W. Cheng, "Disruptive, soft, wearable sensors," *Advanced Materials*, vol. 32, no. 18, p. 1904664, 2020, doi: https://doi.org/10.1002/adma.201904664.

5 A. S. Dahiya, J. Thireau, J. Boudaden, S. Lal, U. Gulzar, Y. Zhang, T. Gil, N. Azemard, P. Ramm, T. Kiessling, C. O'Murchu, F. Sebelius, J. Tilly, C. Glynn, S. Geary, C. O'Dwyer, K. M. Razeeb, A. Lacampagne, B. Charlot, and A. Todri-Sanial, "Review—energy autonomous

wearable sensors for smart healthcare: a review," *Journal of the Electrochemical Society*, vol. 167, no. 3, p. 037516, 2020, doi: https://doi.org/10.1149/2.0162003jes.

6 A. S. Dahiya, T. Gil, J. Thireau, N. Azemard, A. Lacampagne, B. Charlot, and A. Todri-Sanial, "1D nanomaterial-based highly stretchable strain sensors for human movement monitoring and human–robotic interactive systems," *Advanced Electronic Materials*, vol. 6, no. 10, p. 2000547, 2020, doi: https://doi.org/10.1002/aelm.202000547.

7 T. Someya and M. Amagai, "Toward a new generation of smart skins," *Nature Biotechnology*, vol. 37, no. 4, pp. 382–388, 2019, doi: https://doi.org/10.1038/s41587-019-0079-1.

8 R. Dahiya, N. Yogeswaran, F. Liu, L. Manjakkal, E. Burdet, V. Hayward, and H. Jörntell, "Large-area soft e-skin: the challenges beyond sensor designs," *Proceedings of the IEEE*, vol. 107, no. 10, pp. 2016–2033, 2019, doi: https://doi.org/10.1109/JPROC.2019.2941366.

9 R. Dahiya, D. Akinwande, and J. S. Chang, "Flexible electronic skin: from humanoids to humans [scanning the issue]," *Proceedings of the IEEE*, vol. 107, no. 10, pp. 2011–2015, 2019, doi: https://doi.org/10.1109/JPROC.2019.2941665.

10 R. Dahiya, "E-skin: from humanoids to humans [point of view]," *Proceedings of the IEEE*, vol. 107, no. 2, pp. 247–252, 2019, doi: https://doi.org/10.1109/JPROC.2018.2890729.

11 S. Ma, Y. Kumaresan, A. S. Dahiya, and R. Dahiya, "Ultra-thin chips with printed interconnects on flexible foils," *Advanced Electronic Materials*, p. 2101029, 2022, doi: https://doi.org/10.1002/aelm.202101029.

12 O. Ozioko, H. Nassar, and R. Dahiya, "3D printed interdigitated capacitor based tilt sensor," *IEEE Sensors Journal*, vol. 21, no. 23, pp. 26252–26260, 2021, doi: https://doi.org/10.1109/JSEN.2021.3058949.

13 O. Ozioko and R. Dahiya, "Smart tactile gloves for haptic interaction, communication and rehabilitation," *Advanced Intelligent Systems*, vol. 4, p. 2100091, 2021, doi: https://doi.org/10.1002/aisy.202100091.

14 F. Nikbakhtnasrabadi, H. El Matbouly, M. Ntagios, and R. Dahiya, "Textile-based stretchable microstrip antenna with intrinsic strain sensing," *ACS Applied Electronic Materials*, vol. 3, no. 5, pp. 2233–2246, 2021, doi: https://doi.org/10.1021/acsaelm.1c00179.

15 Y. Kumaresan, O. Ozioko, and R. Dahiya, "Multifunctional electronic skin with a stack of temperature and pressure sensor arrays," *IEEE Sensors Journal*, vol. 21, no. 23, pp. 26243–26251, 2021, doi: https://doi.org/10.1109/JSEN.2021.3055458.

16 Y. Kumaresan, G. Min, A. S. Dahiya, A. Ejaz, D. Shakthivel, R. Dahiya, "Kirigami and Mogul patterned ultra-stretchable high-performance ZnO nanowires based photodetector," *Advanced Materials Technologies*, vol. 7, no. 1, p. 2100804, 2022, doi: https://doi.org/10.1002/admt.202100804.

17 P. Karipoth, A. Christou, A. Pullanchiyodan, and R. Dahiya, "Bioinspired inchworm and earthworm like soft robots with intrinsic strain sensing," *Advanced Intelligent Systems*, vol. 4, p. 2100092, 2021, doi: https://doi.org/10.1002/aisy.202100092.

18 A. Zumeit, W. T. Navaraj, D. Shakthivel, and R. Dahiya, "Nanoribbon-based flexible high-performance transistors fabricated at room temperature," *Advanced Electronic Materials*, vol. 6, no. 4, p. 1901023, 2020, doi: https://doi.org/10.1002/aelm.201901023.

19 P. Escobedo, M. Ntagios, D. Shakthivel, W. T. Navaraj, and R. Dahiya, "Energy generating electronic skin with intrinsic touch sensing," *IEEE Transactions on Robotics*, vol. 37, no. 2, pp. 683–690, 2021, doi: https://doi.org/10.1109/TRO.2020.3031264.

20 O. Ozioko, P. Karipoth, M. Hersh, and R. Dahiya, "Wearable assistive tactile communication interface based on integrated touch sensors and actuators," *IEEE Transactions on Neural Systems and Rehabilitation Engineering*, vol. 28, no. 6, pp. 1344–1352, 2020, doi: https://doi.org/10.1109/TNSRE.2020.2986222.

21 O. Ozioko, P. Karipoth, P. Escobedo, M. Ntagios, A. Pullanchiyodan, and R. Dahiya, "SensAct: the soft and squishy tactile sensor with integrated flexible actuator," *Advanced Intelligent Systems*, vol. 3, p. 1900145, 2021, doi: https://doi.org/10.1109/JIOT.2020.3048282.

22 T. R. Ray, J. Choi, A. J. Bandodkar, S. Krishnan, P. Gutruf, L. Tian, R. Ghaffari, and J. A. Rogers, "Bio-integrated wearable systems: a comprehensive review," *Chemical Reviews*, vol. 119, no. 8, pp. 5461–5533, 2019, doi: https://doi.org/10.1021/acs.chemrev.8b00573.

23 S. Niu, N. Matsuhisa, L. Beker, J. Li, S. Wang, J. Wang, Y. Jiang, X. Yan, Y. Yun, W. Burnett, A. S. Y. Poon, J. B.-H. Tok, X. Chen, and Z. Bao, "A wireless body area sensor network based on stretchable passive tags," *Nature Electronics*, vol. 2, no. 8, pp. 361–368, 2019, doi: https://doi.org/10.1038/s41928-019-0286-2.

24 F. Liu, S. Deswal, A. Christou, Y. Sandamirskaya, M. Kaboli, and R. Dahiya, "Neuro-inspired e-Skin for Robots," *Science Robotics*, Vol. 7, abl7344, 2022.

25 J. C. Yang, J. Mun, S. Y. Kwon, S. Park, Z. Bao, and S. Park, "Electronic skin: recent progress and future prospects for skin-attachable devices for health monitoring, robotics, and prosthetics," *Advanced Materials*, vol. 31, no. 48, p. 1904765, 2019, doi: https://doi.org/10.1002/adma.201904765.

26 W. Gao, H. Ota, D. Kiriya, K. Takei, and A. Javey, "Flexible electronics toward wearable sensing," *Accounts of Chemical Research*, vol. 52, no. 3, pp. 523–533, 2019, doi: https://doi.org/10.1021/acs.accounts.8b00500.

27 H. U. Chung, B. H. Kim, J. Y. Lee, J. Lee, Z. Xie, E. M. Ibler, K. Lee, A. Banks, J. Y. Jeong, J. Kim, C. Ogle, D. Grande, Y. Yu, H. Jang, P. Assem, D. Ryu, J. W. Kwak, M. Namkoong, J. B. Park, Y. Lee, D. H. Kim, A. Ryu, J. Jeong, K. You, B. Ji, Z. Liu, Q. Huo, X. Feng, Y. Deng, Y. Xu, K.-I. Jang, J. Kim, Y. Zhang, R. Ghaffari, C. M. Rand, M. Schau, A. Hamvas, D. E. Weese-Mayer, Y. Huang, S. M. Lee, C. H. Lee, N. R. Shanbhag, A. S. Paller, S. Xu, and J. A. Rogers, "Binodal, wireless epidermal electronic systems with in-sensor analytics for neonatal intensive care," *Science*, vol. 363, no. 6430, p. eaau0780, 2019, doi: https://doi.org/10.1126/science.aau0780.

28 F. Liu, S. Deswal, A. Christou, R. Chirila, M. S. Baghini, D. Shakthivel, M. Chakraborty, and R. Dahiya, "Printed synaptic transistor–based electronic skin for robots to feel and learn," *Science Robotics*, vol. 7, eabl7286, 2022.

29 W. T. Navaraj, S. Gupta, L. Lorenzelli, and R. Dahiya, "Wafer scale transfer of ultrathin silicon chips on flexible substrates for high performance bendable systems," *Advanced Electronic Materials*, vol. 4, no. 4, p. 1700277, 2018, doi: https://doi.org/10.1002/aelm.201700277.

30 C. Linghu, S. Zhang, C. Wang, and J. Song, "Transfer printing techniques for flexible and stretchable inorganic electronics," *npj Flexible Electronics*, vol. 2, no. 1, p. 26, 2018, doi: https://doi.org/10.1038/s41528-018-0037-x.

31 C. García Núñez, F. Liu, W. T. Navaraj, A. Christou, D. Shakthivel, and R. Dahiya, "Heterogeneous integration of contact-printed semiconductor nanowires for high-performance devices on large areas," *Microsystems & Nanoengineering*, vol. 4, no. 1, p. 22, 2018, doi: https://doi.org/10.1038/s41378-018-0021-6.

32 C. G. Núñez, W. T. Navaraj, E. O. Polat, and R. Dahiya, "Energy-autonomous, flexible, and transparent tactile skin," *Advanced Functional Materials,* vol. 27, no. 18, p. 1606287, 2017, doi: https://doi.org/10.1002/adfm.201606287.

33 A. S. Dahiya, C. Opoku, G. Poulin-Vittrant, N. Camara, C. Daumont, E. G. Barbagiovanni, G. Franzò, S. Mirabella, and D. Alquier, "Flexible organic/inorganic hybrid field-effect transistors with high performance and operational stability," *ACS Applied Materials & Interfaces*, vol. 9, no. 1, pp. 573–584, 2017, doi: https://doi.org/10.1021/acsami.6b13472.

34 S. Khan, L. Lorenzelli, and R. S. Dahiya, "Technologies for printing sensors and electronics over large flexible substrates: a review," *IEEE Sensors Journal*, vol. 15, no. 6, pp. 3164–3185, 2015, doi: https://doi.org/10.1109/JSEN.2014.2375203.

35 R. Dahiya, W. T. Navaraj, S. Khan, and E. O. Polat, "Developing electronic skin with the sense of touch," *Information Display*, vol. 31, no. 4, pp. 6–10, 2015, doi: https://doi.org/10.1002/j.2637-496X.2015.tb00824.x.

36 S. Ma, Y. Kumaresan, A. S. Dahiya, L. Lorenzelli, and R. Dahiya, "Flexible tactile sensors using AlN and MOSFETs based ultra-thin chips," *IEEE Sensors Journal*, p. 1, 2022, doi: https://doi.org/10.1109/JSEN.2022.3140651.

37 M. Shojaei Baghini, A. Vilouras, and R. Dahiya, "Ultra-thin chips with ISFET array for continuous monitoring of body fluids pH," *IEEE Transactions on Biomedical Circuits and Systems*, vol. 15, no. 6, pp. 1174–1185, 2021, doi: https://doi.org/10.1109/TBCAS.2022.3141553.

38 A. Zumeit, A. S. Dahiya, A. Christou, D. Shakthivel, and R. Dahiya, "Direct roll transfer printed silicon nanoribbon arrays based high-performance flexible electronics," *npj Flexible Electronics*, vol. 5, no. 18, 2021, doi: https://doi.org/10.1038/s41528-021-00116-w.

39 A. Zumeit, A. S. Dahiya, A. Christou, and R. Dahiya, "High performance p-channel transistors on flexible substrate using direct roll transfer stamping," *Japanese Journal of Applied Physics*, vol. 61, SC1042, 2022. http://iopscience.iop.org/article/10.35848/1347-4065/ac40ab.

40 A. S. Dahiya, D. Shakthivel, Y. Kumaresan, A. Zumeit, A. Christou, and R. Dahiya, "High-performance printed electronics based on inorganic semiconducting nano to chip scale structures," *Nano Convergence*, vol. 7, no. 1, p. 33, 2020, doi: https://doi.org/10.1186/s40580-020-00243-6.

41 U. Zschieschang, J. W. Borchert, M. Giorgio, M. Caironi, F. Letzkus, J. N. Burghartz, U. Waizmann, J. Weis, S. Ludwigs, H. Klauk, "Roadmap to gigahertz organic transistors," *Advanced Functional Materials*, vol. 30, no. 20, p. 1903812, 2020, doi: https://doi.org/10.1002/adfm.201903812.

42 P. C. Y. Chow and T. Someya, "Organic photodetectors for next-generation wearable electronics," *Advanced Materials*, vol. 32, no. 15, p. 1902045, 2020, doi: https://doi.org/10.1002/adma.201902045.

43 N. L. Vaklev, J. H. G. Steinke, and A. J. Campbell, "Gravure printed ultrathin dielectric for low voltage flexible organic field-effect transistors," *Advanced Materials Interfaces*, vol. 6, no. 11, p. 1900173, 2019, doi: https://doi.org/10.1002/admi.201900173.

44 C. Koutsiaki, T. Kaimakamis, A. Zachariadis, and S. Logothetidis, "Electrical performance of flexible OTFTs based on slot-die printed dielectric films with different thicknesses," *Materials Today: Proceedings*, vol. 19, no. Part 1, pp. 58–64, 2019, doi: https://doi.org/10.1016/j.matpr.2019.07.657.

45 M. Fattori, J. Fijn, P. Harpe, M. Charbonneau, S. Lombard, K. Romanjek, D. Locatelli, L. Tournon, C. Laugier, and E. Cantatore, "A gravure-printed organic TFT technology for active-matrix addressing applications," *IEEE Electron Device Letters*, vol. 40, no. 10, pp. 1682–1685, 2019, doi: https://doi.org/10.1109/LED.2019.2938852.

46 K. Fukuda and T. Someya, "Recent progress in the development of printed thin-film transistors and circuits with high-resolution printing technology," *Advanced Materials*, vol. 29, no. 25, p. 1602736, 2017, doi: https://doi.org/10.1002/adma.201602736.

47 P. Forouzandeh, P. Ganguly, R. Dahiya, and S. C. Pillai, "Supercapacitor electrode fabrication through chemical and physical routes," *Journal of Power Sources*, vol. 519, no. 31 January, p. 230744, 2022.

48 T. Sekitani and T. Someya, "Stretchable, large-area organic electronics," *Advanced Materials*, vol. 22, no. 20, pp. 2228–2246, 2010, doi: https://doi.org/10.1002/adma.200904054.

49 E. Ozer, J. Kufel, J. Myers, J. Biggs, G. Brown, A. Rana, A. Sou, C. Ramsdale, and S. White, "A hardwired machine learning processing engine fabricated with submicron metal-oxide thin-film transistors on a flexible substrate," *Nature Electronics*, vol. 3, no. 7, pp. 419–425, 2020, doi: https://doi.org/10.1038/s41928-020-0437-5.

50 F. Liu, A. S. Dahiya, and R. Dahiya, "A flexible chip with embedded intelligence," *Nature Electronics*, vol. 3, no. 7, pp. 358–359, 2020, doi: https://doi.org/10.1038/s41928-020-0446-4.

51 M. S. Baghini, A. Vilouras, M. Douthwaite, P. Georgiou, R. Dahiya, "Ultra-thin ISFET based sensing systems," *Electrochemical Science Advances*, p. e2100202, 2021, doi: https://doi.org/10.1002/elsa.202100202.

52 A. Liu, H. Zhu, and Y.-Y. Noh, "Solution-processed inorganic p-channel transistors: recent advances and perspectives," *Materials Science and Engineering Reports*, vol. 135, no. January, pp. 85–100, 2019, doi: https://doi.org/10.1016/j.mser.2018.11.001.

53 Y. Kumaresan, H. Kim, Y. Pak, P. K. Poola, R. Lee, N. Lim, H. C. Ko, G. Y. Jung, and R. Dahiya, "Omnidirectional stretchable inorganic-material-based electronics with enhanced performance," *Advanced Electronic Materials*, vol. 6, no. 7, p. 2000058, 2020, doi: https://doi.org/10.1002/aelm.202000058.

54 X. Han, K. J. Seo, Y. Qiang, Z. Li, S. Vinnikova, Y. Zhong, X. Zhao, P. Hao, S. Wang, and H. Fang, "Nanomeshed Si nanomembranes," *npj Flexible Electronics*, vol. 3, no. 1, p. 9, 2019, doi: https://doi.org/10.1038/s41528-019-0053-5.

55 A. S. Dahiya, A. Zumeit, A. Christou, and R. Dahiya, "High-performance n-channel printed transistors on biodegradable substrate for transient electronics," *Advanced Electronic Materials*, 2022, 2200098. https://doi.org/10.1002/aelm.202200098.

56 S. Gupta, W. T. Navaraj, L. Lorenzelli, and R. Dahiya, "Ultra-thin chips for high-performance flexible electronics," *npj Flexible Electronics*, vol. 2, no. 1, p. 8, 2018, doi: https://doi.org/10.1038/s41528-018-0021-5.

57 B. Guilhabert, A. Hurtado, D. Jevtics, Q. Gao, H. H. Tan, C. Jagadish, and M. D. Dawson, "Transfer printing of semiconductor nanowires with lasing emission for controllable nanophotonic device fabrication," *ACS Nano*, vol. 10, no. 4, pp. 3951–3958, 2016, doi: https://doi.org/10.1021/acsnano.5b07752.

58 A. S. Dahiya, A. Christou, J. Neto, A. Zumeit, D. Shakthivel, and R. Dahiya, "In tandem contact-transfer printing for high-performance transient electronics," *Advanced Electronic Materials*, 2022, 2200170. https://doi.org/10.1002/aelm.202200170.

59 Z. Fan, J. C. Ho, Z. A. Jacobson, H. Razavi, and A. Javey, "Large-scale, heterogeneous integration of nanowire arrays for image sensor circuitry," *Proceedings of the National Academy of Sciences of the United States of America*, vol. 105, no. 32, pp. 11066–11070, 2008, doi: https://doi.org/10.1073/pnas.0801994105.

60 D. Shakthivel, A. S. Dahiya, R. Mukherjee, and R. Dahiya, "Inorganic semiconducting nanowires for green energy solutions," vol. 34, December 2021, 100753 https://doi.org/10.1016/j.coche.2021.100753.

61 M. C. McAlpine, H. Ahmad, D. Wang, and J. R. Heath, "Highly ordered nanowire arrays on plastic substrates for ultrasensitive flexible chemical sensors," *Nature Materials*, vol. 6, no. 5, pp. 379–384, 2007, doi: https://doi.org/10.1038/nmat1891.

62 A. Javey, S. Nam, R. S. Friedman, H. Yan, and C. M. Lieber, "Layer-by-layer assembly of nanowires for three-dimensional, multifunctional electronics," *Nano Letters*, vol. 7, no. 3, pp. 773–777, 2007, doi: https://doi.org/10.1021/nl063056l.

63 Y. Sun, H.-S. Kim, E. Menard, S. Kim, I. Adesida, and J. A. Rogers, "Printed arrays of aligned GaAs wires for flexible transistors, diodes, and circuits on plastic substrates," *Small*, vol. 2, no. 11, pp. 1330–1334, 2006, doi: https://doi.org/10.1002/smll.200500528.

64 A. Christou, F. Liu, and R. Dahiya, "Development of a highly controlled system for large-area, directional printing of quasi-1D nanomaterials," *Microsystems & Nanoengineering*, vol. 7, no. 82, 2021.

65 Y. Oshima, A. Nakamura, and K. Matsunaga, "Extraordinary plasticity of an inorganic semi-conductor in darkness," *Science*, vol. 360, no. 6390, pp. 772–774, 2018, doi: https://doi.org/10.1126/science.aar6035.

66 W. Wu, "Inorganic nanomaterials for printed electronics: a review," *Nanoscale*, vol. 9, no. 22, pp. 7342–7372, 2017, doi: https://doi.org/10.1039/C7NR01604B.

67 M. A. Rose, T. P. Vinod, and S. A. Morin, "Microscale screen printing of large-area arrays of microparticles for the fabrication of photonic structures and for optical sorting," *Journal of Materials Chemistry C*, vol. 6, no. 44, pp. 12031–12037, 2018, doi: https://doi.org/10.1039/C8TC02978D.

68 F. Zheng, Z. Wang, J. Huang, and Z. Li, "Inkjet printing-based fabrication of microscale 3D ice structures," *Microsystems & Nanoengineering*, vol. 6, no. 1, p. 89, 2020, doi: https://doi.org/10.1038/s41378-020-00199-x.

69 J. B. Park, W. S. Choi, T. H. Chung, S. H. Lee, M. K. Kwak, J. S. Ha, and T. Jeong, "Transfer printing of vertical-type microscale light-emitting diode array onto flexible substrate using biomimetic stamp," *Optics Express*, vol. 27, no. 5, pp. 6832–6841, 2019, doi: https://doi.org/10.1364/OE.27.006832.

70 J. F. Salmerón, F. Molina-Lopez, D. Briand, J. J. Ruan, A. Rivadeneyra, M. A. Carvajal, L. F. Capitán-Vallvey, N. F. de Rooij, and A. J. Palma, "Properties and printability of inkjet and screen-printed silver patterns for RFID antennas," *Journal of Electronic Materials*, vol. 43, no. 2, pp. 604–617, 2014, doi: https://doi.org/10.1007/s11664-013-2893-4.

71 S. Emamian, B. B. Narakathu, A. A. Chlaihawi, B. J. Bazuin, and M. Z. Atashbar, "Screen printing of flexible piezoelectric based device on polyethylene terephthalate (PET) and paper for touch and force sensing applications," *Sensors and Actuators A: Physical*, vol. 263, no. 15 August, pp. 639–647, 2017, doi: https://doi.org/10.1016/j.sna.2017.07.045.

72 X. Cao, C. Lau, Y. Liu, F. Wu, H. Gui, Q. Liu, Y. Ma, H. Wan, M. R. Amer, and C. Zhou, "Fully screen-printed, large-area, and flexible active-matrix electrochromic displays using carbon nanotube thin-film transistors," *ACS Nano*, vol. 10, no. 11, pp. 9816–9822, 2016, doi: https://doi.org/10.1021/acsnano.6b05368.

73 A. J. Bandodkar, I. Jeerapan, J.-M. You, R. Nuñez-Flores, and J. Wang, "Highly stretchable fully-printed CNT-based electrochemical sensors and biofuel cells: combining intrinsic and design-induced stretchability," *Nano Letters*, vol. 16, no. 1, pp. 721–727, 2016, doi: https://doi.org/10.1021/acs.nanolett.5b04549.

74 F. Nikbakhtnasrabadi, E. S. Hosseini, S. Dervin, D. Shakthivel, and R. Dahiya, "Smart bandage with inductor-capacitor resonant tank based printed wireless pressure sensor on electrospun poly-L-lactide nanofibers," *Advanced Electronic Materials*, 2101348, 2022.

75 C. García Núñez, F. Liu, S. Xu, and R. Dahiya, *Integration Techniques for Micro/Nanostructure-based Large-Area Electronics*, Cambridge: Cambridge University Press, 2018.

76 D. Shakthivel, W. T. Navaraj, S. Champet, D. H. Gregory, and R. S. Dahiya, "Propagation of amorphous oxide nanowires via the VLS mechanism: growth kinetics," *Nanoscale Advances*, vol. 1, no. 9, pp. 3568–3578, 2019, doi: https://doi.org/10.1039/C9NA00134D.

77 D. Shakthivel, M. Ahmad, M. R. Alenezi, R. Dahiya, and S. R. P. Silva, *1D Semiconducting Nanostructures for Flexible and Large-Area Electronics: Growth Mechanisms and Suitability*, Cambridge: Cambridge University Press, 2019.

78 C. García Núñez, W. T. Navaraj, F. Liu, D. Shakthivel, and R. Dahiya, "Large-area self-assembly of silica microspheres/nanospheres by temperature-assisted dip-coating," *ACS Applied Materials & Interfaces*, vol. 10, no. 3, pp. 3058–3068, 2018, doi: https://doi.org/10.1021/acsami.7b15178.

79 E. O. Polat, O. Balci, N. Kakenov, H. B. Uzlu, C. Kocabas, and R. Dahiya, "Synthesis of large area graphene for high performance in flexible optoelectronic devices," *Scientific Reports*, vol. 5, no. 1, p. 16744, 2015, doi: https://doi.org/10.1038/srep16744.

80 A. S. Dahiya, C. Opoku, D. Alquier, G. Poulin-Vittrant, F. Cayrel, O. Graton, L.-P. T. H. Hue, and N. Camara, "Controlled growth of 1D and 2D ZnO nanostructures on 4H-SiC using Au catalyst," *Nanoscale Research Letters*, vol. 9, no. 1, p. 379, 2014, doi: https://doi.org/10.1186/1556-276X-9-379.

81 Y. Wei, W. Wu, R. Guo, D. Yuan, S. Das, and Z. L. Wang, "Wafer-scale high-throughput ordered growth of vertically aligned ZnO nanowire arrays," *Nano Letters*, vol. 10, no. 9, pp. 3414–3419, 2010, doi: https://doi.org/10.1021/nl1014298.

82 A. C. Ford, J. C. Ho, Z. Fan, O. Ergen, V. Altoe, S. Aloni, H. Razavi, and A. Javey, "Synthesis, contact printing, and device characterization of Ni-catalyzed, crystalline InAs nanowires," *Nano Research*, vol. 1, no. 1, pp. 32–39, 2008, doi: https://doi.org/10.1007/s12274-008-8009-4.

83 G. Zhu, R. Yang, S. Wang, and Z. L. Wang, "Flexible high-output nanogenerator based on lateral ZnO nanowire array," *Nano Letters*, vol. 10, no. 8, pp. 3151–3155, 2010, doi: https://doi.org/10.1021/nl101973h.

84 Y.-K. Chang and F. C.-N. Hong, "The fabrication of ZnO nanowire field-effect transistors by roll-transfer printing," *Nanotechnology*, vol. 20, no. 19, p. 195302, 2009, doi: https://doi.org/10.1088/0957-4484/20/19/195302.

85 K. Takei, T. Takahashi, J. C. Ho, H. Ko, A. G. Gillies, P. W. Leu, R. S. Fearing, and A. Javey, "Nanowire active-matrix circuitry for low-voltage macroscale artificial skin," *Nature Materials*, vol. 9, no. 10, pp. 821–826, 2010, doi: https://doi.org/10.1038/nmat2835.

86 C. G. Núñez, A. Vilouras, W. T. Navaraj, F. Liu, and R. Dahiya, "ZnO nanowires-based flexible UV photodetector system for wearable dosimetry," *IEEE Sensors Journal*, vol. 18, no. 19, pp. 7881–7888, 2018, doi: https://doi.org/10.1109/JSEN.2018.2853762.

87 J. Yoon, S. Jo, I. S. Chun, I. Jung, H.-S. Kim, M. Meitl, E. Menard, X. Li, J. J. Coleman, U. Paik, and J. A. Rogers, "GaAs photovoltaics and optoelectronics using releasable multilayer epitaxial assemblies," *Nature*, vol. 465, no. 7296, pp. 329–333, 2010, doi: https://doi.org/10.1038/nature09054.

88 X. Feng, B. D. Yang, Y. Liu, Y. Wang, C. Dagdeviren, Z. Liu, A. Carlson, J. Li, Y. Huang, and J. A. Rogers, "Stretchable ferroelectric nanoribbons with wavy configurations on elastomeric substrates," *ACS Nano*, vol. 5, no. 4, pp. 3326–3332, 2011, doi: https://doi.org/10.1021/nn200477q.

89 L. Mai, Y. Gu, C. Han, B. Hu, W. Chen, P. Zhang, L. Xu, W. Guo, and Y. Dai, "Orientated Langmuir−Blodgett assembly of VO_2 nanowires," *Nano Letters*, vol. 9, no. 2, pp. 826–830, 2009, doi: https://doi.org/10.1021/nl803550k.

90 G. Yu, A. Cao, and C. M. Lieber, "Large-area blown bubble films of aligned nanowires and carbon nanotubes," *Nature Nanotechnology*, vol. 2, no. 6, pp. 372–377, 2007, doi: https://doi.org/10.1038/nnano.2007.150.

91 X. Zhou, Y. Zhou, J. C. Ku, C. Zhang, and C. A. Mirkin, "Capillary force-driven, large-area alignment of multi-segmented nanowires," *ACS Nano*, vol. 8, no. 2, pp. 1511–1516, 2014, doi: https://doi.org/10.1021/nn405627s.

92 T. Takahashi, K. Takei, E. Adabi, Z. Fan, A. M. Niknejad, and A. Javey, "Parallel array InAs nanowire transistors for mechanically bendable, ultrahigh frequency electronics," *ACS Nano*, vol. 4, no. 10, pp. 5855–5860, 2010, doi: https://doi.org/10.1021/nn1018329.

93 X. Liang, Z. Fu, and S. Y. Chou, "Graphene transistors fabricated via transfer-printing in device active-areas on large wafer," *Nano Letters,* vol. 7, no. 12, pp. 3840–3844, 2007, doi: https://doi.org/10.1021/nl072566s.

94 F. Liu, W. T. Navaraj, N. Yogeswaran, D. H. Gregory, and R. Dahiya, "van der Waals contact engineering of graphene field-effect transistors for large-area flexible electronics," *ACS Nano*, vol. 13, no. 3, pp. 3257–3268, 2019, doi: https://doi.org/10.1021/acsnano.8b09019.

95 S. Haque, M. Alexandre, M. J. Mendes, H. Águas, E. Fortunato, and R. Martins, "Design of wave-optical structured substrates for ultra-thin perovskite solar cells," *Applied Materials Today*, vol. 20, no. September, p. 100720, 2020, doi: https://doi.org/10.1016/j.apmt.2020.100720.

96 M. Wu, Z. Gong, A. J. C. Kuehne, A. L. Kanibolotsky, Y. J. Chen, I. F. Perepichka, A. R. Mackintosh, E. Gu, P. J. Skabara, R. A. Pethrick, and M. D. Dawson, "Hybrid GaN/organic microstructured light-emitting devices via ink-jet printing," *Optics Express*, vol. 17, no. 19, pp. 16436–16443, 2009, doi: https://doi.org/10.1364/OE.17.016436.

97 X. Li, L. V. Nguyen, Y. Zhao, H. Ebendorff-Heidepriem, and S. C. Warren-Smith, "High-sensitivity Sagnac-interferometer biosensor based on exposed core microstructured optical fiber," *Sensors and Actuators B: Chemical*, vol. 269, no. 15 September, pp. 103–109, 2018, doi: https://doi.org/10.1016/j.snb.2018.04.165.

98 Y. Qi, J. Kim, T. D. Nguyen, B. Lisko, P. K. Purohit, and M. C. McAlpine, "Enhanced piezo-electricity and stretchability in energy harvesting devices fabricated from buckled PZT ribbons," *Nano Letters*, vol. 11, no. 3, pp. 1331–1336, 2011, doi: https://doi.org/10.1021/nl104412b.

99 Y. Huang, N. Zheng, Z. Cheng, Y. Chen, B. Lu, T. Xie, and X. Feng, "Direct laser writing-based programmable transfer printing via bioinspired shape memory reversible adhesive," *ACS Applied Materials & Interfaces*, vol. 8, no. 51, pp. 35628–35633, 2016, doi: https://doi.org/10.1021/acsami.6b11696.

100 S. Cho, N. Kim, K. Song, and J. Lee, "Adhesiveless transfer printing of ultrathin microscale semiconductor materials by controlling the bending radius of an elastomeric stamp," *Langmuir*, vol. 32, no. 31, pp. 7951–7957, 2016, doi: https://doi.org/10.1021/acs.langmuir.6b01880.

101 M. Choi, B. Jang, W. Lee, S. Lee, T. W. Kim, H.-J. Lee, J.-H. Kim, and J.-H. Ahn, "Stretchable active matrix inorganic light-emitting diode display enabled by overlay-aligned roll-transfer printing," *Advanced Functional Materials*, vol. 27, no. 11, p. 1606005, 2017, doi: https://doi.org/10.1002/adfm.201606005.

102 N. Mahmoud, W. Walravens, J. Kuhs, C. Detavernier, Z. Hens, and G. Roelkens, "Micro-transfer-printing of Al_2O_3-capped short-wave-infrared PbS quantum dot photoconductors," *ACS Applied Nano Materials*, vol. 2, no. 1, pp. 299–306, 2019, doi: https://doi.org/10.1021/acsanm.8b01915.

103 L. Shen, H. Li, X. Meng, and F. Li, "Transfer printing of fully formed microscale InGaP/GaAs/InGaNAsSb cell on Ge cell in mechanically-stacked quadruple-junction architecture," *Solar*

Energy, vol. 195, no. 1 January, pp. 6-13, 2020, doi: https://doi.org/10.1016/j.solener.2019.11 .046.

104 T.-H. Kim, A. Carlson, J.-H. Ahn, S. M. Won, S. Wang, Y. Huang, and J. A. Rogers, "Kinetically controlled, adhesiveless transfer printing using microstructured stamps," *Applied Physics Letters*, vol. 94, no. 11, p. 113502, 2009, doi: https://doi.org/10.1063/1.3099052.

105 C. Wang, C. Linghu, S. Nie, C. Li, Q. Lei, X. Tao, Y. Zeng, Y. Du, S. Zhang, K. Yu, H. Jin, W. Chen, and J. Song, "Programmable and scalable transfer printing with high reliability and efficiency for flexible inorganic electronics," *Science Advances*, vol. 6, no. 25, p. eabb2393, 2020, doi: https://doi.org/10.1126/sciadv.abb2393.

106 D. S. Wie, Y. Zhang, M. K. Kim, B. Kim, S. Park, Y.-J. Kim, P. P. Irazoqui, X. Zheng, B. Xu, and C. H. Lee, "Wafer-recyclable, environment-friendly transfer printing for large-scale thin-film nanoelectronics," *Proceedings of the National Academy of Sciences of the United States of America*, vol. 115, no. 31, pp. E7236–E7244, 2018, doi: https://doi.org/10.1073/pnas .1806640115.

107 R. S. Dahiya and S. Gennaro, "Bendable ultra-thin chips on flexible foils," *IEEE Sensors Journal*, vol. 13, no. 10, pp. 4030–4037, 2013, doi: https://doi.org/10.1109/JSEN.2013.2269028.

108 D. Shahrjerdi and S. W. Bedell, "Extremely flexible nanoscale ultrathin body silicon integrated circuits on plastic," *Nano Letters*, vol. 13, no. 1, pp. 315–320, 2013, doi: https://doi.org/10.1021/ nl304310x.

109 S. W. Bedell, K. Fogel, P. Lauro, D. Shahrjerdi, J. A. Ott, and D. Sadana, "Layer transfer by controlled spalling," *Journal of Physics D: Applied Physics*, vol. 46, no. 15, p. 152002, 2013, doi: https://doi.org/10.1088/0022-3727/46/15/152002.

110 Y. Kumaresan, R. Lee, N. Lim, Y. Pak, H. Kim, W. Kim, and G.-Y. Jung, "Extremely flexible indium-gallium-zinc oxide (IGZO) based electronic devices placed on an ultrathin poly(methyl methacrylate) (PMMA) substrate," *Advanced Electronic Materials*, vol. 4, no. 7, p. 1800167, 2018, doi: https://doi.org/10.1002/aelm.201800167.

111 H.-J. Kim-Lee, A. Carlson, D. S. Grierson, J. A. Rogers, and K. T. Turner, "Interface mechanics of adhesiveless microtransfer printing processes," *Journal of Applied Physics*, vol. 115, no. 14, p. 143513, 2014, doi: https://doi.org/10.1063/1.4870873.

112 R. Dahiya, G. Gottardi, and N. Laidani, "PDMS residues-free micro/macrostructures on flexible substrates," *Microelectronic Engineering*, vol. 136, no. 25 March, pp. 57-62, 2015, doi: https://doi.org/10.1016/j.mee.2015.04.037.

113 A. Chortos, J. Liu, and Z. Bao, "Pursuing prosthetic electronic skin," *Nature Materials*, vol. 15, no. 9, pp. 937–950, 2016, doi: https://doi.org/10.1038/nmat4671.

114 X. Wang, L. Dong, H. Zhang, R. Yu, C. Pan, and Z. L. Wang, "Recent progress in electronic skin," *Advancement of Science*, vol. 2, no. 10, p. 1500169, 2015, doi: https://doi.org/10.1002/ advs.201500169.

115 R. S. Dahiya, G. Metta, M. Valle, and G. Sandini, "Tactile sensing—from humans to humanoids," *IEEE Transactions on Robotics*, vol. 26, no. 1, pp. 1-20, 2010, doi: https://doi .org/10.1109/TRO.2009.2033627.

116 A. Polishchuk, W. T. Navaraj, H. Heidari, and R. Dahiya, "Multisensory smart glove for tactile feedback in prosthetic hand," *Procedia Engineering*, vol. 168, pp. 1605-1608, 2016, doi: https://doi.org/10.1016/j.proeng.2016.11.471.

117 Z. Wang, C. Dong, X. Wang, M. Li, T. Nan, X. Liang, H. Chen, Y. Wei, H. Zhou, M. Zaeimbashi, S. Cash, and N.-X. Sun, "Highly sensitive integrated flexible tactile sensors with piezoresistive $Ge_2Sb_2Te_5$ thin films," *npj Flexible Electronics*, vol. 2, no. 1, p. 17, 2018, doi: https://doi.org/10.1038/s41528-018-0030-4.

118 H. Kawasaki, T. Komatsu, K. Uchiyama, and T. Kurimoto, "Dexterous anthropomorphic robot hand with distributed tactile sensor: Gifu hand II," in *IEEE SMC'99 Conference Proceedings. 1999 IEEE International Conference on Systems, Man, and Cybernetics (Cat. No.99CH37028)*, 12–15 October 1999, vol. 2, pp. 782–787, doi: https://doi.org/10.1109/ICSMC.1999.825361.

119 H. Kawasaki, T. Komatsu, and K. Uchiyama, "Dexterous anthropomorphic robot hand with distributed tactile sensor: Gifu hand II," *IEEE/ASME Transactions on Mechatronics*, vol. 7, no. 3, pp. 296–303, 2002, doi: https://doi.org/10.1109/TMECH.2002.802720.

120 M. Shimojo, A. Namiki, M. Ishikawa, R. Makino, and K. Mabuchi, "A tactile sensor sheet using pressure conductive rubber with electrical-wires stitched method," *IEEE Sensors Journal*, vol. 4, no. 5, pp. 589–596, 2004, doi: https://doi.org/10.1109/JSEN.2004.833152.

121 N. Wettels, V. J. Santos, R. S. Johansson, and G. E. Loeb, "Biomimetic tactile sensor array," *Advanced Robotics*, vol. 22, no. 8, pp. 829–849, 2008, doi: https://doi.org/10.1163/156855308X314533.

122 R. L. Truby, R. K. Katzschmann, J. A. Lewis, and D. Rus, "Soft robotic fingers with embedded ionogel sensors and discrete actuation modes for somatosensitive manipulation," in *2019 2nd IEEE International Conference on Soft Robotics (RoboSoft)*, 14–18 April 2019, pp. 322–329, doi: https://doi.org/10.1109/ROBOSOFT.2019.8722722.

123 M. Ntagios, P. Escobedo, and R. Dahiya, "3D printed robotic hand with embedded touch sensors," *2020 IEEE International Conference on Flexible and Printable Sensors and Systems (FLEPS 2020), Manchester, UK, 16–19 August 2020, (Accepted for Publication)*, 2020, doi: https://doi.org/10.1109/FLEPS49123.2020.9239587.

124 Y. Gao, G. Yu, T. Shu, Y. Chen, W. Yang, Y. Liu, J. Long, W. Xiong, and F. Xuan, "3D-printed coaxial fibers for integrated wearable sensor skin," *Advanced Materials Technologies*, vol. 4, no. 10, p. 1900504, 2019, doi: https://doi.org/10.1002/admt.201900504.

125 M. Ntagios, H. Nassar, A. Pullanchiyodan, W. T. Navaraj, and R. Dahiya, "Robotic hands with intrinsic tactile sensing via 3D printed soft pressure sensors," *Advanced Intelligent Systems*, vol. 2, no. 6, p. 1900080, 2020, doi: https://doi.org/10.1002/aisy.201900080.

126 W. Navaraj and R. Dahiya, "Fingerprint-enhanced capacitive-piezoelectric flexible sensing skin to discriminate static and dynamic tactile stimuli," *Advanced Intelligent Systems*, vol. 1, no. 7, p. 1900051, 2019, doi: https://doi.org/10.1002/aisy.201900051.

127 S. Herat, "Sustainable management of electronic waste (e-waste)," *CLEAN – Soil Air Water*, vol. 35, no. 4, pp. 305–310, 2007, doi: https://doi.org/10.1002/clen.200700022.

128 M. Chakraborty, J. Kettle, and R. Dahiya, "Electronic waste reduction through devices and printed circuit boards designed for circularity," *IEEE Journal on Flexible Electronics*, vol. 1, no. 1, pp. 4–23, 2022, doi: https://doi.org/10.1109/JFLEX.2022.3159258.

129 E. S. Hosseini, S. Dervin, P. Ganguly, and R. Dahiya, "Biodegradable materials for sustainable health monitoring devices," *ACS Applied Bio Materials*, vol. 4, no. 1, pp. 163–194, 2021, doi: https://doi.org/10.1021/acsabm.0c01139.

130 N. Yogeswaran, E. S. Hosseini, and R. Dahiya, "Graphene based low voltage field effect transistor coupled with biodegradable piezoelectric material based dynamic pressure sensor," *ACS Applied Materials & Interfaces*, vol. 12, no. 48, pp. 54035–54040, 2020, doi: https://doi.org/10.1021/acsami.0c13637.

131 M. A. Kafi, A. Paul, A. Vilouras, E. S. Hosseini, and R. S. Dahiya, "Chitosan-graphene oxide-based ultra-thin and flexible sensor for diabetic wound monitoring," *IEEE Sensors Journal*, vol. 20, no. 13, pp. 6794–6801, 2020, doi: https://doi.org/10.1109/JSEN.2019.2928807.

132 E. S. Hosseini, L. Manjakkal, D. Shakthivel, and R. Dahiya, "Glycine–chitosan-based flexible biodegradable piezoelectric pressure sensor," *ACS Applied Materials & Interfaces*, vol. 12, no. 8, pp. 9008–9016, 2020, doi: https://doi.org/10.1021/acsami.9b21052.

133 S.-W. Hwang, J.-K. Song, X. Huang, H. Cheng, S.-K. Kang, B. H. Kim, J.-H. Kim, S. Yu, Y. Huang, and J. A. Rogers, "High-performance biodegradable/transient electronics on biodegradable polymers," *Advanced Materials,* vol. 26, no. 23, pp. 3905–3911, 2014, doi: https://doi.org/10.1002/adma.201306050.

134 H. Heidari, N. Wacker, and R. Dahiya, "Bending induced electrical response variations in ultra-thin flexible chips and device modeling," *Applied Physics Reviews,* vol. 4, no. 3, p. 031101, 2017, doi: https://doi.org/10.1063/1.4991532.

135 A. Vilouras, H. Heidari, S. Gupta, and R. Dahiya, "Modeling of CMOS devices and circuits on flexible ultrathin chips," *IEEE Transactions on Electron Devices*, vol. 64, no. 5, pp. 2038–2046, 2017, doi: https://doi.org/10.1109/TED.2017.2668899.

136 S. Gupta, H. Heidari, A. Vilouras, L. Lorenzelli, and R. Dahiya, "Device modelling for bendable piezoelectric FET-based touch sensing system," *IEEE Transactions on Circuits and Systems I: Regular Papers*, vol. 63, no. 12, pp. 2200–2208, 2016, doi: https://doi.org/10.1109/TCSI.2016.2615108.

137 A. Vilouras, A. Christou, L. Manjakkal, and R. Dahiya, "Ultrathin ion-sensitive field-effect transistor chips with bending-induced performance enhancement," *ACS Applied Electronic Materials*, vol. 2, no. 8, pp. 2601–2610, 2020, doi: https://doi.org/10.1021/acsaelm.0c00489.

138 X. Liao, L. Xiao, C. Yang, and Y. Lu, "MilkyWay-2 supercomputer: system and application," *Frontiers of Computer Science*, vol. 8, no. 3, pp. 345–356, 2014, doi: https://doi.org/10.1007/s11704-014-3501-3.

139 M. Soni and R. Dahiya, "Soft eSkin: distributed touch sensing with harmonized energy and computing," *Philosophical Transactions of the Royal Society of London, Series A: Mathematical, Physical and Engineering Sciences*, vol. 378, no. 2164, p. 20190156, 2020, doi: https://doi.org/10.1098/rsta.2019.0156.

140 Z. Wang, L. Wang, M. Nagai, L. Xie, M. Yi, and W. Huang, "Nanoionics-enabled memristive devices: strategies and materials for neuromorphic applications," *Advanced Electronic Materials*, vol. 3, no. 7, p. 1600510, 2017, doi: https://doi.org/10.1002/aelm.201600510.

141 W. Taube Navaraj, C. García Núñez, D. Shakthivel, V. Vinciguerra, F. Labeau, D. H. Gregory, and R. Dahiya, "Nanowire FET based neural element for robotic tactile sensing skin," (in English), *Frontiers in Neuroscience,* Original Research, vol. 11, no. 501, 2017, doi: https://doi.org/10.3389/fnins.2017.00501.

10

Hybrid Systems-in-Foil

Mourad Elsobky

Institut für Mikroelektronik Stuttgart, Sensor Systems, Stuttgart, Germany

10.1 Introduction

Plastic electronics, including flexible, ultrathin, printed, and organic, is an emerging field in which conformable electronic systems are designed to introduce intelligence and connectivity to almost everything. Consequently, flexible electronics is considered one of the main enablers of the Internet of Things (IoT). When combined with new materials and advanced fabrication processes, flexible electronics offer unique characteristics such as mechanical flexibility, thin-form factor, large-area scaling feasibility, and adaptability to nonplanar surfaces. In contrast to the dense and highly integrated devices on silicon chips, large-area electronics include devices that can be widely spread across a large substrate [1].

Hybrid Systems-in-Foil (HySiF) is a branch of flexible electronics and defined as an approach for a promising technology direction, which describes the fabrication process and design rules for integrating electronic components into polymeric foil. HySiF serves as a platform, which complements the merits of the very large-scale integration (VLSI) of silicon ICs with the large-area electronics by combining mechanically flexible System-in-Packages and System-on-Packages (SiPs and SoPs) concepts.

Figure 10.1 graphically illustrates one variant of HySiF as it includes, large-area and on-chip sensors, silicon-based and nonsilicon-based ICs, antennas, passives, display, and energy storage elements [1–3]. Here, ultrathin chips (UTCs) are main building blocks, which are either directly embedded in the polymeric substrate or indirectly assembled via an interposer package.

10.1.1 System-Level Concept

Figure 10.2 shows a block diagram for a generic model reflecting the envisioned HySiF concept in which main electronic subsystems are defined. By closely considering a target application, this block diagram can be customized and certain blocks can be omitted (removal of flexible display for economic implementation).

In this HySiF concept, a centralized System-on-Chip (SoC) UTC performs accurate sensor read-out and data conversion. In addition, high-speed digital signal processing, wired and/or wireless communication, and power management are incorporated. In this way, the complex electronic

M. Elsobky, (2021), Hybrid Systems-in-Foil, Wiley Semicon Ref, 2021;xx:x–x.

Advances in Semiconductor Technologies: Selected Topics Beyond Conventional CMOS, First Edition. Edited by An Chen.

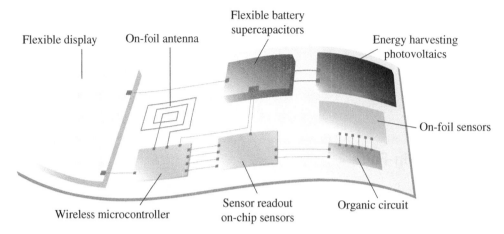

Figure 10.1 Concept of a Hybrid Systems-in-Foil (HySiF) [1–3] in which multiple ultrathin silicon ICs are combined with large-area electronic components and integrated into flexible polymeric foils.

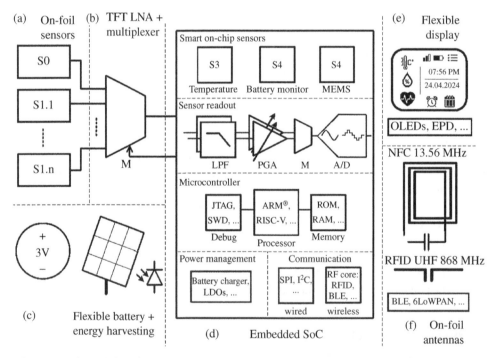

Figure 10.2 Block diagram for a flexible generic model of a smart sensor system reflecting the targeted HySiF concept [4]. a) Large-area off-chip sensor arrays are distributed on the foil as they complement the operation of the miniaturized on-chip smart sensors. b) Thin-film transistors can be employed to preamplify and multiplex the analog signals. c) Flexible batteries or energy haversting, which requires specific power management circuitry in the SoC. d) A centralized SoC fabricated by UTC performs accurate sensor readout and data conversion. Here, the complex electronic tasks are assigned to the well-established CMOS technology and the number of embedded silicon UTCs is limited to 2 or 3 ICs per HySiF. e) Flexible display. f) On-foil flexible anntenas.

tasks are assigned to the well-established CMOS technology and the number of embedded silicon UTCs is limited to two or three ICs per HySiF.

Moreover, large-area off-chip sensor arrays are distributed on the foil as they complement the operation of the miniaturized on-chip smart sensors. Here, on-foil sensors require signal-conditioning circuits, which eventually bring the signal flow back to the centralized SoC UTCs. Optionally, thin-film transistors can be employed to preamplify and multiplex the analog signals. Therefore, signal integrity is preserved and the number of on-foil sensors can be extended beyond the limits of the number of SoC pads. Wireless microcontroller UTCs are a good candidate for combining high-speed computational power with multistandard wireless connectivity. For ultralow power applications, bespoke microcontrollers on plastic substrates are good alternatives due to their minimalistic approach [5].

After this introduction to HySiF section, few emerging applications for hybrid flexible electronics are reviewed in Section 10.2. In Section 10.3, different integration technologies are discussed. Next, in Section 10.4, state-of-the-art active and passive electronic components that are used in ultrathin flexible sensor systems are listed with examples from recent literature. In Section 10.5, a quick overview of a testing methodology for HySiF is presented.

10.2 Emerging Applications

10.2.1 Smart Labels for Logistics Tracking

According to the Organic and Printed Electronics Association (OE-A) [6], RFID technology is the second driver of growth for organic and printed electronics after OLED displays. In fact, RFID technology is rapidly replacing the conventional barcode readers in supply chain management and logistics tracking. It provides secure authentication, tamper control, and real-time monitoring on item-level. Additionally, RFID readers are able to wirelessly scan through multiple tags at a time and do not require careful positioning of the label's antenna relative to that of the reader, which results in time and cost savings.

The simple concept and small count of electronic components in an RFID tag, mainly a low-frequency antenna and an IC with two to four pads, allowed the RFID technology to be a key application field for hybrid electronics. The hybrid integration of silicon UTCs and silver-ink printed antennas into flexible packaging is widely utilized nowadays in the manufacturing of RFID tags. For example, the company PragmatIC mass produces ultra-low-cost RFID and NFC tags, where UTCs are embedded in spin-coated polymeric substrates resulting in a reliable FlexICs packaging (Figure 10.3a) [7]. The FlexIC interposer is then assembled to another thicker flexible board, where the low-frequency antenna and circuit interconnects are printed. Alternatively, either organic or amorphous InGaZnO TFT-based NFC circuitry are proposed to replace conventional Si ICs (Figure 10.3b) [8, 9]. This allows the ultrathin RFID tags to be inherently flexible and disposable, which reduces its environmental impact. However, TFT low-yield, high-temperature, and air stability are main challenges that are regularly addressed in literature.

The next generation of RFID tags requires not only simple communication and memory blocks but also signal processing core(s), support for various communication protocols (e.g. BLE, 6LoWPAN), and integrated readout for off-chip/on-foil sensors [12].

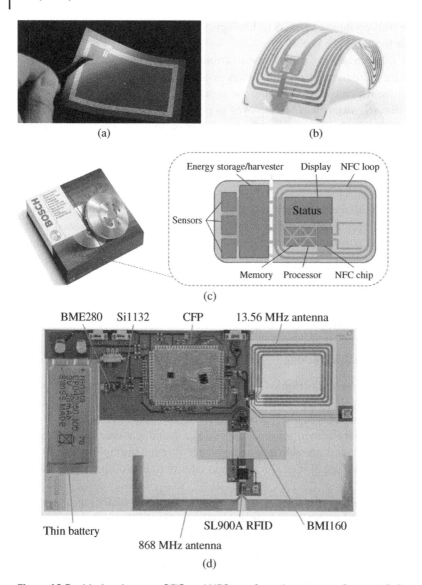

Figure 10.3 (a) ultra-low-cost RFID and NFC tags from the company PragmatIC. Source: From [6, 7] / with permission of IEEE. (b) TFT-based RFID. (c) Conceptual diagram of the ParsiFAL4.0 project. Source: From [10]. (d) Flexible HySiF for smart packaging. The smart label integrates an inertial measurement unit (BMI160, Bosch), an environmental sensor (BME280, Bosch), a 868-MHz UHF (SL900A, ams), and 13.56-MHz NFC (NF4, em Microelectronics) ICs. The NFC UTC is embedded together with an ultralow power microcontroller unit (MCU) (Apollo, Ambiq) using the CFP process. Source: From [11] /with permission of IEEE.

Figure 10.3c,d illustrates a use case and the conceptual diagram of the ParsiFAL4.0 project [10], in which a flexible HySiF is developed for smart packaging. The smart label includes an inertial measurement unit (BMI160, Bosch) in order to detect and log sudden or irregular handling of sensitive parcels. In addition, an environmental sensor (BME280, Bosch) is used to continuously monitor the surrounding environmental parameters, such as air temperature, humidity, and pressure of the parcel. Furthermore, two RFID standards are supported, namely 868-MHz

UHF (SL900A, ams) and 13.56-MHz NFC (NF4, em Microelectronics). The NFC UTC (thickness = 30 μm) is embedded using the polyimide (PI)-based ChipFilm Patch (CFP) (thickness less than 100 μm) together with an ultralow power microcontroller unit (MCU) (Apollo, Ambiq) [11]. The CFP is then used as an interposer, which is assembled to the flexible motherboard (liquid crystal polymer [LCP] Würth). An ultrathin 3-V 28-mAh lithium battery (thickness = 0.42 mm, bending radius = 25 mm, Renata SA, a subsidiary of Swatch Group), which can be recharged from the NFC field, provides the required energy for the label. Another variant of this demonstrator uses a BLE wireless MCU (TI, CC2640), which is capable of communicating with a remote central base-station.

10.2.2 Electronic Skin

Emulating the human sensory system is the ultimate goal for advancing the fields of robotics and prosthetics. Hybrid flexible electronics combines novel materials and high-speed computing to bring about such revolution. Different sensors (e.g. temperature and pressure), sensor arrays (for spatial resolution), smart sensors (i.e. power efficient digital sensors), and thin bendable sensors should be integrated into the e-skin system. In fact, the research in advancing electronic skin (e-skin) focuses on designing robust bendable materials that are suitable to be attached to the sturdy hull of the robot and at the same time allow for the integration of different electronic components.

E-skin can be realized either by soldering off-the-shelf components, or directly printing of sensors or embedding sensing UTCs into a flexible substrate [13]. The following examples demonstrate the level of innovation and the endless possibilities for realizing e-skin systems. They also prove that no single material, set of sensors, or sensing technique is a clear winner.

In [14], multiple flexible CMOS stress sensors are embedded in polymeric foil and attached to a robotic gripper.

Figure 10.4c shows the e-skin, which senses the object shape by measuring the in-plane stress. This demonstrates that flexible electronics can benefit from the characteristics of silicon, such as the piezoresistive effect.

Another technology demonstrator which implements a hybrid e-skin by combining printing and embedding of UTCs was presented in [4, 15] and shown in Figure 10.4a,b. Figure 10.4e shows a 3D-printed glove, which presents an unconventional method of using optical fibers for realizing e-skin [16]. Here, stretchable fiber-optic sensors combined with low-cost LEDs and dyes, can detect bending and determine its location.

The limited energy supply in e-skin is addressed in literature [17–19]. For instance, Figure 10.4d shows an array of solar cells distributed on soft elastomeric substrate is used as energy harvesters (about 10 mW cm^{-2}) as well as touch sensors [18]. Alternatively, a flexible polymeric array of biofuel cells that uses human sweat to generate energy (about 3.5 mW cm^{-2}) was developed [17]. Using this limited power budget, the presented e-skin is able to power several integrated sweat, temperature, and strain sensors, in addition to wirelessly transmitting sensory data.

10.2.2.1 E-Skin with Embedded AI

The human skin covers an area of about 2 mm^2 [20] with millions of touch receptors. Therefore, the capability to process large amount of sensor data and provide feedback in real-time is key enabler for e-skin. Several algorithms have been developed to handle large-amount of data given the complex constraints enforced by e-skin [21–23].

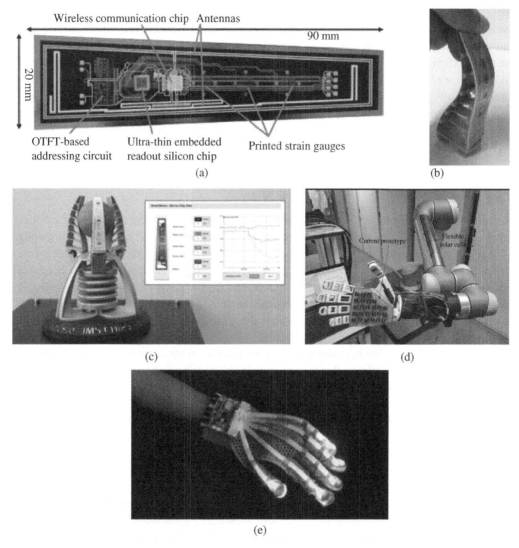

Figure 10.4 Several e-skin demonstrators, where in (a) on-foil printing of strain gauges, OTFT-based addressing circuit, and embedding of UTC is combined. Source: From [15]/with permission of John Wiley & Sons. (b) The e-skin in (a) is attached to a 3D printed robotic finger. (c) Another variant of (a) in which multiple UTCs are embedded and used as wireless stress sensors. Source: [14]. (d) An array of solar cells distributed on soft elastomeric substrate used for energy harvesting and touch sensing. Source: [18].

In [22], a methodology for realizing large-area e-skin system that efficiently handles event-driven information is presented. Here, a number of 1260 multimodal skin cells are demonstrated, which emphasize the need for local processing of large amount of data across distributed sensing points. Alternatively, AI trained using large and diverse datasets can be utilized to provide e-skin with complex human-like properties. In [24], a deep convolutional neural network (CNN) is used to teach robots how to see by using low-cost touch sensors. The presented CNN is trained using hundred thousands of tactile frames, which corresponds to 26 objects in addition to the frames from an empty hand.

10.2.3 Biomedical

A clear advantage for hybrid flexible electronics is the ability to be integrated in and adapted to complex nonplanar surfaces and at the same time enables high-speed smart sensor systems. Therefore, it is possible to use smart flexible biomedical sensors to monitor, in real-time and non-intrusive manner, various human-body biomarkers, such as levels of glucose [25–27], pH [28–30], and cholesterol [27] in addition to physiological parameters, such as temperature [31–34], pulse rate [34], and respiration [35, 36].

As an example of nonintrusive health monitoring, a stress monitoring patches, which integrates multimodal temperature, skin-conductance, and arterial pulsewave sensors, have been developed [34].

The patch, shown in Figure 10.5b can measure sensitivities upto $0.31\,\Omega/°C$, $14\,\mu V\,\mu S^{-1}$, and 70 ms for the skin temperature, skin conductance, and pulsewave sensors, respectively. Wearable Lab on Body is another novel approach for health monitoring, which aims at integrating digital sensors, such as GPS and IMU for location and activity recognition, with conventional paper-based biochemical sensors in a hybrid flexible sensor system (Figure 10.5a) [37].

For vision impairment correction and potential use in virtual and augmented reality applications, development of smart low-power contact lenses is needed [38–41]. Particularly, researchers have recently reported an artificial iris (Figure 10.5c), which is embedded in a smart contact lens that utilizes liquid crystal cells as an active electro-optical component [41]. The iris aperture is tunable through concentric rings offering an innovative solution for people who suffer from human eye iris deficiencies, e.g. light sensitivity or photophobia.

We now switch from devices that are placed on the surface of the human body to implanted devices. Implantable biosensors have important applications in precision medicine, which targets the prevention, investigation, and treatment of diseases while taking individual variability into account [42]. In general, implanted devices can be categorized into two main categories based on their rated lifespan, biodegradable, and long-lived systems. Here, the materials involved in the manufacturing of the flexible sensor/actuator systems primarily determine how long the implanted device can function in a human body. In fact, Si, Ge, and ZnO have been proven to be biodegradable, with Si being the most common choice of bioresorbable electronics. A 100 nm thick Si-based TFT and CMOS circuits using Tungsten interconnects can gradually disappear from the human body after about month of implantation [43, 44]. Figure 10.5d shows a bioresorbable intracranial pressure sensor that uses membranes of monocrystalline silicon. Such sensor can operate up to three weeks [45].

Finally, intracellular and extracellular interfaces benefits from the significant improvements in precision readout electronics on ultrathin substrates. Ultrathin probes minimize tissue damage and cause minimal bleeding. As an example, the company Neuralink has recently presented advances toward a scalable high-bandwidth brain-machine interface system using an array of 3072 polymeric microelectrodes [46]. Despite the bendability and precision of the microelectrodes, shown in Figure 10.5e, and the subsequent low-power readout, the readout electronics is manufactured in bulky package with a thickness of $1.65\,mm^2$, which limits the stability and overall form factor of the system. On-foil TFT devices can be employed to filter and preamplify the voltage signals when polymeric probes are used. Silicon probes are also used for neural recorders as readout electronics can be directly integrated at the pixel site. Here, a 20 μm thick 960-channel 130 nm CMOS probe called neuropixels is presented in Figure 10.5f [47]. The reported channel-to-channel crosstalk is less than 5% and a reliable rms noise of about 6 μV can be achieved.

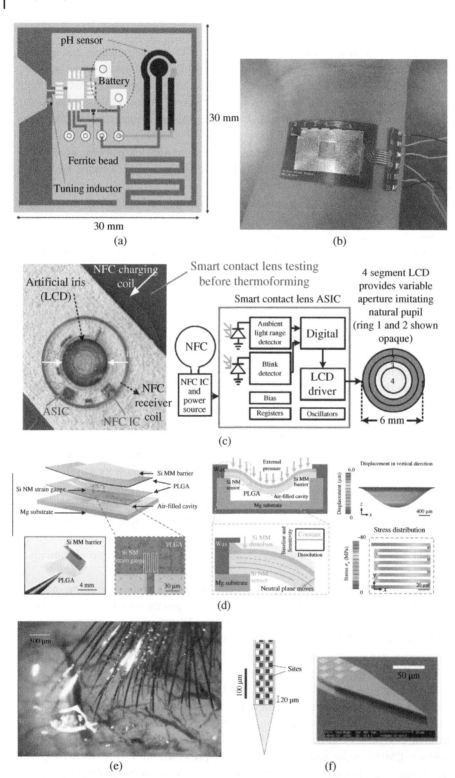

Figure 10.5 (a) Compact wireless flexible on-skin printed pH sensor is combined with the energy harvesting antenna of a UHF-band RFID. Source: [29]/IEEE. (b) Stress-monitoring patch integrates multimodal temperature, skin-conductance, and arterial pulsewave sensors. Source: [34]. Sunghyun Yoon, et. al., 2016, Springer Nature. (c) Artificial iris where the aperture is tunable through concentric rings. Source: [41] Andrés Vásquez Quintero et al., 2020, Springer Nature. (d) Bioresorbable multisensor platforms for pressure monitoring where a multilayer membrane is sealed over an air-filled cavity. Source: [45]/ John Wiley & Sons. (e) Scalable high-bandwidth brain-machine interface system using an array of 3072 polymeric microelectrodes from the company Neuralink. Source: From [46] / with permission of JMIR Publications. (f) Neuropixels are a 20 μm thick 960 channel 130 nm CMOS Silicon probe used for neural recorders. Source: [47] James J Jun et al., 2017, Springer Nature.

10.3 Integration Technologies

10.3.1 Substrate and Interconnect Materials

In hybrid flexible systems exploiting metallic foil substrates, wire interconnects are already available, and their properties are coupled to the substrate properties. Metal foil substrates offer the capability of higher process temperature, form factor stability with minimal shrinkage, and almost no moisture uptake [48]. Drawbacks of using metallic foils as flexible substrates include higher parasitic capacitances and poorer surface smoothness which degrades the performance of some flexible electronic components, e.g. TFT and nanosheet sensors.

A diverse range of flexible polymeric materials are available for HySiF applications including PI, poly(ethylene napthalate) (PEN), and polyethylene terephthalate (PET). In systems employing such insulating substrates, metallic films are deposited or printed to electrically connect different electronic components. Conventional CMOS-compatible lithography could be used to trade-off smaller-area coverage and lower-throughput with fine-pitch interconnects. Fortunately, the thinner the material, the more flexible it becomes. This is true in the case of thin-film metals, such as gold (Au), silver (Ag), and aluminum (Al), which normally are used to manufacture flexible electronic components. However, the thinner the metal, the higher the sheet resistance as well as the possibility of cracking. As an example of this trade-off, multiple overpasses are performed during metal inkjet printing to tailor the sheet resistance to that of a printed single metal layer. Long wire interconnects that are often used in large-area flexible electronics affect signal integrity as they limit the precision and speed of the embedded systems by adding line delays (higher parasitic RLC).

The chosen plastic substrate despite being inherently mechanically flexible, needs to offer low-moisture uptake, smooth surface with minimal topographies, excellent dimensional and thermal stability, and low coefficient of thermal expansion (CTE) mismatch with other HySiF components [48, 49].

The earlier mentioned properties of metallic and insulating substrates are general and related to physical properties of each material. However, there are application-related properties that can lead to new materials or the usage of hybrid combination of different materials. For example, hydrogels are more than 70% water, making them very compatible with human skin tissues [50]. Inside the hydrogel, an UTC in PI package can be embedded. In [1, 3], CMOS UTCs are embedded in an interposer polymeric fanout package, which is then assembled to a flex-PCB (Figure 10.7d).

10.3.2 Flex-PCB

The well-established Printed Circuit Board (PCB) manufacturing has inspired the realization of various integration and packaging technologies for ultrathin and flexible electronic components.

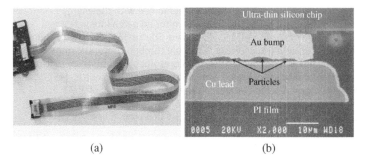

(a)   (b)

Figure 10.6   (a) Ultrathin Chip Package (UTCP) technology realizing a symmetrical substrate sandwich. Polyimide (PI) is used as a substrate material. Source: Würth Elektronik Group. (b) Flex Silicon-on-Polymer. Source: [57] / IEEE. (c) ChipFilm Patch (CFP) with two function ICs embedded in PI. Source: From Burghartz et al. [1], Albrecht et al. [11], and Hassan et al. [56]/with permission of IEEE. (d) The CFP interposer is then assembled to a flex-PCB motherboard.

The natural extension of the rigid PCBs are the flexible PCBs, in which the brittle FR4 substrate is replaced with flexible sheets of polymers, such as PI-based resin [51] and LCP (Figure 10.6a) [52]. Board thicknesses from 1 mm down to 100 μm are achieved depending on the number of metal layers and substrate materials. Since flexible PCBs integrate conventional through-hole and surface mount components in a similar fashion to the rigid PCBs, their flexibility and form factor are limited by the utilized active and passive components.

Another variant of flexible PCBs uses bare die UTCs (shown in Figure 10.6b), which are either embedded in the substrate sheets or mounted on the substrate surface using flip-chip bonding [53]. The form factor is significantly reduced but another challenge remains, which is to achieve fine-pitch circuit interconnects, particularly the fan-out interconnects of UTCs with

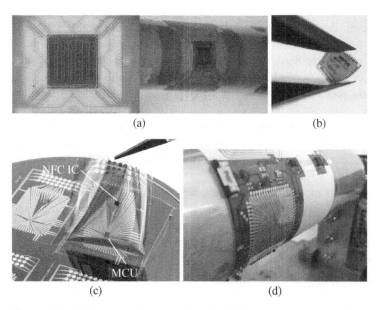

(a)   (b)

(c)   (d)

Figure 10.7   (a) Bendable biocompatible flex-PCB that can be integrated into textile (TWINflex-Stretch technology from the company Würth Elektronik). Source: From [52] / with permission of IEEE. (b) Ultrathin chip-on-flex (UTCOF) interconnects formed using anisotropic conductive adhesive (ACA). American Semiconductor, Inc.

large I/O count. Current flexible PCB manufacturing technologies provide a minimum pitch in the range of about 100 μm, which is limited by the metal structuring and drilling processing steps. This, in turn, restricts the pin-count for the utilized ICs and reflects the need for an alternative integration technology.

10.3.3 Wafer-Based Processing

In order to fill this gap, UTC embedding technologies, such as Ultrathin Chip Package (UTCP) [54, 55], CFP [1, 11, 56], and Flex Silicon-on-Polymer [57], have emerged to provide a conformable UTC package with finepitch wire interconnects. Figure 10.7 highlights such UTC embedding technologies, and in the next Section 10.3.3, the CFP processing steps are shortly reviewed.

10.3.3.1 ChipFilm Patch
The CFP is an embedding technology, which provides a flexible fan-out package for single or multiple UTCs. The fine-pitch interconnects (<10 μm) of the CFP enable the integration of high-performance pad-limited ICs into flexible electronic systems. Moreover, several forms of the HySiF concept are achieved in part using CFP technology. In fact, the CFP-compatibly is a principle aim for the ultrathin active and passive components that are implemented in this thesis. Therefore, it is reasonable to overview the process sequence of one version of the CFP technology.

Figure 10.8 shows the cross section of the CFP processing steps [1, 11, 56]. The flexible substrate is a stack of two spin-on polymers, namely BCB and PI. The embedding polymer is BCB, which is

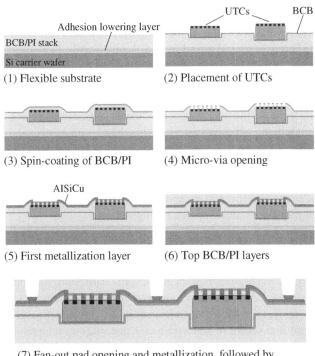

Figure 10.8 Fabrication process steps for the CFP embedding technology [49, 56].

CMOS-compatible and has fine-pitch patterning capability. On the other side, PI is the reinforcing polymer, which is highly bendable and stable and is widely used as a substrate material in flexible electronics. The temporary carrier wafer is coated with an adhesion lowering and stress-relief layer. Next, successive spin-coating of BCB/PI layers defines the thickness of the flexible substrate, which is designed considering the embedded UTC thickness. A BCB layer is spin-coated in which a cavity resembling the UTC is patterned and etched. Another adhesive layer of BCB is spin-coated on which the UTC is placed in face-up orientation.

After curing of the BCB, additional BCB layers are spin-coated to completely encapsulate the UTC. Dry etching using BCB or AlSiCu as a hard mask enables the realization of micro-vias on the UTC pads. Chip warpage, nonuniform planarization, process-induced stress, and air bubbles are nonidealities, which are directly linked to the quality and shape of the micro-vias. To counteract these nonidealities, the idea of standard via sizes is borrowed from the CMOS technology and introduced here as a design rule for other CFP-compatible components.

Conventional AlSiCu sputtering is used to fabricate the fan-out interconnects on the smooth BCB surface. A superstrate layer, with a thickness similar to the substrate thickness, is fabricated using successive spin-coating of BCB/PI layers. Finally, coarse fan-out pads are opened and optionally covered with a noble metal. When the CFP is used as an interposer, the CFP is mechanically released from the rigid carrier either using a manual or laser cutter. Alternatively, further processing steps are followed (e.g. sensors or antenna fabrication), when the CFP is utilized as a substrate for HySiF integration.

10.3.4 Challenges

Several challenges are faced during the processing and integration of HySiF. The robustness and reliability of HySiF can be affected after several stretching or bending cycles, which can cause delamination or cracking and ultimately failure of several ultrathin HySiF components. In medical wearables, even the quality of the electrical signals is coupled with the adhesion properties of the flexible electrodes. Other HySiF integration challenges can be related to interfacing different types of metal, biocompatibility of the used metallization and substrate materials, the effect of ink viscosity on the adhesion of printed metallic films. As the thickness of the flexible substrate decreases, it becomes negligible compared to the underlying adhesive material. This calls for innovative and stable adhesives with minimal hysteresis and thickness, properties that are closely related to the utilized substrate.

The quality of integrated components are directly coupled to the underlying substrate. Surface roughness and macroscopic topographic properties, thermal stress due to CTE mismatch, and other assembly inaccuracies within the HySiF framework are discussed in this section.

10.3.4.1 Surface Topography

Most polymers suffer from poor planarization behavior due to their high viscosity. In general, uniformity of polymeric foils varies with process parameters such as the spin-coating and the temperature profile during the soft and hard bake [35]. This leads to several challenges during UTC embedding (e.g. during opening of microvias) [11, 35, 49, 56, 58, 59], in addition to mismatch resulting from the deposition or printing of thin-film sensor or semiconductor materials on nonplanar surfaces [3]. In fact, the inherent surface properties of the polymer are more relevant than the size and orientation of underlying features [60].

However, in special cases, the underlying features can introduce varying topographies that can propagate to the foil surface. Figure 10.9a shows the topography of spin-coated PI on top of an embedded chip during CFP process, where the chip edges are highlighted [35]. Variations of only

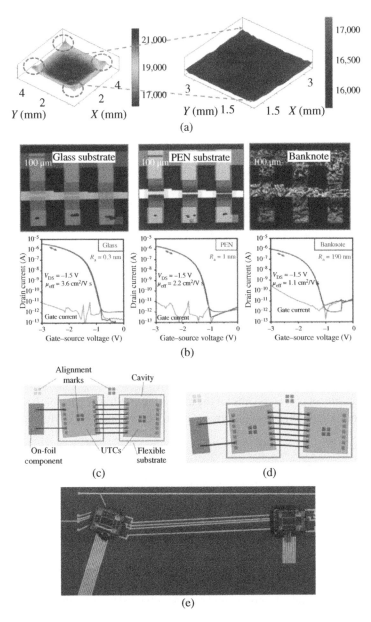

Figure 10.9 (a) Topography of spin-coated PI on top of an embedded chip during CFP process where the chip edges are highlighted. Source: [35]/IEEE. (b) Photographs and transfer characteristics of p-channel DNTT TFTs fabricated on glass, PEN, and a banknote. Source: [61] Ulrike Kraft et al., 2018, John Wiley & Sons. (c) Illustration of inaccuracies during chip placement. (d) Application of adaptive layout technique in which the first metallization layer and all other consequent layers are adjusted to compensate the displacement and rotational errors [58, 59]. (e) Micrograph of two logic chips, integrated into CFP using the adaptive layout technique. Source: [59] Golzar Alavi et al., 2018, IEEE.

700 nm is achieved in the inner region on top of the embedded UTC. However, due to chip warpage and the fact that a height difference equal to the chip thickness must be overcome, increased PI topography is observed at the chip edges. This can degrade subsequent process steps such as microvia opening especially when the chip I/O pads are located near the chip edges.

One solution is using thicker PI layers or multiple layers, which can degrade the fine-pitch interconnection property of CFP. Another solution is to extend the chip edges by cutting a larger chip area so that the I/O pads lie in the planar region. However, this wastes area and reduces yield.

Furthermore, surface roughness directly affects the electrical properties of devices manufactured on the foil surface. It has been shown that the mobility of TFTs and consequently the maximum operational speed of TFT-based circuits is a function of surface quality [4, 15, 49, 58, 61, 62]. Figure 10.9b shows photographs and transfer characteristics of p-channel DNTT TFTs fabricated on glass, PEN, and a banknote [61]. It is clear that the extracted mobility degrades toward poorer surface quality. Also, the sensitivity and mismatch of thin-film nanosheet- and resistive sensors are related to the quality of the substrate surface [3, 15, 63].

10.3.4.2 Thermal Stress

The CTE describes how the volume of a layer changes with temperature. As an example, unwanted wafer warpage is caused by the deposition of polymers (CTE of Si, BCB, PI2611 is 2.5, 42, 3 ppm/°C, respectively). Stoney's formula [64] is then used to calculate the stress resulting from the Si-polymer, two-material system. During the embedding of UTC in polymeric foils, the CTE mismatch between the polymers and the embedded UTCs causes internal residual stress inside the film and on the chips. Using low CTE polymers and placing the embedded UTC in the neutral plan of stress are steps to counteract process-induced thermal stress.

Systematic substrate shrinkage can be taken into account during the early design phases, for example if the spacing between sensor electrodes or in cases where TFT channel length and width need to be accurately defined. This technique is particularly used during the fabrication of OLED backplanes [48].

The low thermal dissipation capacity of polymers calls for the utilization of ultrathin flexible heat sinks or substructuring trench formation [58] as a countermeasure in case self-heating of HySiF is an issue.

10.3.4.3 Assembly and Positioning Errors

During the assembly of rigid ultra-thin HySiF components, such as UTCs, onto the surface of the polymeric substrate and due to the limited accuracy of the placement tool, unwanted rotation and positioning errors occurs due to the component movement on top of the foil during the curing and hardening step [58]. Figure 10.9c depicts such inaccuracies during UTC placement during CPF processing.

Adaptive layout technique can be utilized to resolve this situation [58, 59]. The coordinates of each HySiF component are measured by utilizing alignment marks and a maskless laser direct writer. Layouts for the first metallization and all consequent layers are updated in order to preserve the original circuit interconnects, as shown in Figure 10.9d,e. The alignment of on-foil components, such as sensors, TFT circuitry, and antennas, is either adjusted to the orientation of the associated chip or is left unadjusted provided that the adjustments in the layout of the respective wire interconnects are sufficient. Note, that the overall accuracy of the CFP is now linked to the structuring as well as to the overlay accuracy of the lithography tool.

10.4 State-of-the-Art Components

In this section, design and fabrication considerations for commonly used active and passive components are presented to enable their integration into and compatibility with HySiF.

10.4.1 Active Electronics

Table 10.1 compares different TFT technologies that are used to fabricate flexible sensor frontends and digital processors, namely single-crystalline UTCs, poly-crystalline silicon (poly-Si), amorphous silicon (a-Si), OTFTs, and metal-oxide TFTs (e.g. InGaZnO).

10.4.1.1 Microcontrollers

Computation and digital signal processing are continuously pushed to edge devices (i.e. locally in the sensor node) as user privacy and communication availability, bandwidth, and latency are main concerns [65]. For achieving high-performance battery-operated flexible sensor nodes, low-power microcontrollers ICs are needed (Figure 10.10c). A flexible microcontroller, which is implemented using organic TFTs on a plastic substrate, is operated using a 10-V supply voltage and consumes about 90 µW [66]. In addition, bespoke microcontrollers on plastic substrates are good alternatives since their implementation is optimized for the target application [5]. However, the low operating speed and high-supply voltage are main concerns.

Alternatively, silicon-based commercial microcontrollers have been thinned to less than 50 µm and embedded in polymeric fan-out packages, which were presented in various IoT-driven demonstrators, such as health monitoring, e-skin, and logistics tracking [1, 3, 4, 63, 67].

Figure 10.10a shows a commercial microcontroller that have been thinned down to less than 30 µm and then embedded in CFP (cf. Figure 10.7c) [3, 4]. Further, the company American Semiconductors have released its Flex-MCU (NHS3100 from NXP) that integrates NFC temperature logging IC with arm M0+ (Figure 10.10b) [57].

10.4.1.2 Sensor Frontends

Most hybrid flexible sensors presented in literature are either demonstrated as standalone sensors or combined with bulky readout and digital signal processing electronics. In HySif, the target is to enable the realization of smart flexible sensor systems with digital outputs, which require the integration of the whole sensor frontend into the same sensor substrate or using an interposer subsystem.

Figure 10.11a shows an example of complete inherently flexible smart sensor systems, in which an RFID-ready ADC is demonstrated together with a printed NTC sensor fabricated using InGaZnO TFTs on ultra-thin PI foil [69]. The sensor system enables the next-generation RFID-tags with integrated sensors. The same technology is used in [70] and realizes a charge-sense amplifier for 1 megapixel in-panel sensor imager applications. The amplifier achieves 31.17 dB gain, 140 kHz GBW, and 87 µW power consumption at 15 V supply. A low-voltage light-sensing oscillator using active-inductors have been also demonstrated in a similar TFT technology [71]. Several standalone analog building blocks based on TFTs can also be found in the literature [72–76].

Table 10.1 Comparison of different TFT technologies used to fabricate flexible sensor frontends.

Technology	Si-UTC	Poly-Si	a-Si	OTFT	Metal-oxide
Resolution	■■■	■■□	■□□	■□□	■□□
Speed	■■■	■■□	■■□	■□□	■□□
CMOS availability	☺	☺	☹	☺	☹
Room-temperature processing	☹	☺	☺	☺	☺
Large-area coverage Roll-to-roll	☹	☺	☺	☺	☺

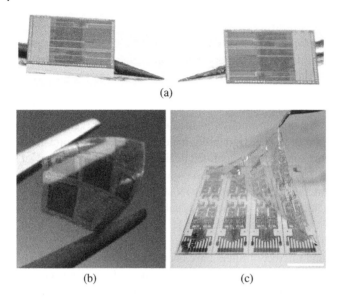

Figure 10.10 Several e-skin demonstrators, where in (a) Source: From [3] / with permission of IEEE (b) Source: American Semiconductor, Inc. [57]. (c) Photograph of organic CMOS logic circuit on ultrathin substrate. Source: [68] Yasunori Takeda et al., 2016, Springer Nature.

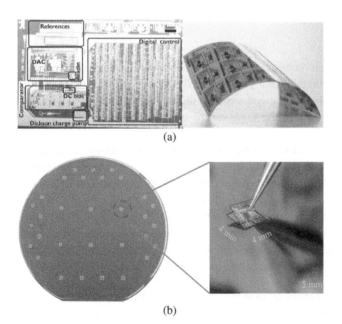

Figure 10.11 (a) RFID-ready ADC is with integrated NTC sensor fabricated using InGaZnO TFTs on ultrathin PI foil. Source: [69] Nikolas P. Papadopoulos et al., 2018, IEEE. (b) Readout ASIC implement on CFP for a wireless smart breath sensor developed for newborn. Source: [35] Ulrike Passlack et al., 2021, IEEE.

Fabricated on UTCs, several sensor frontends that are HySiF compatible or already into HySiF have been demonstrated. In [14], stress-insensitive current readout followed by 10-bit SAR ADC are used to interface with on-chip CMOS stress sensors. The ASICs are embedded in a polymeric foil and attached to a robotic gripper to monitor its bending activity. A similar system that uses on-foil strain gauges uses differential difference amplifier for voltage readout and offset compensation of nonidealities of the sensor printing process [15]. In [35], a readout ASIC for resistive bridge breathing sensor is characterized after back-thinning and embedding into the CFP (Figure 10.11b). The chip-to-chip performance variation is within the standard CMOS process variations and is not related to stress-induced errors.

Flexible Analog-to-Digital Converters ADCs are at the heart of every smart sensor system and notably required for achieving high-performance flexible electronic systems. Although ADCs integrated into UTCs represent the natural extension for the silicon-based semiconductor technology to the flexible electronics world, ADCs fabricated using other technologies, notably organic and inorganic metal-oxide semiconductors, are on the rise.

Figure 10.7b shows a photograph of the Silicon-on-Polymer package developed by the company American Semiconductor in which the industry's first bendable CMOS ADC is embedded [57]. The 8-channel 8-bit 100-kS/s SAR ADC is fabricated using a 180-nm CMOS technology and its die size is about $2.5 \times 2.5\,mm^2$. The so-called "FleX-ICs" are essentially UTCs with thickness of $25\,\mu m$ and are embedded in $50\,\mu m$ thick polymeric packages using spin-on polymers.

Alternatively, Figure 10.11a shows a photograph of the 6-bit SAR ADC fabricated on PI substrate using unipolar dual-gate metal-oxide (InGaZnO) TFTs [69]. The ADC occupies an area of $27.5\,mm^2$ and consumes about $73\,\mu W$ when operated using a supply voltage of $15\,V$ at a conversion rate of $26.67\,S/s$. Other examples include $\Delta\Sigma$-based ADCs fabricated on polymeric foils utilizing OTFTs [77] and InGaZnO TFTs [78] achieving FoMW of 3.45 and $0.39\,\mu J/$conv.-step, respectively.

Circuit Design Challenges The availability of device models and process design kits (PDKs) aides the customization and wide adoption of the HySiF approach. In fact, more PDKs are being introduced for TFT technologies and foundries are promising a less than two-week manufacturing cycle. One expects the codesign of TFT and conventional CMOS circuits to enable optimal task and block allocation for each technology. By coupling the mechanical, thermal, and electrical properties of the substrate, as well as, each HySiF component into a single software simulator, more mechanically compliant and electrical reliable HySiF can be achieved.

In circuit design for flexible electronics, new challenges are faced, not only concerning the impact of mechanical stress on circuit blocks, but also due to other factors, such as parasitic RLC and cross-talk, which increase as long thin-film wire interconnects and thin substrates are used. TFT amplifiers and repeaters can be used to ensure analog and digital signal integrity across large-area systems. Additionally, the sensor readout should be able to handle elevated offset in the sensor signal, which traces back to higher mismatch and resolution of simple printing technologies.

Furthermore, shielding the HySiF polymeric foil against electromagnetic interference is another challenge. Employing fully differential signaling directly reduces the effect of environmental and mechanical disturbances. Moreover, the high internal resistance of flexible batteries limits peak currents, which are needed in high-speed circuits. Therefore, the developing of thin-film supercapacitors can be particularly important for mitigating nonidealities of flexible batteries.

As more UTCs are integrated into flexible electronic applications, the modeling and incorporation of stress effects, e.g. piezoresistive and piezojunction effects of Si, in PDKs are of paramount importance. The characterization of high-resolution ADCs during elevated levels of mechanical stress can provide critical feedback for design improvements. Finally, noise analysis of stressed devices and the investigation of noise propagation in thin chips can improve the performance of ultrathin microsystems.

10.4.1.3 Sensor Addressing and Multiplexing

For economic reasons, UTC embedding in hybrid flexible electronic systems can be limited to a maximum number of two to three chips per foil [1, 3]. The interface required between the large-area components and silicon UTCs limits the number of accessible on-foil sensors [79].

Figure 10.12a shows a schematic illustration of a core-limited pad ring UTC, which is embedded, for example in CFP foil, having larger perimeter fan-out on-foil pads. Here, the chip doesn't require high I/O count and the UTC-foil interconnection profits from such arrangement by spreading the UTC pads to the perimeter foil pads using a low-resistance fan-out [49]. On the contrary, due to CMOS-scaling and increased SoC integration, the pad-limited pad rings are increasingly used. Consequently, the number of I/O pads increases, which demand more on-foil fan-out pads, as illustrated in Figure 10.12b. In this situation, the foil area in close proximity to the embedded UTC is utilized and minimum width of the wire interconnects are used for routing.

For both pad ring variants and in order to address more on-foil sensors using the limited UTC access points, the chip I/O count can be extended using active TFT multiplexing circuitry. Different system architectures for large-area acquisition and I/O count extension are available, such as active matrices, binary addressing circuits [15], and frequency-hopping oscillators [79].

10.4.2 Passive Components and Sensors

10.4.2.1 Flexible Antennas

Several flexible antennas have been reported, and their performance is steadily approaching their rigid counterparts [80, 81].

Figure 10.13a shows an inkjet-printed multiband antenna (1.2–3.4 GHz), which is fabricated on a Kapton substrate with a near-constant performance during bending [80]. Another example is a stretchable 915 MHz antenna that is fabricated by integrating conductive fibers in a PDMS substrate [82]. Note that, the electrical behavior of antennas, flexible antennas in particular, is strongly coupled to the electromagnetic properties of the substrate, as well as the passivation superstrate. Low-electrical permittivity and dielectric loss in the vicinity of the antenna tend to improve its efficiency.

Off-chip/on-foil inductors are inevitably used in wireless battery charging and NFC applications [83, 84]. In [85], various on-foil inductor loops, shown in Figure 10.13c, are fabricated on a

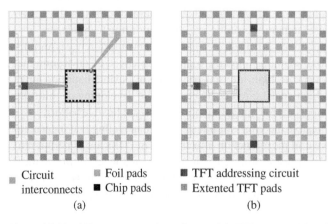

  ▪ Circuit   ▪ Foil pads   ▪ TFT addressing circuit
     interconnects ▪ Chip pads   ▪ Extented TFT pads

         (a)   (b)

Figure 10.12 I/O count extension using on-foil TFT sensor addressing circuits. Source: Elsobky [4]. (a) Core-limited and (b) pad-limited I/O pad ring design for embedded chips with the associated fan-out perimeter and area pads, respectively.

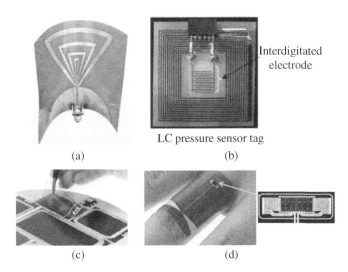

Interdigitated electrode

LC pressure sensor tag

(a) (b)

(c) (d)

Figure 10.13 Several flexible antennas, where (a) is an inkjet-printed multiband antenna (from 1.2 to 3.4 GHz) fabricated on Kapton substrate. Source: [80] S. Ahmed et al., 2015, IEEE. (b) Wireless passive LC pressure sensor tag to measure pressure from 0 Pa to 60 kPa. Source: [90] M. R. R. Khan, et. al., 2020, IEEE. (c) On-foil inductors for NFC applications fabricated on BCB/PI foil, which is used for UTC-embedding. Source: [85] Mourad Elsobky et al., 2018, MDPI / CC BY 4.0. (d) Flexible 5.5-GHz antenna and RF transmitter thin ASIC integrated on BCB/PI substrate. Source: [92] S. Özbek, et. al., 2018, IEEE.

BCB/PI flexible substrate targeting NFC applications have been demonstrated. A demonstration of hybrid flexible electronic sensor systems is when flexible inductors are combined with on-foil sensors for chipless remote passive sensing applications (e.g. wireless passive LC pressure sensor tag Figure 10.13b) [86–90]. Flexible inductors can also be used in harvesting energy from the movement magnets in robotic applications [91].

Figure 10.13d shows an ultrathin and conformal antenna and RF transmitter ASIC [92, 93]. The presented system targets the 5.5-GHz frequency range and has been integrated into a flexible BCB/PI substrate. The processing steps of the chip-embedding and antenna fabrication follow the CFP UTC-embedding process. The low-loss tangent (tan $\delta = 0.0008$ at 1 GHz) and low electrical permittivity ($\epsilon_r = 2.65$) of BCB, with its stable dielectric properties over a wide range of frequencies, is very attractive due to its CMOS-compatibility [94]. Since the combination of bendable high-frequency communication ICs with flexible antennas are rarely presented in literature, this demonstration unlocks an important milestone for high-speed hybrid flexible electronics.

10.4.2.2 HySiF-Compatible Sensors

Several standalone bendable sensors, which can be considered HySiF-compatible, are available in the literature, and there are few review articles that summarizes the state-of-the-art and can be found elsewhere, e.g. temperature [95–97], humidity and gas [98–100], and tactile [101–104]. Here, few points regarding the codesign of HySiF and sensor elements are discussed.

Sensors are integrated into HySiF while considering cumbersome boundary conditions regarding sensor protection, thermal, and moisture uptake of the next processing steps, providing low RC interconnects, and smooth substrate surface. For instance, the electrical properties (i.e. electron/hole, ionic, or proton conductivity) of resistive thin-film sensors are affected by the physical properties, in addition to the mechanical activities, of the flexible substrate. This correlation can be mitigated by either employing bridge configurations, special layout techniques [105, 106], or more effectively using capacitive sensors that benefit from the stress-insensitivity of the parallel plate electrode design.

Table 10.2 Comparison between the concepts of on-chip vs. off-chip/on-foil sensors

	On-chip sensors	On-foil sensors
Array formation	Complex	Simple
Interconnect parasitics	Low	High
Readout electronics	Integrated	Remote
Signal integrity	☺	☹
Area coverage, footprint	Small	Large
Thickness	$10\,\mu m$ to few mm	$<1\,\mu m$
Thermal contact resistance	High when in package	Low
Lithography	Complex	Simple
Fabrication temperature	High	Low
Power consumption	Low	High
Reliable lifetime	Years	Weeks to months
Biodegradable	☹	☺
Substrate	Mainly silicon	PI, BCB, LCP, paper, etc.
Package	Ceramic, plastic, etc.	PI, BCB, LCP, …

In order to simplify the HySiF fabrication process and increase the information per unit area, multimodal sensors (i.e. a single sensor cell that is able to sense different physical parameters) can be utilized. For instance, a ferroelectric gate insulator is integrated into the gate oxide of an organic TFT [107]. Since ferroelectricity includes both pyroelectricity and piezoelectricity, and by establishing a tensor relation for the TFT electrical parameters, the temperature, and mechanical strain can be simultaneously detected. Besides, 3D integration of thin-film sensors by vertical stacking is another possibility of achieving multimodal sensors as long as cross sensitivity is minimized [108, 109].

On-Chip vs. Off-Chip/On-Foil Components HySiF components can either be integrated into thin silicon chips or be fabricated directly on the foil substrate [1]. For sensing elements, Table 10.2 summarizes key differences between the concepts of integrated on-chip vs. large-area off-chip/on-foil sensors. It is clear that matrix or array formation is simpler and more economic for the on-foil sensors in contrast to the complexity of handling and assembling several silicon chips. Furthermore, signal integrity and readout speed are negatively impacted when long wire interconnects are used to connect the large-area sensors with a central readout IC.

10.5 HySiF Testing

For the characterization of UTC integrated stress sensors or checking stress sensitivity of readout and microcontroller ICs, UTCs can be placed on a tape and bent to different radii with exchangeable metal cylinders over which the tape is wrapped and strapped, as shown in Figure 10.14a [110–112]. For extracting the piezoresistive coefficient, Si strips can be placed in conventional four-point bending tool, shown in Figure 10.14b. Although the applied bending

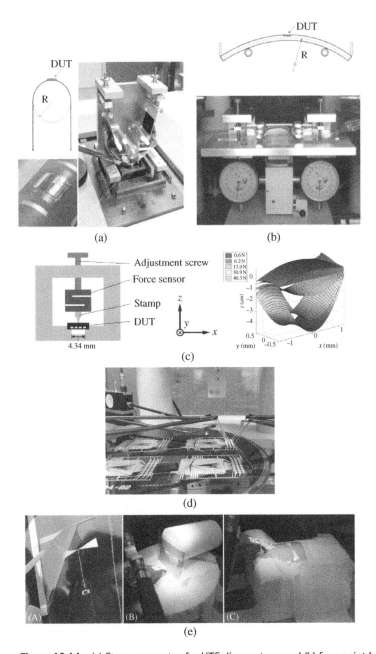

Figure 10.14 (a) Stress apparatus for UTC dies on tape and (b) four-point beam bending tool for bulk Si strips. Source: [110] Mahadi-Ul Hassan, et. al., 2009, IEEE. (c) Nonmagnetic load cell that bends the package of hall plates. Source: [113]/IEEE. The stamp is applied to the top face of the package. Measured deformation of the die surface due to vertical forces applied. (d) Electrical characterization of the UTCs embedded in CFP. Source: M. Elsobky et al., 2018, IEEE. (e) Illustration of bowtie antenna measurement on the polymer foil using VNA in straight and bent status. Source: [58]. Golzar Alavi, et. al., 2019, IEEE.

strain can be tracked accurately using such techniques, the long wire interconnects involved in these setups can degrade the signal integrity specially if the purpose is the characterization of high-precision or high-speed sensor frontends. Custom-made testing methods are developed for the specific use-case, e.g. stress-compensation of hall sensors, depicted in Figure 10.14c [113].

Figure 10.14d shows the measurement setup, which consists of a manual prober and source measurement units (SMUs), used for the continuity test [3] performed on the CFP with two embedded ICs (Figure 10.7c). Here, electrostatic discharge (ESD) devices at the I/O, supply and ground pins are used to check the basic functionality of the embedded UTC(s) after critical CFP processing steps [15]. Such preliminary test checks the overall state of the chip supply and ground pins connections to the fanout pads but does not say much about the performance of the embedded UTC or the connection of other I/O pads. Higher leakage current of ESD diodes can be a hint toward high process-induced stressors.

The electrical characterization of flexible antennas is usually not straightforward. Figure 10.14e shows the measurement setup used to characterize a bowtie antenna with operating frequency of 5.5 GHz. Antennas can be measured before releasing from the rigid carrier substrate; however, their properties might change significantly afterward due to the absorption or distortion of the high-frequency signal by the lossy substrate in case of Si and also due to variations in the trace dimensions (shrinkage or relaxation depending on the initial process-induced stress). As shown in Figure 10.14e, the antenna, after release from carrier wafer, is measured during the straight and the bent status above the white cylinders with the permittivity close to air.

10.6 Conclusion and Future Directions

HySiF is increasingly adopted in practical applications, including e-skin for prosthetics and robots and smart catheters and bandages. However, these applications represent niche markets, thereby call for more research and development in order to introduce the HySiF approach into mass production. To this end, improving the reliability, process yield, and air-stability of on-foil components is of prime importance.

Furthermore, the availability of accurate device models and PDKs aides the customization and wide adoption of the HySiF approach. In fact, more PDKs are being introduced for TFT technologies, and foundries are promising a less than two-week manufacturing cycle. One expects the codesign of TFT and conventional CMOS circuits to enable optimal task and block allocation for each technology. By coupling the mechanical, thermal, and electrical properties of the substrate, as well as, each HySiF component into a single software simulator, more mechanically compliant and electrical reliable HySiF can be achieved.

Besides, as more sensors are crammed into the foil, big data is generated and needs to be locally or remotely processed and communicated. Therefore, high-speed microprocessors and communication ICs are inevitably needed. Many questions arise regarding the economical feasibility of utilizing UTCs embedded in flexible packages or one can still utilize conventional rigid packages. By adapting the existing UTC embedding methods to fit into R2R manufacturing, high-speed UTCs can claim a central role in the mass production of flexible electronics.

References

1 J. N. Burghartz, G. Alavi, B. Albrecht, T. Deuble, M. Elsobky, S. Ferwana, C. Harendt, Y. Mahsereci, H. Richter, and Z. Yu, *IEEE Journal of the Electron Devices Society*, vol. 7, pp. 776–783, 2019.

2 J. N. Burghartz, W. Appel, C. Harendt, H. Rempp, H. Richter, and M. Zimmermann, in *2009 Proceedings of the European Solid State Device Research Conference*, pp. 29–36, 2009.

3 M. Elsobky, T. Deuble, S. Ferwana, B. Albrecht, C. Harendt, A. Ottaviani, M. Alomari, and J. N. Burghartz, *IEEE Sensors Journal*, vol. 20, no. 14, pp. 7595–7604, 2020.

4 M. Elsobky, Ph.D. thesis, University of Stuttgart, 2021, submitted for publishing.

5 E. Ozer, J. Kufel, J. Biggs, G. Brown, J. Myers, A. Rana, A. Sou, and C. Ramsdale, in *2019 IEEE International Conference on Flexible and Printable Sensors and Systems (FLEPS)*, 2019, pp. 1–3.

6 *Organic and Printed Electronics Association*, 2021, Available from: http://oe-a.org/ [last accessed May 2021].

7 G. Dou, A. S. Holmes, B. Cobb, S. Devenport, A. Jeziorska-Chapman, J. Meeth, and R. Price, in *2018 7th Electronic System-Integration Technology Conference (ESTC)*, pp. 1–6, 2018.

8 E. Cantatore, T. C. T. Geuns, G. H. Gelinck, E. van Veenendaal, A. F. A. Gruijthuijsen, L. Schrijnemakers, S. Drews, and D. M. de Leeuw, *IEEE Journal of Solid-State Circuits*, vol. 42, no. 1, pp. 84–92, 2007.

9 P. Heremans, N. Papadopoulos, A. de Jamblinne de Meux, M. Nag, S. Steudel, M. Rockelé, G. Gelinck, A. Tripathi, J. Genoe, and K. Myny, in *2016 IEEE International Electron Devices Meeting (IEDM)*, pp. 6.3.1–6.3.4, 2016.

10 T. Deuble, ParsiFal4.0 project report, Verbundprojekt ParsiFal40, Teilvorhaben HySiF: Schlussbericht Sachbericht: Förderprogramm IKT 2020 - Forschung für Innovationen des Bundesministeriums für Bildung und Forschung BMBF: Verbundprojekt Produktfähige autarke und sichere Foliensysteme für Automatisierungslösungen in Industrie 4.0 (ParsiFal40): Teilvorhaben Hybride Systeme in Folie,. Stuttgart: Institut für Mikroelektronik Stuttgart, 2019. https://www.tib.eu/de/suchen/id/TIBKAT%3A1678568309.

11 B. Albrecht, G. Alavi, M. Elsobky, S. Ferwana, U. Passlack, C. Harendt, and J. N. Burghartz, in *2018 7th Electronic System-Integration Technology Conference (ESTC)*, pp. 1–5, 2018.

12 H. Reinisch, M. Wiessflecker, S. Gruber, H. Unterassinger, G. Hofer, M. Klamminger, W. Pribyl, and G. Holweg, *IEEE Journal of Solid-State Circuits*, vol. 46, no. 12, pp. 3075–3088, 2011.

13 R. Dahiya, *Proceedings of the IEEE*, vol. 107, no. 2, pp. 247–252, 2019.

14 Y. Mahsereci, S. Saller, H. Richter, and J. N. Burghartz, *IEEE Journal of Solid-State Circuits*, vol. 51, no. 1, pp. 273–280, 2016.

15 M. Elsobky, Y. Mahsereci, Z. Yu, H. Richter, J. N. Burghartz, J. Keck, H. Klauk, and U. Zschieschang, *Electronics Letters*, vol. 54, 2018.

16 H. Bai, S. Li, J. Barreiros, Y. Tu, C. R. Pollock, and R. F. Shepherd, *Science*, vol. 370, no. 6518, pp. 848–852, 2020.

17 S. Thomas, *Nature Electronics*, vol. 3, p. 235, 2020.

18 P. Escobedo, M. Ntagios, D. Shakthivel, W. T. Navaraj, and R. Dahiya, *IEEE Transactions on Robotics*, vol. 37, no. 2, pp. 1–8, 2020.

19 N. Garcia, C. Manjakkal, and R. Dahiya, *npj Flexible Electronics*, vol. 3, 2019.

20 N. Mosteller R.D. *New England Journal of Medicine*, vol. 317 (17), p. 1098, 1987. PMID: 3657876.

21 P. Mittendorfer, E. Yoshida, and G. Cheng, *Advanced Robotics*, vol. 29, no. 1, pp. 51–67, 2015.

22 F. Bergner, E. Dean-Leon, and G. Cheng, *Sensors*, vol. 20, no. 7, 2020.

23 A. Moin, A. Zhou, A. Rahimi, A. Menon, S. Benatti, G. Alexandrov, S. Tamakloe, J. Ting, N. Yamamoto, Y. Khan, F. Burghardt, L. Benini, A. Arias, and J. M. Rabaey, *Nature Electronics*, vol. 4, pp. 1–10, 2021.

24 S. Sundaram, P. Kellnhofer, Y. Li, J.-Y. Zhu, A. Torralba, and W. Matusik, *Nature*, vol. 569, pp. 698–702, 2019.

25 H. Lee, Y. Hong, S. Baik, T. Hyeon, and D.-H. Kim, *Advanced Healthcare Materials*, vol. 7, p. 1701150, 2018.

26 J. Kim, A. S. Campbell, and J. Wang, *Talanta*, 177, pp. 163–170, 2018, Special issue dedicated to Professor Gary Christian's 80th Birthday.

27 X.-T. Sun, Y. Zhang, D.-H. Zheng, S. Yue, C.-G. Yang, and Z.-R. Xu, *Biosensors and Bioelectronics*, vol. 92, pp. 81–86, 2017.

28 S. Nakata, M. Shiomi, Y. Fujita, T. Arie, S. Akita, and K. Takei, *Nature Electronics*, 1, pp. 596–603, 2018.

29 S. Nappi, V. Mazzaracchio, L. Fiore, F. Arduini, and G. Marrocco, in *2019 IEEE International Conference on Flexible and Printable Sensors and Systems (FLEPS)*, pp. 1–3, 2019.

30 E. Scarpa, V. Mastronardi, F. Guido, L. Algieri, A. Qualtieri, R. Fiammengo, F. Rizzi, and M. Vittorio, *Scientific Reports*, vol. 10, 2020.

31 S. Konishi and A. Hirata, *Scientific Reports*, 9, pp. 15634, 2019.

32 S. Ali, J. Bae, and A. Bermak, in *2019 IEEE International Conference on Flexible and Printable Sensors and Systems (FLEPS)*, pp. 1–3, 2019.

33 P. Descent, R. Izquierdo, and C. Fayomi, in *2018 IEEE International Symposium on Circuits and Systems (ISCAS)*, pp. 1–4, 2018.

34 S. Yoon, J. Sim, and Y.-H. Cho, *Scientific Reports*, vol. 6, p. 23468, 2016.

35 U. Passlack, M. Elsobky, Y. Mahsereci, C. Scherjon, C. Harendt, and J. N. Burghartz, *IEEE Sensors Journal*, vol. 21 (23), pp. 26345–26354, 2021.

36 D. Lo Presti, C. Massaroni, J. D'Abbraccio, L. Massari, M. Caponero, U. G. Longo, D. Formica, C. M. Oddo, and E. Schena, *IEEE Sensors Journal*, vol. 19 (17), pp. 7391–7398, 2019.

37 P. Pataranutaporn, A. Jain, C. M. Johnson, P. Shah, and P. Maes, in *2019 41st Annual International Conference of the IEEE Engineering in Medicine and Biology Society (EMBC)*, pp. 3327–3332, 2019.

38 H. Liu, Y. Huang, and H. Jiang, *Proceedings of the National Academy of Sciences of the United States of America*, vol. 113, p. 201517953, 2016.

39 F. De Roose, S. Steudel, K. Myny, M. Willegems, S. Smout, M. Ameys, P. E. Malinowski, R. Gehlhaar, R. Poduval, X. Chen, J. De Smet, A. V. Quintero, H. De Smet, W. Dehaene, and J. Genoe, in *2016 IEEE International Electron Devices Meeting (IEDM)*, 2016, pp. 32.1.1–32.1.4.

40 J.-C. Chiou, S.-H. Hsu, Y.-C. Huang, G.-T. Yeh, W.-T. Liou, and C.-K. Kuei, *Sensors*, vol. 17, no. 1, 2017.

41 A. V. Quintero, P. Pérez-Merino, and H. De Smet, *Scientific Reports*, vol. 10, no. 1, pp. 2045–2322, 2020.

42 M. Gray, J. Meehan, C. Ward, S. P. Langdon, I. H. Kunkler, A. Murray, and D. Argyle, *The Veterinary Journal*, vol. 239, pp. 21–29, 2018.

43 L. Yin, C. Bozler, D. V. Harburg, F. Omenetto, and J. A. Rogers, *Applied Physics Letters*, vol. 106, p. 014105, 2015.

44 H.-P. Phan, *Micromachines*, vol. 12, p. 157, 2021.

45 Q. Yang, S. Lee, Y. Xue, Y. Yan, T. L. Liu, S.-K. Kang, Y. J. Lee, S. H. Lee, M. Seo, D. Lu, J. Koo, M. MacEwan, R. T. Yin, W. Z. Ray, Y. Huang, and J. Rogers, *Advanced Functional Materials*, vol. 30, p. 1910718, 2020.

46 E. Musk, *Journal of Medical Internet Research*, vol. 21, 2019.

47 J. J. Jun, N. A. Steinmetz, J. H. Siegle, D. J. Denman, M. Bauza, B. Barbarits, A. K. Lee, C. A. Anastassiou, A. Andrei, Ç. Aydın, M. Barbic, T. J. Blanche, V. Bonin, J. Couto, B. Dutta,

S. L. Gratiy, D. A. Gutnisky, M. Häusser, B. Karsh, P. Ledochowitsch, C. M. Lopez, C. Mitelut, S. Musa, M. Okun, M. Pachitariu, J. Putzeys, P. D. Rich, C. Rossant, W.-L. Sun, K. Svoboda, M. Carandini, K. D. Harris, C. Koch, J. O'Keefe, and T. D. Harris, *Nature*, vol. 551, no. 7679, pp. 232–236, 2017.

48 J. Chen, W. Cranton, and M. Fihn (eds.), *Flexible Displays: TFT Technology: Substrate Options and TFT Processing Strategies*. Berlin, Heidelberg: Springer-Verlag, pp. 897–932, 2012.

49 M.-UI Hassan, Ph.D. thesis, University of Stuttgart, Shaker Verlag, 2017.

50 Y. Cai, J. Shen, C.-W. Yang, Y. Wan, H.-L. Tang, A. A. Aljarb, C. Chen, J.-H. Fu, X. Wei, K.-W. Huang, Y. Han, S. J. Jonas, X. Dong, and V. Tung, *Science Advances*, vol. 6, p. 48, 2020.

51 J. R. Ganasan, *IEEE Transactions on Electronics Packaging Manufacturing*, vol. 23, no. 1, pp. 28–31, 2000.

52 Würth Elektronik Group, 2021. Available from: https://www.we-online.de/web/de/ leiterplatten/produkte/ [last accessed May 2021].

53 S.-T. Lu and W.-H. Chen, *IEEE Transactions on Advanced Packaging*, vol. 33, no. 3, pp. 702–712, 2010.

54 J. Govaerts, W. Christiaens, E. Bosman, and J. Vanfleteren, *IEEE Transactions on Advanced Packaging*, vol. 32, no. 1, pp. 77–83, 2009.

55 W. Christiaens, E. Bosman, and J. Vanfleteren, *IEEE Transactions on Components and Packaging Technologies*, vol. 33, no. 4, pp. 754–760, 2010.

56 M. Hassan, C. Schomburg, C. Harendt, E. Penteker, and J. N. Burghartz, in *2013 Eurpoean Microelectronics Packaging Conference (EMPC)*, pp. 1–6, 2013.

57 American Semiconductor Homepage, 2021, Available from: www.americansemi.com/flex-ics .html [last accessed May 2021].

58 G. Alavi, Ph.D. thesis, University of Stuttgart, 2010.

59 G. Alavi, H. Sailer, B. Albrecht, C. Harendt, and J. N. Burghartz, *IEEE Transactions on Components, Packaging and Manufacturing Technology*, vol. 8, no. 5, pp. 802–810, 2018.

60 P. Chiniwalla, R. Manepalli, K. Farnsworth, M. Boatman, B. Dusch, P. Kohl, and S. A. Bidstrup-Alen, *IEEE Transactions on Advanced Packaging*, vol. 24, no. 1, pp. 41–53, 2001.

61 U. Kraft, T. Zaki, F. Letzkus, J. N. Burghartz, E. Weber, B. Murmann, and H. Klauk, *Advanced Electronic Materials*, vol. 5, no. 2, p. 1800453, 2019.

62 U. Zschieschang and H. Klauk, *Journal of Materials Chemistry C*, vol. 7, pp. 5522–5533, 2019.

63 M. Elsobky, B. Albrecht, H. Richter, J. N. Burghartz, P. Ganter, K. Szendrei, and B. V. Lotsch, in *2017 IEEE SENSORS*, pp. 1–3, 2017.

64 G. G. Stoney, *Proceedings of the Royal Society of London Series A*, vol. 82, pp. 172–175, 1909.

65 D. Bankman, L. Yang, B. Moons, M. Verhelst, and B. Murmann, in *2018 IEEE International Solid - State Circuits Conference - (ISSCC)*, 2018, pp. 222–224.

66 K. Myny, E. van Veenendaal, G. H. Gelinck, J. Genoe, W. Dehaene, and P. Heremans, in *2011 IEEE International Solid-State Circuits Conference*, pp. 322–324, 2011.

67 T. Sterken, J. Vanfleteren, T. Torfs, M. O. de Beeck, F. Bossuyt, and C. Van Hoof, in *2011 Annual International Conference of the IEEE Engineering in Medicine and Biology Society*, pp. 6886–6889, 2011.

68 Y. Takeda, K. Hayasaka, R. Shiwaku, K. Yokosawa, T. Shiba, M. Mamada, D. Kumaki, K. Fukuda, and S. Tokito, *Scientific Reports*, vol. 6, p. 25714, 2016.

69 N. P. Papadopoulos, F. De Roose, J.-L. P. J. van der Steen, E. C. P. Smits, M. Ameys, W. Dehaene, J. Genoe, and K. Myny, *IEEE Journal of Solid-State Circuits*, vol. 53, no. 8, pp. 2263–2272, 2018.

70 N. Papadopoulos, S. Steudel, F. De Roose, D. M. Eigabry, A. J. Kronemeijer, J. Genoe, W. Dehaene, and K. Myny, in ESSCIRC 2018 - IEEE 44th European Solid State Circuits Conference (ESSCIRC), pp. 194–197, 2018.

71 T. Meister, K. Ishida, S. Knobelspies, G. Cantarella, N. Münzenrieder, G. Tröster, C. Carta, and F. Ellinger, *IEEE Journal of Solid-State Circuits*, vol. 54, no. 8, pp. 2195–2206, 2019.

72 T. Zaki, F. Ante, U. Zschieschang, J. Butschke, F. Letzkus, H. Richter, H. Klauk, and J. N. Burghartz, in 2011 IEEE International Solid-State Circuits Conference, pp. 324–325, 2011.

73 H. Marien, M. S. J. Steyaert, E. van Veenendaal, and P. Heremans, *IEEE Journal of Solid-State Circuits*, vol. 47, no. 7, pp. 1712–1720, 2012.

74 S. Elsaegh, C. Veit, U. Zschieschang, M. Amayreh, F. Letzkus, H. Sailer, M. Jurisch, J. N. Burghartz, U. Würfel, H. Klauk, H. Zappe, and Y. Manoli, *IEEE Journal of Solid-State Circuits*, vol. 55, no. 9, pp. 2553–2566, 2020.

75 G. H. Ibrahim, U. Zschieschang, H. Klauk, and L. Reindl, *IEEE Transactions on Electron Devices*, vol. 67, no. 6, pp. 2365–2371, 2020.

76 M. Seifaei, D. Schillinger, M. Kuhl, M. Keller, U. Zschieschang, H. Klauk, and Y. Manoli, *IEEE Solid-State Circuits Letters*, vol. 2, no. 10, pp. 219–222, 2019.

77 H. Marien, M. S. J. Steyaert, E. van Veenendaal, and P. Heremans, *IEEE Journal of Solid-State Circuits*, vol. 46, no. 1, pp. 276–284, 2011.

78 C. Garripoli, J.-L. P. J. van der Steen, E. Smits, G. H. Gelinck, A. H. M. Van Roermund, and E. Cantatore, in 2017 IEEE International Solid-State Circuits Conference (ISSCC), pp. 260–261, 2017.

79 Y. Afsar, T. Moy, N. Brady, S. Wagner, J. C. Sturm, and N. Verma, *IEEE Journal of Solid-State Circuits*, vol. 53, no. 1, pp. 297–308, 2018.

80 S. Ahmed, F. A. Tahir, A. Shamim, and H. M. Cheema, *IEEE Antennas and Wireless Propagation Letters*, vol. 14, pp. 1802–1805, 2015.

81 M. Tang, T. Shi, and R. W. Ziolkowski, *IEEE Transactions on Antennas and Propagation*, vol. 63, no. 12, pp. 5343–5350, 2015.

82 A. Kiourti and J. L. Volakis, *IEEE Antennas and Wireless Propagation Letters*, vol. 13, pp. 1381–1384, 2014.

83 S. Bandyopadhyay, P. P. Mercier, A. C. Lysaght, K. M. Stankovic, and A. P. Chandrakasan, in 2014 IEEE International Solid-State Circuits Conference Digest of Technical Papers (ISSCC), pp. 396–397, 2014.

84 S. Ha, C. Kim, J. Park, S. Joshi, and G. Cauwenberghs, *IEEE Journal of Solid-State Circuits*, vol. 51, no. 11, pp. 2664–2678, 2016.

85 M. Elsobky, G. Alavi, B. Albrecht, T. Deuble, C. Harendt, H. Richter, Z. Yu, and J. N. Burghartz, *Proceedings*, vol. 2, no. 13, 2018.

86 W. Deng, B. Zhou, M. Xie, L. Wang, L. Dong, Q. Huang, and Z. Yi, in 2019 IEEE SENSORS, pp. 1–4, 2019.

87 Y. Chiu, Y. Chen, C. Hsieh, and H. Hong, in 2019 IEEE SENSORS, pp. 1–4, 2019.

88 M. A. Carvajal, P. Escobedo, A. Martínez-Olmos, and A. J. Palma, *IEEE Sensors Journal*, vol. 20, no. 2, pp. 885–891, 2020.

89 M. Xie, L. Wang, L. Dong, W. Deng, and Q. Huang, *IEEE Sensors Journal*, vol. 19, no. 12, pp. 4717–4725, 2019.

90 M. R. R. Khan, T. K. An, and H. S. Lee, *IEEE Sensors Journal*, vol. 21, no. 2, pp. 2184–2193, 2021.

91 T. Deuble, et al., Produktfähige autarke und sichere Foliensysteme für Automatisierungslö-sungen in Industrie 4.0, 2021, Available from: http://www.parsifal40.de/ [last accessed March 2021].

92 S. Özbek, M. Grözing, G. Alavi, J. N. Burghartz, and M. Berroth, in 2018 48th European Microwave Conference (EuMC), pp. 1241–1244, 2018.

93 G. Alavi, S. özbek, M. Rasteh, M. Grözing, M. Berroth, J. Hesselbarth, and J. N. Burghartz, in 2018 7th Electronic System-Integration Technology Conference (ESTC), pp. 1–5, 2018.

94 G. Di Massa, S. Costanzo, A. Borgia, F. Venneri, and I. Venneri, in 2010 Conference Proceedings ICECom, 20th International Conference on Applied Electromagnetics and Communications, pp. 1–4, 2010.

95 B. A. Kuzubasoglu and S. K. Bahadir, *Sensors and Actuators A: Physical* vol. 315, p. 112282, 2020.

96 S. Yi, M. Chunsheng, C. Jing, H. Wu, W. Lio, Y. Peng, Z. Luo, L. Li, Y. Tan, O.M. Omisore, Z. Zhu, L. Wang, H. Li, *Nanoscale Research Letters*, vol. 15, 2020.

97 Q. Li, L.-N. Zhang, X. Tao, and X. Ding, *Advanced Healthcare Materials*, vol. 6, 2017.

98 R. Liang, A. Luo, Z. Zhang, Z. Li, C. Han, and W. Wu, *Sensors*, vol. 20, no. 19, 2020.

99 H. He, Y. Fu, S. Liu, J. Cui, and W. Xu, Research Progress and Application of Flexible Humidity Sensors for Smart Packaging: A Review, pp. 429–435, 2020.

100 Z. Chen and C. Lu, *Sensor Letters*, vol. 3, 2005.

101 D. Barmpakos and G. Kaltsas, *Sensors*, vol. 21, no. 3, 2021.

102 S. Stassi, C. Valentina, G. Canavese, and C. Pirri, *Sensors (Basel, Switzerland)*, vol. 14, pp. 5296–5332, 2014.

103 M. Cheng, G. Zhu, F. Zhang, W.-l. Tang, S. Jianping, J.-q. Yang, and L. Zhu, *Journal of Advanced Research*, vol. 26, pp. 53–68, 2020.

104 K. S. Kumar and P.-Y. Chen, *Research*, vol. 2019, 2019.

105 T. Widlund, S. Yang, Y.-Y. Hsu, and N. Lu, *International Journal of Solids and Structures*, vol. 51, pp. 4026–4037, 2014.

106 H. Hocheng and C.-M. Chen, *Sensors (Basel, Switzerland)*, vol. 14, pp. 11855–11877, 2014.

107 T. T. Nguyen, S. Jeon, D.-I. Kim, T. Q. Trung, M. Jang, B.-U. Hwang, K.-E. Byun, J. Bae, E. Lee, J. Tok, Z. Bao, N.-E. Lee, and J.-J. Park, *Advanced Materials (Deerfield Beach, Fla.)*, vol. 26, pp. 796–804, 2014.

108 Q. Hua, J. Sun, H. Liu, R. Bao, R. Yu, C. Pan, and Z. Wang, *Nature Communications*, vol. 9, 2018.

109 K. Kim, M. Jung, B. Kim, J. Kim, K. Shin, O.-S. Kwon, and S. Jeon, *Nano Energy*, vol. 41, 2017.

110 M.-U. Hassan, H. Rempp, T. Hoang, H. Richter, N. Wacker, and J. N. Burghartz, in 2009 IEEE International Electron Devices Meeting (IEDM), pp. 1–4, 2009.

111 M. Poliks, J. Turner, K. Ghose, Z. Jin, M. Garg, Q. Gui, A. Arias, Y. Kahn, M. Schadt, and F. Egitto, in 2016 IEEE 66th Electronic Components and Technology Conference (ECTC), pp. 1623–1631, 2016.

112 A. Davila-Frias, O. P. Yadav, and V. Marinov, *IEEE Transactions on Components, Packaging and Manufacturing Technology*, vol. 10, no. 11, pp. 1902–1912, 2020.

113 H. Husstedt, U. Ausserlechner, and M. Kaltenbacher, *IEEE Sensors Journal*, vol. 11, no. 11, pp. 2993–3000, 2011.

11

Optical Detectors

Lis Nanver and Tihomir Knežević

Faculty of Electrical Engineering Mathematics & Computer Science, MESA+ Institute of Technology, University of Twente, Enschede, The Netherlands

11.1 Introduction

In the drive toward More than Moore (MtM) micro/nanoelectronic systems that interact with the analog world, there is an increasing demand for nondigital functionality. This has pushed the R&D of optical detectors for a broad spectral range that can be integrated with electronic readout circuits as well as circuits for signal amplification and speed enhancement. The electronics world itself is also embracing the advantages of light, turning more and more to solutions that adapt features from photonics integrated circuits (PICs) [1]. Today, *optical detectors* form part of *optical sensors* used in a wide range of technologies including cameras, fiber optics, laser, remote sensing, data storage, and optical communication systems. More specific applications include monitoring/diagnostic systems in medicine, flame detection, radiation detection, biosensors, as well as specialized industrial and military applications.

An impressive growth in photodetector application in all abovementioned areas has long been interlinked with the development of CMOS technology which now offers inexpensive, yet powerful, analog and digital signal processing capabilities [2]. In addition, silicon is in itself a potent photosensitive semiconductor. Si photodiodes are readily made with good sensitivity in the visual range, and imaging sensors fabricated in CMOS are found in almost every digital camera, medical imaging devices, and security systems to name just a few of a multitude of examples. A limitation of Si is that it only absorbs photons with an energy above the bandgap energy, $E_{gap} = 1.12\,eV$, that gives a cutoff wavelength above which Si becomes transparent, $\lambda_{gap} = hc/E_{gap} \approx 1100\,nm$, where h is the Planck constant and c is the speed of light. For detection of photons with wavelengths above 1100 nm, other semiconductor materials with smaller bandgaps are then employed such as *germanium* (Ge) with a cutoff at 1600 nm [3]. In Figure 11.1a, a number of commonly used semiconductors are compared with respect to their ability to absorb light: the *absorption length*, which is the distance the light penetrates into the material before the intensity has dropped by $1/e$ (equivalent to absorption of 63% of the photons), is plotted as a function of wavelength.

Besides the basic absorption properties of the photosensitive material, when incorporated into an optical detector, many aspects of the total device design become important for the applicability. A number of performance metrics, also called figures of merit (FOMs), are used to characterize and compare detectors. Due to the diversity of detector types and their long development history, different ways of defining the FOMs have subsisted. This list gives a selection of definitions commonly used for photodetectors, and they will be applied in this chapter:

Advances in Semiconductor Technologies: Selected Topics Beyond Conventional CMOS, First Edition. Edited by An Chen.
© 2023 The Institute of Electrical and Electronics Engineers, Inc. Published 2023 by John Wiley & Sons, Inc.

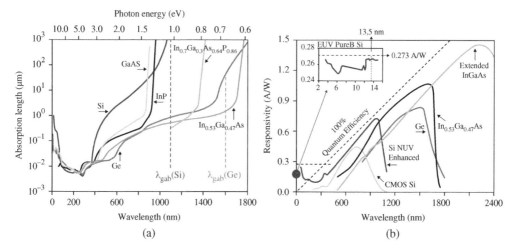

Figure 11.1 (a) Absorption length as a function of wavelength/photon energy for different semiconductor materials commonly used in optical detectors. The cutoff wavelengths (λ_{gap}) for Si and Ge are also indicated. (b) Typical spectral response of commercially available photodiode detectors made in Si, Ge, or InGaAs. The 100% quantum efficiency ($\eta = 1$) line indicates the ideal responsivity in the long-wavelength region, where one electron per absorbed photon is created. In the short-wavelength region, the horizontal dashed line indicates the ideal responsivity in Si, where 3.66 eV is needed to create one electron–hole pair. This means that the responsivity is constant with a value ≈ 0.273 A/W. The inset shows the almost ideal responsivity of PureB photodiodes developed for detection at the extreme-ultraviolet (EUV) wavelength of 13.5 nm. Source: Adapted from Nanver et al. [4].

Spectral response: The response of a photodetector as a function of photon frequency.

Quantum efficiency: η, is a basic property of the material used to quantify the processes that absorb and convert light to other measurable quantities. In semiconductors, absorbed photons generate charge carriers, and η is defined as the number of primary electron–hole pairs generated, divided by the number of absorbed photons.

External quantum efficiency: EQE, depends on η but takes into account the limitation of specific detector implementations. EQE is defined as the number of photogenerated charge carriers (coming out as photocurrent from the device) measured across the detector divided by the number of photons incident on the device.

Responsivity: R (A/W), is the photocurrent per unit optical power incident on a photodetector.

Noise-equivalent power: *NEP* (W), is the input power to a detector which produces the same signal output power as the internal noise of the device.

Detectivity: D (1/W), is the inverse of the noise equivalent power.

Specific detectivity: D^* (cm·Hz$^{1/2}$/W = Jones), is the detectivity normalized to a unit detector area and detection bandwidth.

Dark current: I_{dark} (A), is the current flowing through a photodetector in the absence of light.

Gain: G, is the output current of a photodetector divided by the current directly produced by the photons incident on the detector, i.e. produced in a biasing region where there is no built-in current gain.

Response time: (s), is the time needed for a photodetector to go from 10% to 90% of final output.

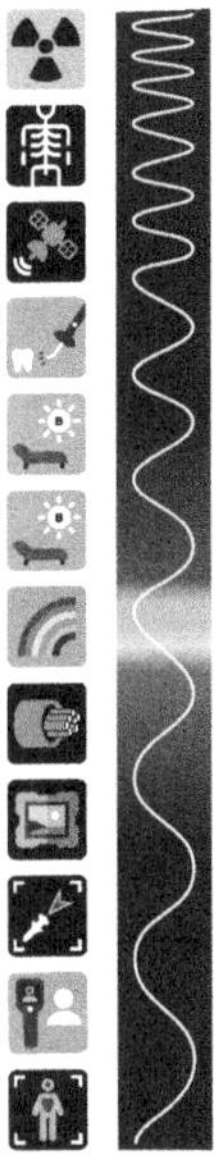

Abbr.	Type of electromagnetic radiation	Wavelength range (nm)
GR	Gamma radiation	< 0.01
XRAY	X-ray radiation	0.01–10
VUV	Vacuum ultraviolet	10–200
FUV	Far ultraviolet	200–280
MUV	Medium ultraviolet	280–315
NUV	Near ultraviolet	315–400
RGB	Visual	400–700
NIR	Near infrared	700–1100
SWIR	Short wavelength infrared	1100–3000
MWIR	Medium wavelength infrared	3000–6000
LWIR	Long wavelength infrared	6000–15000
FIR	Far infrared	$15000–10^6$

Figure 11.2 List of the terms used in this chapter for the different wavelength ranges of electromagnetic radiation that can be monitored by optical detectors.

Typical values of the spectral response of commercially available detectors made in Si, Ge, or InGaAs are shown in Figure 11.1b. Above about 400 nm, the responsivity ideally increases linearly with wavelength because one electron–hole pair is created per absorbed photon. In Si, below this wavelength, the number of electrons created increases linearly with the photon energy and the ideal responsivity becomes constant. In an actual detector design, several factors can lower the responsivity such as reflection, absorption within the nonactive layers of the device, and carrier recombination at (surface) defects. In addition, the cost and manufacturability of the detector have a large influence on the final choice of technology. This has led to a great variety of detector types, and also the different application fields have often developed a specific classification terminology. In this chapter, we use the division of the electromagnetic spectrum given in Figure 11.2.

11.2 Si Photodiodes Designed in CMOS

Among the semiconductor materials suitable for detector fabrication, Si has a uniquely prominent position. The Si integrated-circuit (IC) industry has over the last 60 years solved a multitude of material and fabrication problems and developed equipment that allows almost atomic precision in both vertical and horizontal device layer positioning. Although other group IV and also compound semiconductors have many inherent advantages with respect to detector realization, such as significantly lower pair-creation energies or very high material strength, the material processing is still under development to solve difficulties with obtaining low-defect material growth, effective surface passivation, localized doping methods, and low-ohmic contacts that are both stable and uniform.

Si has also been blessed with a spectral responsivity that makes photodiodes fabricated in standard CMOS very suitable for visible light detection as well as for part of the near-ultraviolet

(NUV) and near-infrared (NIR), up to 1100 nm ranges. Pivotal for the impressive proliferation of CMOS imagers in sectors such as mobile communication, has been the fact that the photodiodes can be integrated on-chip with the read-out electronics [2]. The development was set off in the late 1960s by the invention of *charge-coupled devices* (CCDs) that used gate-controlled potential wells to collect and read-out the photo-generated carriers. CCDs developed into a major digital imaging technology, but in the 1990s, *active pixel CMOS sensors* combining imaging and processing functions in one circuit, appeared as an alternative imaging solution that has several advantages. CCDs are commendable for their very high-quality imaging in large-area, high-resolution cameras with low noise, excellent uniformity and linearity, and near-theoretical sensitivity. For these reasons, they are still often preferred for scientific instrumentation and astronomy imaging. However, for commercial applications, CMOS imagers offer faster response times and lower power requirements than CCDs, and the required high-charge transfer efficiency of CCDs is an extra critical fabrication concern that cannot be met in Si (CMOS) foundries not specialized in this respect.

The design of photodiode imager pixels in CMOS has seen intensive development that continues to this day. Not only can the basic p–n junction photodiode be operated in different ways, with and without internal gain, but read-out electronics are also of critical importance for the final sensitivity and speed of the imager. In addition to the basic detector FOMs discussed in the introduction, there are several other important parameters that must be considered for optimal pixel imager performance. The photodiode *capacitance* and *series/load resistance* both influence the response time that is decisive for the imager speed. For some applications, the *nonlinearity* of the photodetector can be a limitation, as for example, for scientific imaging systems where precision is achieved through linear response to the incident light. The *dynamic range* of the photodetector determines how low a light level can be distinguished despite the detector noise, and also sets the upper limit of detection before the light-generated carriers overload the system and are no longer accurately monitored. Minimizing the level of read-out noise and dark current is critical for maintaining a high dynamic range.

The Si p–n photodiodes are integrated into CMOS with light-entrance windows covered with oxide passivation/isolation layers and interconnect metal is avoided. Some of the most basic structures are illustrated in Figure 11.3. The light generates electron–hole pairs that can give a measurable voltage or current. A current output is often preferred for good linearity, offset, and bandwidth, and can be converted to voltage in the electronic read-out circuit. The photodiode can be operated without biasing or with a reverse bias that may be used to widen the junction depletion

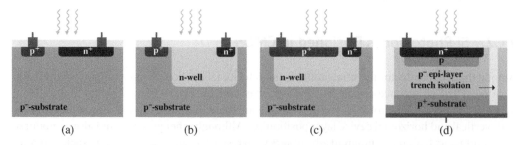

Figure 11.3 Schematic cross sections of photodiode designs in standard CMOS where the photosensitive depletion region is formed (a) at an n⁺/p-substrate junction, (b) at an n-well/p-substrate junction for a wider depletion, (c) with a pinned n-well/p-substrate photodiode that gives an increased depletion at the p⁺/n-well junction and also moves the depletion away from the defected Si surface, and (d) in an p–i–n design with an extra p-region adjacent to the n⁺-region to set the avalanche field for operation as an avalanche photodiode (APD).

and thus reduce capacitance. These two modes of operation are referred to as "photoconductive" (biased) and "photovoltaic" (unbiased) modes. These terms should not be confused with the classification of photodetectors in the two types: *photovoltaic detectors* which make use of diode junctions as the ones being discussed here, and *photoconductive detectors* that basically are light-dependent resistors as will be discussed in Section 11.4.

The photo-generated signal is mainly amplified via the electronic circuit but lower noise and power consumption can in some cases be achieved if the photodiode is operated with internal gain. This is implemented for detection of very low light levels, as indicated in Figure 11.4, where the biasing applied when going from linear mode to avalanche mode operation is indicated. For these *avalanche photodiodes* (APDs), a reverse biasing is applied to create a high electric field that is strong enough to accelerate photo-generated carriers and give them the energy necessary for generating secondary carriers. In this *avalanche multiplication region*, as the voltage is increased, the multiplication factor M can go from 1 to about 1000. As shown in Figure 11.3d, the p-doping just under the anode n-doping is increased. This locally increases the electric field strength, thus increasing avalanche multiplication while maintaining a low dark current at the perimeter and in the wide lightly p-doped absorber region. The detector still has a linear response to the light if it is operated with a gain in the range of 1–100. Therefore, less signal amplification is needed in the read-out electronics which means less electronic noise is introduced.

The high gain of APDs marks a strong increase in responsivity of the photodiode but at the cost of reliability due to the statistical and defect-sensitive nature of the avalanching process. The gain is strongly dependent on the reverse voltage and may vary significantly even between devices processed adjacently on the same wafer. The quantum efficiency in APD-mode may also be lower than in photodiode-mode due to increasing recombination as the reverse voltage increases. The reliability of APDs can be improved by careful design of the absorbing and multiplication regions, for example by placing the high electric-field region away from the oxide interfaces, as is shown in Figure 11.3c,d. The exact APD design will also impact the detection bandwidth that can be high, but there is an inherent trade-off between bandwidth and gain that must be taken into account.

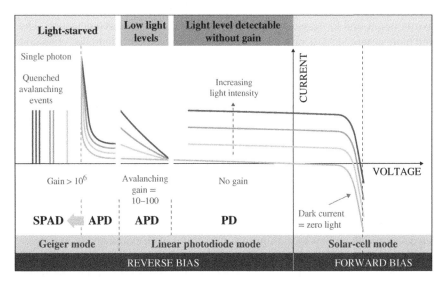

Figure 11.4 Modes of operation of a generic p–n-junction Si photodiode seen through typical optoelectrical current–voltage characteristics.

The avalanche process may also deteriorate the response time but due to the enhanced sensitivity, operation with a smaller shunt resistor may compensate this.

APDs with gain above 1000 and operated in so-called *Geiger mode* are nowadays routinely made in CMOS, and arrays with a large active area, sometimes made by stitching several chips together, are marketed as *silicon photomultipliers* (SiPMs) that have replaced photomultiplier tubes for many applications. SiPMs are cheaper and much more compact and robust. The photomultiplier tubes do, however, mainly have lower amplification noise despite a lower quantum efficiency. The noise of APDs is expressed in the *excess noise factor*, F, and takes into account the quantum and amplification noise that are absent when operating the diode in photodiode-mode. This factor increases with increasing internal amplification, i.e. increasing reverse voltage, so settings are often chosen to minimize the overall noise by accepting about equal noise from the multiplication process and the electronic amplifier.

When APDs are operated in Geiger mode, the electronics are designed to stop the avalanche events by a passive or active quenching mechanism. Near breakdown, such avalanching would otherwise cause current runaway. Photodiodes can be developed to have extremely low-defect densities near and around the photosensitive depletion region, so that it becomes possible to bias the diode past the breakdown voltage without actually causing breakdown in a brief period of time. In this case, a single absorbed photon generating an electron–hole pair can give a measurable avalanche event, thus enabling single photon counting. These devices are called *single-photon avalanche diodes*, SPADs. Carrier generation events can also be caused by thermal fluctuations without the presence of light, and they will be more numerous the more imperfections are present in the Si crystal. These are characterized by a *dark count rate* (*DCR*) that is measured as the frequency of dark counts. It can be as low as a few hertz in Si devices.

The recovery time (*dead time*) after the quenching of an event limits the bandwidth of SPADs making it difficult to surpass the MHz range, whereas APDs in linear mode reach GHz bandwidth operation [5]. It is also a drawback for some applications, among others quantum optics for quantum information technology such as cryptography, that not every incident photon is able to trigger an avalanche. Much research is therefore focused on increasing the quantum efficiency and speed of SPADs. Today, large arrays of SPADs integrated in CMOS are available, e.g. for imaging in light-starved environments, and for single-photon 3D imaging via time-resolved detection. The development of even more potent SPAD arrays is being pushed by areas such as light detection and ranging (LIDAR) and medical imaging.

Many detector developments are aimed at increasing the sensitivity of optical detectors by increasing the fraction of light that actually reaches the photosensitive diode region. In Si, the reflection can be as high as 30% in the visual range. Methods to mitigate this include the application of anti-reflection coatings, surface roughening or periodic nanopatterning of the Si surface, Si surface treatments to create "black silicon," or deposition of nanospheres [6].

An important FOM for imaging arrays is also the *fill factor*, i.e. the fraction of photosensitive area with respect to the total surface area. One solution that has won popularity over the last decade is the use of *backside illumination* (BSI) as illustrated in Figure 11.5. The fill-factor is significantly increased by moving the light-entrance window to the back of the wafer, while the metal wiring of the photodiode and electronics remains on the front [8]. To expose the photosensitive layer, processes have been developed that allow local thinning of the wafer in an extremely uniform, precise manner without introducing damage that still allows reliable wafer handling. In BSI imagers, *optical crosstalk* due to light scattering to neighboring pixels can become a more serious problem. Hence, development of pixel isolation trenches with reflecting metal is also undertaken. With oxide isolation as shown in Figure 11.3d, reflections will give some reduction of optical crosstalk.

(a) CMOS imager

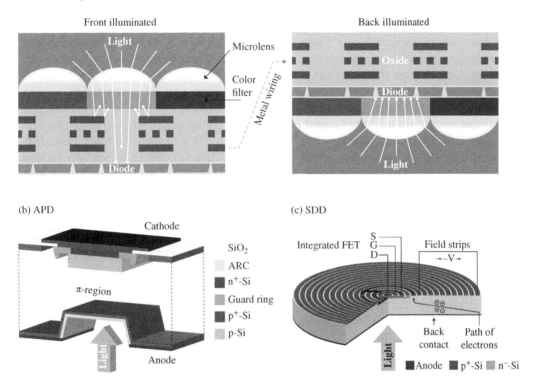

Figure 11.5 Examples of silicon photodiode detectors that make use of backside illumination (BSI). (a) CMOS imager where the fill-factor is increased by BSI because the metal wiring contacting the photodiodes and associated electronics does not block the light as with front-illuminaton. (b) An avalanche-photodiode (APD) high-sensitivity detector with an almost intrinsic fully-depleted p-region. A micro-machined cavity is etched to reduce the required reverse bias. (c) A silicon drift detector (SDD) with fully depleted Si substrate where an electric field is set by the p^+ rings, sweeping the light-generated electrons to the anode. This structures makes high sensitivity possible as well as energy and position-dependent detection. Source: Modified from Metzger et al. [7].

11.3 Ultraviolet Photodetectors

The development of BSI CMOS imagers has also been strongly motivated by applications in particularly medical and scientific instrumentation fields that require detection below 450 nm. Below this wavelength, the spectral sensitivity of CMOS imaging sensors significantly drops since the electrodes and oxide isolation layers absorb NUV wavelengths. Phosphorus scintillation layers are also used to increase short wavelength sensitivity, and today there are CMOS and CCD imagers that allow deep ultraviolet (UV) imaging down to 200 nm.

Wavelengths in the 10–200 nm range are strongly absorbed in air so detector operation in vacuum is necessary, hence, the term *vacuum-ultraviolet light* (VUV light). For almost a century, *photomultiplier tubes* (PMTs) have been used for sensitive detection in this wavelength range and for the even shorter wavelengths of X-ray and gamma radiation. This changed in the course of the last two decades when the need to replace the bulky and expensive PMTs with compact Si-based detectors was spurred by the electronics industry that became dependent on development of lithography systems using, in first instance, 193 nm light for deep ultraviolet (DUV) lithography, and later 13.5 nm soft X-rays for extreme ultraviolet (EUV) lithography.

The very low penetration depth of VUV radiation in Si becomes especially critical at 193 nm where the absorption length is only 4.5 nm. For p–n diodes, this means that the photosensitive depletion region needs to be placed very close to the Si surface. This, combined with the high-energy of DUV photons, exposed short-comings in the robustness and stability of conventional Si detector configurations that, for detection of the even higher energy 13.5-nm EUV photons, became a serious bottle-neck. A solution for both problems was found in PureB technology where p–n photodiodes have anode regions of robust pure boron thin-films deposited at temperatures that restrict the B-dopant diffusion into the Si [4]. In contrast to other methods for fabricating ultra-shallow p^+-type regions such as low-energy B implants, B-doped Si epitaxy, or Schottky barriers made with thin metal-silicide layers, PureB combines low dark currents and high responsivity with high robustness/stability, as well as front-end and back-end CMOS-compatible process modules. The latter has also stimulated application as a p-type coating for the light-entrance surface of BSI CMOS imagers for the NUV range. As seen in the inset of Figure 11.1b, PureB Si detectors achieve near-ideal responsivity even below the EUV wavelength.

In addition to lithography systems, PureB is used commercially in VUV instrumentation and electron microscopes for low-energy electron detection that use PureB-only light entrance windows. More long-standing applications of VUV light are found in space exploration, solar radiation monitoring, and high-energy physics. The high sensitivity and stability of PMTs were instrumental in advancing the studies in all these areas, but nowadays low-cost semiconductor devices with performance approaching that of PMTs have extended interest to areas such as spectroscopy for biomedical, chemistry, and physics research. In addition, many of the analysis techniques that are used to monitor the environment or (bio)medical conditions, profit from the absolute accuracy of radiant power detection that can be achieved with high-performance VUV Si photodetectors [9].

Good quality c-Si is in itself very robust with respect to high-dose exposure to VUV, X-ray, and γ radiation, but interfaces with the isolation/passivation layers, other than, for example PureB layers, can be very vulnerable, so protection of these interfaces must be included in the detector design. *Wide-bandgap semiconductors* are, in the case of ideal materials, more radiation hard than Si and have lower dark currents even at high temperatures. Several of these have also been adopted for detection in the VUV/NUV wavelength range which, however, is to a large degree because they are *solar-blind* materials. Even though there are excellent Si detectors spanning the whole VUV range, their good sensitivity in the NUV-visual range becomes a drawback when natural background light is an unwanted disturbance. Si detectors can be rendered solar blind to a reasonable degree by reducing the thickness of the active Si layer and/or by capping with wavelength-selective filter/absorber layers that reduce responsivity. The wide-bandgap materials may also unwantedly respond to wavelengths above their cut-off frequency if the material is defect-rich or if high optical intensities lead to two-photon absorption.

For example, SiC with bandgaps just above 3 eV, is used commercially for detectors blocking wavelengths above about 350–400 nm, called *visible-blind* detectors. SiC detectors are particularly attractive for their exceptional stability under high-dose irradiation and high-temperature operation to above 200 °C combined with very low dark currents. Based on the bandgap, GaN and ZnO are also binary compound semiconductors that could be suitable for developing visible-blind UV detectors [10]. The cutoff wavelength of these two materials can also be tuned to meet the criteria of solar-blind photodetectors with a cutoff below 280 nm, by forming ternary alloys such as $Al_xGa_{1-x}N$ and $Mg_xZn_{1-x}O$ to tune the bandgap up to about 4.4 eV, some of which are suitable for making *focal plane arrays* for solar-blind cameras. For even lower cutoff wavelength near 200 nm (6.2 eV), ultrawide bandgap materials such as AlN, BN, and *diamond* are under development.

Recent years have also seen a rapid development of β-Ga$_2$O$_3$ with cutoff wavelengths around 255 nm [11]. This material offers high carrier mobility, low cost, and high thermal and chemical stability, but like the other compound materials, high defect densities are a problem at transitions between regions with different material content. Particularly, the fabrication of reliable and stable p-type regions is a common issue. These aspects lead to high dark currents and long response times, and high-sensitivity devices like APDs cannot easily be realized.

All these wide-bandgap semiconductors have direct bandgaps so both light sources and detectors working at VUV wavelengths are becoming increasingly available. This has led to developments in solar-blind optical communication that, due to the strong scattering and absorption of the short-wavelength light, can only be realized over short distances which is attractive for secure information transmission.

For shorter wavelengths than soft X-rays, the absorption length in Si becomes longer and longer. *Hard X-rays* with energies up to about 50 keV can be detected directly in *p–i–n photodiodes* where a large part or even the whole thickness of a wafer, normally in the range 300 μm to 1 mm, is depleted. This is illustrated in Figure 11.5b for an APD design that uses local thinning of a high-ohmic Si substrate to achieve full depletion. Special designs such as the *silicon drift detector* (SDD), illustrated in Figure 11.5c, are often used to increase the detector sensitivity as well as to obtain energy and/or position dependent information [12]. These types of detectors are also widely used for the detection of electrons with energies in the range of 200 eV–30 keV. To reduce the noise level of these high-sensitivity photodetectors, the operation temperature is often reduced because the dark current is halved for every 50–100 °C decrease in temperature.

For even shorter X-ray wavelengths, the photons pass through the Si without causing detectable events. *Scintillation layers* are then added to the surface of the photodetectors to convert the incident photon energy into many photons with low energies in the visual spectrum of 2–5 eV range, depending on the type of scintillator. The development of scintillators was mainly driven by the needs of high-energy physics experimentation, originally using PMTs, but increasingly the research is motivated by medical applications such as Positron Emission Tomography (PET) and Single Photon Emission Computed Tomography (SPECT) [13].

11.4   Infrared Optical Detectors

For the NIR wavelength range up to almost 1100 nm, silicon photodiodes function well if the active depletion region is wide enough to absorb the light that starts to penetrate hundreds of micron into the Si. Alternatively, semiconductor materials with a lower bandgap energy and higher absorption coefficients are applied. For short-wavelength infrared (SWIR) wavelengths up to about 1.7 μm, *germanium* (Ge) and *indium gallium arsenide* (InGaAs) are playing prominent roles in MtM developments because monolithic integration in CMOS by epitaxial growth of these materials on the Si substrate, has seen significant progress over the last decade, not the least due to the pursuit of applications in advanced transistor nodes [1, 3, 14, 15]. Development of SWIR detectors was in the past sparked by military detector development, while recent work also focuses on areas such as optical telecommunication systems, quantum communications, LIDAR, 3D imaging, and space imaging and communication. Many of these applications are pushing performance limits with increasing demands on the epitaxial techniques used to grow low-defect crystalline detector materials. This is necessary for minimizing the dark currents that are already much higher than in Si due to the smaller bandgaps.

For longer wavelengths into the medium-wavelength infrared (MWIR) and FIR regions, a large variety of device types and material systems are currently under investigation, many of which

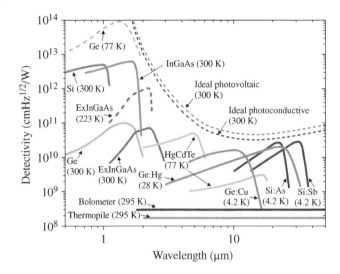

Figure 11.6 Typical values of the specific detectivity for commercially available photodetectors and thermal detectors as a function of wavelength. The reduction of the dark current with decreasing temperature entails an increase in detectivity as can be seen for the Ge and InGaAs photodiodes.

lend themselves to nonepitaxial integration on Si substrates using processes that dispense with the need to form diode junctions [16]. These infrared (IR) detectors are based on the conversion of photon or thermal energy, where the former can be grouped into photodiode and photoconductor detectors. The basic physical mechanism by which they function and the efficiency of the associated detector design are very different. To nevertheless make comparison possible, specific detectivity is an important figure of merit. To enable a comparison of all these different detectors, their performance is often given by the specific detectivity D^*, which in Figure 11.6 is presented for several commonly available detectors types. Performance of ideal infrared detectors is limited by background radiation depending on the detector type, photoconductive or photovoltaic, as shown in Figure 11.6. In background-limited infrared photodetector (BLIP) mode, quantum efficiency is the only detector parameter determining a detector's performance.

11.4.1 Detectors for Photonic Integrated Circuits

The last decade has seen an enormous increase in research directed toward the monolithic integration of photonic devices on a silicon platform, the attraction being the cointegration of photonics with transistor-based electronics. In communication technology, the main motivation is the bottleneck in data exchange rates caused by the limitations of the metal interconnects for data rates beyond 10 Gbit/s. Nowadays, optical interconnects operating in the low-loss window (1.3–1.5 μm) of silica fibers have replaced metal interconnects. However, since silicon is transparent at those wavelengths, it cannot be used at the receiver frontend to convert the optical signal back to the electrical domain, which has pushed the development of photodiode detectors fabricated in either Ge or InGaAs. In addition, many other fields such as spectroscopy for scientific research and medical diagnostics, are positioned to benefit from the development of silicon PICs working in a broader spectrum of wavelengths.

Of all the semiconductors suitable for SWIR detection, Ge is the one that has proven most straightforward to integrate on Si. The realization of sufficiently defect-free crystalline Ge layers went through many years of problematic development due to the lattice mismatch between Ge and Si [14]. Chemical-vapor deposition (CVD) was and still is the main technique for epitaxial Ge film growth on Si, where initial methods used micron-thick SiGe transition (buffer) layers, or Ge-overgrowth of small windows combined with chemical–mechanical polishing (CMP) to obtain a flat surface. In the last decade, there has been a strong development of relatively low-cost methods, where windows are directly filled with buffer-free low-defect Ge islands by first depositing a low-temperature Ge seed layer followed by Ge deposition and annealing at higher temperatures. While CVD enjoys a wide industrialization, research toward even cheaper methods for Ge film growth on Si are receiving attention. They often involve a reduction in processing temperatures and include techniques such as Ge condensation and rapid melting/regrowth of sputtered Ge layers. These techniques also enhance the compatibility with standard CMOS technology.

Besides the growth of low-defect Ge regions, the formation and contacting of the p and n regions of the photodiode are also much more problematic than in Si. They can be grown as layers in Ge CVD films which are then mesa-etched for isolation and contacting purposes. Similar to Si doping, implantation of impurities is also widely used in Ge. However, difficulties with the removal of etch/implantation damage means that these aspects are still the subject of intensive research. Nevertheless, high-performance Ge detectors integrated on Si have been demonstrated for operation in linear and APD mode. For the latter, the Ge island is often used as the absorbing layer, while the underlying Si is the multiplication layer in specially designed separate absorption and multiplication (SAM) structures. SAM structures suitable for operation as SPADs have also been presented.

In Figure 11.7, examples are shown of how, in a Si-based photonic circuit, the light from a wave-guide can be coupled to a Ge photodiode integrated vertically on the Si. In these examples, either bulk Si or silicon-on-insulator (SOI) substrates are used to initiate the crystalline Ge growth. The waveguide is either fabricated in the SOI c-Si or deposited along with the dielectric isolation layers as, for example amorphous Si or silicon nitride. As a very low-cost, low-performance solution for detection, in some integrated electronic/photonic circuits, it suffices to fabricate a photoresistor detector integrated by depositing amorphous Ge as a low-temperature back-end process module [1, 3].

Ge *metal–semiconductor–metal* (MSM) photodiodes also have the advantage of process simplicity as compared to vertically integrated p–i–n diodes because the doped n- and p-type regions are replaced by direct metal contacting. Moreover, the simple structure has enabled continuous improvement of already fast response times and high bit rate operation needed to meet ever-increasing demands on optical fiber communication applications. However, the low metal Schottky barrier height leads to high dark currents that severely degrade device performance and cause high-power consumption. Performance trade-offs are being studied for methods that increase the barrier height.

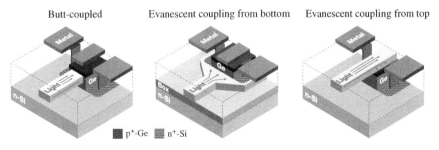

Figure 11.7 Schematic structures of typical waveguide-coupled Ge photodiode detectors: (a) butt-coupled with the waveguide made in α-Si or SiN; evanescent coupling with (b) a Si-on-oxide waveguide under the Ge photodiode, and (c) an α-Si or SiN waveguide on top.

As a drawback compared to the III–V compounds like InGaAs, Ge is an indirect-bandgap semiconductor and therefore not directly suitable for fabricating practical light sources. Nevertheless, Ge does have a direct energy transition at 0.8 eV which is only 0.14 eV above the lower indirect bandgap of 0.66 eV. Therefore, several band engineering methods have been investigated to lower the direct bandgap level. These include tensile strain engineering, n^+-doping, Ge–Sn alloying, and quantum well structures. Particularly, GeSn growth on Ge has for more than a decade received much attention as a low-cost, CMOS compatible option. Despite significant progress in the material quality, the GeSn photdiode performance is still far from being commercially competitive with state-of-the-art InGaAs photodiodes [17].

While the drive to develop Si PICs is relatively new, *indium phosphide* (InP) has for several decades been under research as a platform for photonic circuits where both lasers and photodiodes can be integrated in compound semiconductors comprised of combinations of Ga, As, In, Sb, and P. As a detector, the $In_{0.53}Ga_{0.47}As$ has been a heavily researched material system because it is lattice matched to InP. Good detectivity is achieved up to 1.7 µm, while small alterations of the composition can extend this to about 2.6 µm.

As for Ge-on-Si photodiodes, the sensitivity of InGaAs/InP diodes is increased by designing for operation in avalanche mode, and they have also been commercialized as SPADs. This requires separation of the InGaAs absorption layer and the multiplication layer located in the InP. The compound nature of these materials gives the opportunity to confine the avalanching process to the region where it is most effective and also decrease multiplication noise. An example of such a structure is shown in Figure 11.8, where grading and charge layers have been introduced in a special SAM structure. Discrete SPADs and SPAD imaging arrays in InGaAs/InP technology have

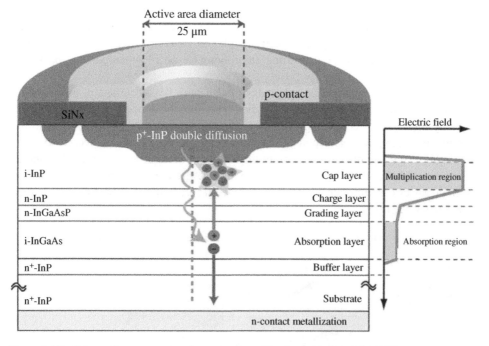

Figure 11.8 Schematic cross section through a front-illuminated InGaAsP/InP APD structure with separate absorption, grading, charge, and avalanche multiplication regions (SAGCM). On the right, the electric field along the center of the active area is shown, indicating the separated avalanche multiplication and absorption regions. Source: Modified from Tosi et al. [15].

been commercialized with better DCRs than comparable Ge SPADs. The main impetus for development of such SPADs is LIDAR and 3D imaging camera systems where faint light in IR is used for eye safety. However, integration of InGaAs/InP SPADs with the supporting circuitry needed for controlling the SPAD operation is still under development, and Ge-on-Si systems show greater promise since the integration complexity is significantly reduced by their CMOS compatibility.

Growing InGaAs on InP rather than Si helps to improve the electrical performance, and diodes can be operated just by using a thermoelectric cooler. For detection of longer wavelengths, the absorbing region of the p–n junction photodiode can be made of InSb which allows detection up to 5 μm, InAs for up to 4 μm, and InAsSb up to 11 μm. For InAsSb ($InAs_{1-x}Sb_x$) and InGaAs ($In_{1-x}Ga_xAs$), not only the cut-off wavelength, but also the electrical properties will depend on the molar composition used in the fabrication of the absorber. For all these materials, an improvement of the electrical performance is achieved by low-temperature operation which can be applied down to cryogenic temperatures. As for Ge, the formation of doped regions is achieved during epitaxial growth or by localized dopant diffusion or implantation/annealing. The latter allows planar diode fabrication, often by using Zn or Be p-dopants, while epitaxy, mainly metal–organic CVD (MOCVD), requires mesa etching and usually results in high perimeter currents due to poor passivation of the junctions.

11.4.2 Infrared Photoconductive Detectors

For detection of infrared wavelengths longer than 3 μm, the MWIR and LWIR regions, photoconductors (photoresistors) are commonly used instead of photodiodes [16]. The devices are no longer fabricated with epitaxial growth aiming at a low-defect material with high quantum efficiency. Instead, it is common to fabricate the photoconductors by depositing polycrystalline films such as in lead-salt detectors containing, for example lead selenide (PbSe) or lead sulfide (PbS). They do not contain a p–n junction, but the incident light induces intraband transitions with the result that there is a reduction in electrical resistance. The EQE is often very high but with the disadvantage that also the dark current is high due to the low bandgap energy and the high defect concentration found in these nonsingle-crystal materials. Therefore, this type of detector can be useful at room temperature, but the detectivity is improved by cooling the devices. This also reduces the bandgap energy which has the advantage that light with longer wavelengths can be detected. The effect of device cooling on different IR detectors can be seen in Figure 11.6, where the specific detectivity is plotted as a function wavelength.

Photoconductors containing mercury cadmium telluride (MCT, $Hg_xCd_{1-x}Te$) are also widely used. This material has several attractive properties: the bandgap can be tailored by varying the Hg content, the absorption coefficient and quantum efficiency are high, and some realizations can also work in photovoltaic mode. Detection at long wavelengths up to about 20 μm is feasible. Most often MCT detectors are cooled, and they can be fabricated in focal plane arrays for thermal imaging.

In the FIR wavelength range, up to and even going a bit beyond 200 μm, detection sensitivity is achieved by highly-doping silicon and germanium layers. The high doping with impurities such as arsenic, copper, gold, or indium not only introduces extra energy levels in the bandgap, allowing intraband transitions, but also causes very high dark current. Therefore, to achieve sufficient sensitivity, these detectors must be operated at cryogenic temperatures as low as those achieved with liquid helium, i.e. in the region of 4 K.

While the InSb- and HgCdTe-based detectors have seen many decades of development and are commercially available for many applications, they do have drawbacks with respect to the material toxicity and fragility. The last decade has therefore seen a marked shift of interest toward developing quantum-well infrared detectors (QWIPs) and type II superlattice (T2SL)-based detectors that also have potential for being integrated in CMOS [18–20]. The most studied material stack for fabrication of QWIPs is lattice-matched GaAs/AlGaAs, but combinations with In and P are also used. A combination of wide and narrow bandgaps create quantum wells that by varying the number of layers, thickness, and composition give a great flexibility in setting the optoelectronic characteristics such as cut-off wavelength, detectivity, and dark currents. QWIPs are usually operated at cryogenic temperatures. The GaAs/AlGaAs-based QWIPs can be fabricated with sensitivity in the 6–20 µm range, but the spectral region will depend on the transitions allowed by the quantum-well configuration and may be relatively narrow. Quantum dot IR photodetectors (QDIPs) are also being investigated, and they may offer more performance enhancements, but research is still in the early stages [21].

T2SL are particularly interesting as MWIR and LWIR detectors since they have the potential to increase the operating temperature close to room temperature (RT) by reducing the dark current. Typical material systems used for formation of type-II superlattice absorbers are InAs/GaSb, InGaAs/GaAsSb, and InAs/InAsSb. A structure is created that is a hybrid between photoconductor and photodiode. Barriers between absorbers are implemented to separately control majority and minority carrier flows, for example in the so-called "nBn structure." Compared to HgCdTe detectors, T2SL detectors have much better material properties with respect to manufacturability, scalability, and stability.

11.4.3 Thermal Infrared Detectors

In thermal detectors, the absorbed infrared photons are first converted to heat before being detected as a temperature increase [16]. While photodetectors have a spectral response that is wavelength-dependent with a cut-off wavelength determined by the specific material, the photosensitivity of thermal detectors is wavelength independent. To increase their thermal resistance, the photosensitive material is often a thin-film material suspended on a plate or as a membrane in a vacuum chamber. This requires a light entrance window, the material of which may limit the spectral response. Thermal detectors have the advantage of working at RT. A minimal cooling is, however, sometimes added by integrating a thermoelectrical cooler. This stabilizes the temperature and gives some improvement in performance.

Bolometer infrared sensors operate by measuring the electrical resistance of thin-films made in materials such as amorphous silicon (α-Si) or vanadium oxide that have a large temperature coefficient. They are widely fabricated as microbolometers that can be used in focal-plane arrays, for example for infrared cameras. To achieve low-cost production, standard IC, and microelectromechanical systems (MEMS) technology are merged to produce imagers based on structures like the example shown in Figure 11.9. These types of IR imagers were primarily developed for military applications, but the favorable cost and performance have made them popular for use in diverse fields such as driver-assistance systems, aircraft flight control, industrial process monitoring, firefighting, night vision, and security systems.

Other types of thermal detectors include pyroelectric detectors, pneumatic cells, and thermopiles that all use different types of heat-generated material responses that can be converted to a measurable electrical signal. For example, pyroelectric detectors and thermopiles generate a voltage in the specific pyroelectric material or a thermocouple, respectively. Common for all these thermal detectors is that there is a fundamental trade-off between sensitivity and bandwidth. The light-absorbing sensor element to be heated will have a thermal resistance that is proportional to the temperature rise and therefore also to the electrical signal to be measured. Thus, a high

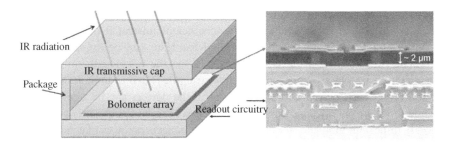

Figure 11.9 Schematic of the basic structure of a vacuum-packaged microbolometer and a cross-sectional scanning electron microscope (SEM) image of the bolometer array. Source: Adapted from [22] Elßner et al., 2015, ELSEVIER.

sensitivity requires a sufficiently high thermal resistance. The smaller the absorber, the higher the thermal resistance, where the size reduction will be limited by the ability to focus the light appropriately. On the other hand, the thermal resistance is proportional to the thermal time constant, which means that a high thermal resistance will make the detector slow. This fact has also motivated the development of thermal imagers rather than scanning systems.

11.5 Emerging Devices

Several movements in physics, chemistry, and materials science, as well as electrical and optical engineering, are coming together to push the development of new materials and device concepts for more potent optical detector technology. From a scientific point of view, the interactions between light and matter are a boundless source of information on the fabric of microscopic to macroscopic environments. The study of these interactions now forms an integral part of intensive research activity directed toward the fabrication, understanding, and potential optical and electronic applications of new materials, very often with focus on 1D and 2D structuring. The application in photodetectors and other sensors is clearly receiving a relatively large part of the attention because, as opposed to the fabrication of advanced transistors, relevant devices can be produced on a small scale using methods unsuited for volume production. Finding routes toward upscaling and integration on Si is at the same time given serious consideration as it is vital for future implementation on a large scale.

Among the 1D structures, silicon-based nanowires with complex structuring have come quite far on the route to usable implementations, particularly with top-down fabrication methods combining nano-patterning techniques with plasma- and wet-etching procedures. Bottom-up techniques using vapor phase epitaxy and other epitaxial growth methods are also applied. Examples are Si nanocolumns and nanopenciles with which the surface to active device volume ratio is increased and, in principle, the photosensitivity, responsivity, and response time of the basic Si p–n junction can be enhanced [23, 24]. This is particularly studied as a low-cost method of enhancing the efficiency of thin-film solar-cells, where also the nanostructuring of metal oxide materials like ZnO is receiving attention. The shape of the nanostructures is developed to control undesirable surface carrier recombination and also achieve features that are advantageous for broadband absorption and light trapping. The latter can also be achieved by adding dielectric/metal layers with suitable nanostructuring for functions such as anti-reflection or light-trapping.

For the development of imaging arrays, nanowires containing III–V or II–VI layer stacks offer unique bandgap engineering possibilities that cannot be realized in larger structures due to the material incompatibilities. When grown on Si substrates, research has shown that nanostructures can allow relief from strain due to lattice and expansion-coefficient mismatch that otherwise would cause material degradation. Therefore, the individual bandgaps can be tuned in many different ways to, among other things, achieve detection in a specific wavelength range. The responsivity can also be increased by using carrier selective structures such as phototransistors. By promoting electron rather than hole injection, the speed of the devices can be significantly increased. At the same time, the energy consumption of each pixel can be significantly reduced. This may become a very important motivator for developing such imagers because, in general, the energy consumption of many emerging systems is seen as a serious limitation, whether it concerns computation and sensing technologies for large-scale systems like advanced machine learning and broadband brain-machine interfaces, or small-scale systems like wearable/implantable devices and self-folding robots. For the latter systems, nanowires also open more options for making flexible devices by enabling growth on flexible substrates.

Since the first modern experiments with graphene more than a decade ago, the compelling potentials of 2D materials have led to research into many other 2D material systems. For optoelectronic photodetectors, graphene is playing a prominent role but transition metal dichalcogenides such as MoS_2 and WSe_2 are also receiving attention [25]. The bandgap of these materials can be tuned by changing the number of monolayers and lateral geometry to enable broadband detection in the visible to far-infrared range, and, despite the present low manufacturability, they are seen as a means of lowering the fabrication costs through simplified fabrication processes, possibly with higher yield than what appears feasible with quantum well/dot detector processes. Ultimately, all these processes aspire to obtain higher operating temperature by using the low dimensionality to reduce dark currents and promote confinement of carriers and photons.

Although the quantum efficiency of the 2D materials can be high, the volume of absorbing matter is small. Therefore, *plasmonic structures* that enhance light–matter interaction are studied as a means of boosting the responsivity [26, 27]. For example, graphene has been investigated for plasmonic-enhanced silicon detectors where NIR light beyond the Si cutoff wavelength is absorbed in the graphene and via an internal photoemission effect is transmitted at a detectable wavelength to a metal-silicon photodetector. Another interesting aspect of plasmonics is the potential for down-scaling PICs that presently have minimum dimensions in the micron range, and thus occupy unattractively large wafer area as compared to Si electronic circuits. Via *surface plasmon polaritons*, the electromagnetic field of the light can couple to charge oscillations in a thin metal and the size of such oscillations can be much smaller than the light wavelength. Thus, plasmonic devices could operate at deep-subwavelength scales.

11.6 Concluding Remarks

In Figure 11.10, an overview is given of many of the optical detector devices and technologies discussed in this chapter in relationship to their potential compatibility with CMOS processing and their present manufacturability level. For the advancement of MtM technology, it is of importance that such sensor functions be directly integrated at the chip level, either as front-end CMOS modules or as back-end add-ons, but integration at the level of the package is also playing an essential role. The funding of optical detector R&D over the last decade has been high and steadily increasing, and it is expected that the introduction of new technologies will proceed at a high pace. Particularly, the drive will continue toward nano-photodetector imagers with single-photon sensitivity, high-speed, and low-energy consumption.

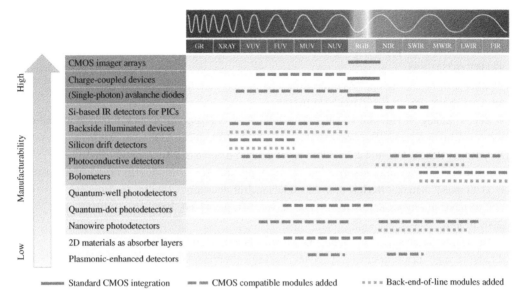

Figure 11.10 Overview of some of the discussed optical detectors giving an impression of their main application wavelengths, present level of manufacturability, and the degree to which they can be integrated with CMOS as modules that can be either both front- and back-end-of-line compatible or only back-end-of-line compatible.

References

1 D. Benedikovic, L. Virot, G. Aubin, J. M. Hartmann, F. Amar, X. Le Roux, C. Alonso-Ramos, É. Cassan, D. Marris-Morini, J. M. Fédéli, and F. Boeuf, "Silicon–germanium receivers for short-wave infrared optoelectronics and communications," *Nano*, vol. 10, no. 3, pp. 1059–1079, 2021, doi: 10.1515/nanoph-2020-0547.

2 D. Durini (ed.), *High Performance Silicon Imaging: Fundamentals and Applications of CMOS and CCD Sensors*, 2nd Edition. UK: Woodhead Publishing, 2019.

3 J. Lui, "Monolithically integrated Ge-on-Si active photonics," *Photonics*, vol. 1, pp. 162–197, 2014, doi: 10.3390/photonics1030162.

4 L. K. Nanver, L. Qi, V. Mohammadi, K. R. M. Mok, W. B. De Boer, N. Golshani, A. Sammak, T. L. Scholtes, A. Gottwald, U. Kroth, and F. Scholze, "Robust UV/VUV/EUV PureB photo-diode detector technology with high CMOS compatibility," *IEEE Journal of Selected Topics in Quantum Electronics*, vol. 20, pp. 306–316, 2014, doi: 10.1109/JSTQE.2014.2319582.

5 F. Ceccarelli, G. Acconcia, A. Gulinatti, M. Ghioni, I. Rech, and R. Osellame, "Recent advances and future perspectives of single-photon avalanche diodes for quantum photonics applications," *Advanced Quantum Technologies*, 2000102, pp. 1–24, 2021, doi: 10.1002/qute.202000102.

6 T. Tsang, A. Bolotnikov, A. Haarahiltunen, and J. Heinonen, "Quantum efficiency of black silicon photodiodes at VUV wavelengths," *Optics Express*, vol. 28, no. 9, pp. 13299–13309, 2020, doi: 10.1364/OE.385448.

7 W. Metzger, J. Engdahl, W. Rossner, O. Boslau, and J. Kemmer, "Large-area silicon drift detectors for new applications in nuclear medicine imaging," *IEEE Transactions on Nuclear Science*, vol. 51, no. 4, pp. 1631–1635, 2004, doi: 10.1109/TNS.2004.832666.

8 A. Lahav, A. Fenigstein, and A. Strum, "Backside illuminated (BSI) complementary metal-oxide-semiconductor (CMOS) image sensors," *High Performance Silicon Imaging, Fundamentals and Applications of CMOS and CCD Sensors*, pp. 98–123, 2014, doi: 10.1533/ 9780857097521.1.98.

9 W. Zheng, L. Jia, F. Huang, "Vacuum-ultraviolet photon detections," *iScience*, vol. 23, no. 6, pp. 1–22, 2020, doi: 10.1016/j.isci.2020.101145.

10 Y. Zou, Y. Zhang, Y. Hu, and H. Gu, "Ultraviolet detectors based on wide bandgap semiconductor nanowire: a review," *Sensors*, vol. 18, no. 7, 2072, 2018, doi:10.3390/s18072072.

11 X. Chen, F. Ren, S. Gu, and J. Ye, "Review of gallium-oxide-based solar-blind ultraviolet photodetectors," *Photonics Research*, vol. 7, no. 4, pp. 381–415, 2019, doi: 10.1364/PRJ.7.000381.

12 E. Gatti and P. Rehak, "Review of semiconductor drift detectors," *Nuclear Instruments and Methods in Physics Research – Section A*, vol. 541, pp. 47–60, 2005, doi: 10.1016/j.nima .2005.01.037.

13 T. Harada, N. Teranishi, T. Watanabe, Q. Zhou, J. Bogaerts, and X. Wang, "High-exposure-durability, high-quantum-efficiency (>90%) backside-illuminated soft-X-ray CMOS sensor," *Applied Physics Express*, vol. 13, no. 016502, 2020, doi: 10.7567/1882-0786/ab5b5e.

14 M. Bosi and G. Attolini, "Germanium: epitaxy and its applications," *Progress in Crystal Growth and Characterization of Materials*, vol. 56, no. 3, 4, pp. 146–174, 2010, doi: 10.1016/j .pcrysgrow.2010.09.002.

15 A. Tosi, F. Acerbi, M. Anti, and F. Zappa, "InGaAs/InP single-photon avalanche diode with reduced afterpulsing and sharp timing response with 30 ps tail," *IEEE Journal of Quantum Electronics*, vol. 48, no. 9, pp. 1227–1232, 2012, doi: 10.1109/JQE.2012.2208097.

16 Photonics Encyclopedia. Infrared Detectors. Available: https://www.rp-photonics.com/infrared_ detectors.html.

17 C. L. Tan and H. Mohseni, "Emerging technologies for high performance infrared detectors," *Nanophotonics*, vol. 7, no. 1, pp. 169–197, 2018, doi: 10.1515/nanoph-2017-0061.

18 D. Z. -Y. Ting, A. Soibel, A. Khoshakhlagh, S. A. Keo, B. Rafol, A. M. Fisher, B. J. Pepper, E. M. Luong, C. J. Hill, and S. D. Gunapala, "Advances in III–V semiconductor infrared absorbers and detectors," *Infrared Physics and Technology*, vol. 97, pp. 210–216, 2019, doi: 10.1016/j.infrared.2018.12.034.

19 A. Rogalski, P. Martyniuk, and M. Kopytko, "Type-II superlattice photodetectors versus HgCdTe photodiodes," *Progress in Quantum Electronics*, vol. 68, no. 100228, 2019, doi: 10.1016/j .pquantelec.2019.100228.

20 D. Z. -Y. Ting, A. Soibel, L. Höglund, J. Nguyen, C. J. Hill, A. Khoshakhlagh, and S. D. Gunapala, "Type-II superlattice infrared detectors," *Semiconductors and Semimetals*, vol. 84, pp. 1–57, 2011, doi: 10.1016/B978-0-12-381337-4.00001-2.

21 A. Ren, L. Yuan, H. Xu, J. Wu, and Z. Wang, "Recent progress of III–V quantum dot infrared photodetectors on silicon," *Journal of Materials Chemistry C*, vol. 7, no. 46, pp. 14441–14453, 2019, doi: 10.1039/C9TC05738B.

22 M. Elßner and H. Vogt, "Reliability of microbolometer thermal imager sensors using chip-scale packaging," *Procedia Engineering*, vol. 120, pp. 1191–1196, 2015, doi: 10.1016/j.proeng .2015.08.784.

23 Z. Li, J. Allen, M. Allen, H. H. Tan, C. Jagadish, and L. Fu, "Review on III–V semiconductor single nanowire-based room temperature infrared photodetectors," *Materials*, vol. 13, no. 6, pp. 1400, 2020, doi: 10.3390/ma13061400.

24 S. K. Ray, A. K. Katiyar, and A. K. Raychaudhuri, "One-dimensional Si/Ge nanowires and their heterostructures for multifunctional applications – a review," *Nanotechnology*, vol. 28, 092001, 2017, doi: 10.1088/1361-6528/aa565c.

25 L. Tao, Z. Chen, H. Fang, X. Li, X. Wang, J. B. Xu, and H. Zhu, "Graphene and related two-dimensional materials: structure-property relationships for electronics and optoelectronics," *Applied Physics Reviews*, vol. 4, 021306, pp. 1–31, 2017, doi: 10.1063/1.4983646.

26 A. Dorodnyy, Y. Salamin, P. Ma, J. V. Plestina, N. Lassaline, D. Mikulik, P. Romero-Gomez, A. F. i Morral, and J. Leuthold, "Plasmonic photodetectors," *IEEE Journal of Selected Topics in Quantum Electronics*, vol. 24, no. 6, pp. 1–13, 2018, doi: 10.1109/JSTQE.2018.2840339.

27 U. Levy, M. Grajower, P. A. D. Gonçalves, N. A. Mortensen, and J. B. Khurgin, "Plasmonic silicon Schottky photodetectors: the physics behind graphene enhanced internal photoemission," *APL Photonics*, vol. 2, 026103, 2017, doi: 10.1063/1.4973537.

12

Environmental Sensing

Tarek Zaki

Munich, Germany

12.1 Motivation

12.1.1 Air Pollution

Air in the Earth's atmosphere consists of 78% nitrogen, 21% oxygen, 0.9% argon, and 0.1% other gases. Trace amounts of carbon dioxide, methane, and water vapor are some of these other gases that make up the remaining 0.1% [1]. The European Environment Agency (EEA) estimates that carbon dioxide, which is one of the abundant trace gases in the atmosphere to be around 733 mg/m^3 (equivalent to 407 ppm) in 2018 [2].

Moreover, there are other gases and particles released into the atmosphere from both natural and man-made sources that pollute the air and are harmful to our environment and health. Such air pollutants can affect quality of life and even reduce life expectancy of humans, animals, plants, water bodies, soils, and even buildings. Examples of natural sources are volcanic eruptions, forest fires, and sandstorms. Some of these substances have even high propensity to interact with other substances to form new ones, forming the so-called "secondary harmful pollutants." Heat, including that from the sun, is usually a catalyst facilitating or triggering such chemical reaction processes [3].

Nowadays, sad to note, the World Health Organization (WHO) estimates that 9 out of 10 people worldwide breathe polluted air. The combined ambient (outdoor) and household (indoor) air pollution has shown correlations to several million premature deaths every year, largely as a result of increased mortality from stroke, heart disease, chronic obstructive pulmonary disease, lung cancer, and acute respiratory infections [4]. This dreadful situation is alarming. Reducing the concentration of such pollutants help not only to improve air quality and public's health but also to combat climate change.

The most common way of quantifying the air pollutants (or air quality) at a given location and at a given time is by using the mass concentration expressed as mass per unit volume of atmospheric air – in grams per cubic meter air or in g/m^3. Since higher concentrations of pollutants can have a big impact on health and the environment, agencies such as the EEA, WHO, the US Environmental Protection Agency (EPA), and the German Federal Environmental Agency (UBA) enforce guiding threshold limits.

This chapter is concerned with environmental sensing, particularly focusing on how to sense and quantify air pollution. This includes design, technology, operating schemes, and manufacturing of air quality sensors. Not only do such devices provide data to scientists and regulators who study air

Advances in Semiconductor Technologies: Selected Topics Beyond Conventional CMOS, First Edition. Edited by An Chen.

pollution and the effectiveness of emissions control strategies, but they are also enabling the public to know precisely what is in the air we breathe. Ultimately, environmental sensors create a plethora of valuable information that helps us to disconnect from wrong habits and to start thoughtfully healing our health, nature, and ecosystem.

12.1.2 Hazardous Pollutants

Air pollutants have different sources and impacts. Among the most harmful air pollutants are **carbon dioxide (CO_2), ozone (O_3), sulfur oxides (SO_x), nitrogen oxides (NO_x)**, and the following [5–8]:

- **Particulate matter (PM)**: PM is microscopic particles consisting of a complex mixture of solid and liquid particles of organic and inorganic substances that are suspended in the air. Due to its significant impact on health, PM is sometimes used as the main indicator for air pollution. PM can originate from adverse sources, for example from energy supply and industrial processes, from metal and steel production, and from natural origin such as soil erosion. While particles with a diameter of 10 μm or less (referred to as PM10) can penetrate and lodge deep inside the lungs, particles with a diameter of 2.5 μm or less (referred to as PM2.5) are more dangerous because they can penetrate even the lung barrier and enter into the blood stream, just like oxygen. This increases risk of developing cardiovascular and respiratory diseases, as well as of lung cancer. Recent findings observed even a link between PM2.5 exposure and sensitivity to viral diseases, including COVID-19 [9]. For the protection of health, the German Federal Environmental Agency enforces limits of 40 μg/m^3 for one-year average PM10 and 25 μg/m^3 for one-year average PM2.5 [5]. See also US EPA's National Ambient Air Quality Standard (NAAQS) in [8].
- **Carbon monoxide (CO)**: CO is a toxic gas that is colorless, nonirritant, odorless, and tasteless. It is produced by the incomplete combustion of carbonaceous fuels such as wood, petrol, coal, and natural gas. For example, one of the dangerous exposure sources of CO is emissions from faulty cooking or heating appliances that burn fossil fuels. Furthermore, cars and other vehicles or machinery that burn fossil fuels contribute to the emissions of CO. CO is dangerous because it affects the oxygen intake of humans and animals and has adverse effects on the central nervous system. It can cause dizziness, unconsciousness, and even death. Accordingly, the German Federal Environmental Agency set an exposure maximum limit of 10 mg/m^3 (daily eight-hour mean) for health protection [5]. See also US EPA's NAAQS in [8].
- **Volatile organic compounds (VOCs)**: VOCs are organic chemicals that are volatile, meaning that they have a high vapor pressure already at room temperature. They are found in the air in gaseous and vaporous forms. Examples of VOCs are hydrocarbons, alcohols, aldehydes, and organic acids. They can be released into the air from a broad range of different sources, for example from residual solvents and building blocks in plastics, from excipients such as plasticizers, solubilizers, antioxidants, stabilizers, and catalysts in production processes, and also from aromatics, flame retardants, and biocides. Materials used in construction and interior decoration, such as paints, varnishes, and adhesives, contribute as well to VOC emissions. Furthermore, bodycare products and cleaning materials also increase VOC concentrations in the air. Among the adverse health effects of VOC are dryness and irritation of the eye, nose, and throat. Therefore, the German Federal Environmental Agency recommends that the Total Volatile Organic Compound (TVOC) concentration of 10–25 mg/m^3 (daily, short residence) and of 1–3 mg/m^3 (daily, long residence) shall not be exceeded [5].

Figure 12.1 illustrates the health impacts associated with hazardous air pollutants [3, 10]. These health impacts result in substantial costs, and thus, the benefits of stringent air quality policies are usually much higher than these costs.

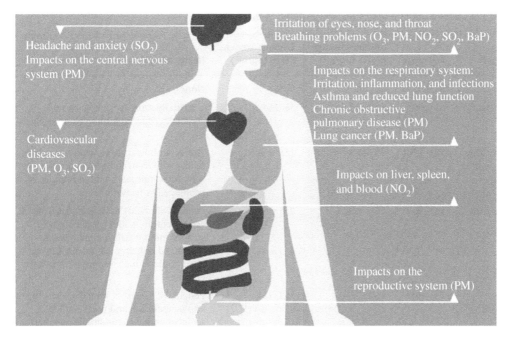

Figure 12.1 Health impacts of hazardous air pollutants [3, 10].

12.1.3 Air Quality Index

Air quality index (AQI) is used as a value to quantify the quality of the air. It is based on the measured concentrations of the hazardous pollutant(s). The higher the AQI, the greater level of air pollution, and thus, the more dangerous the health concern. In order to reflect the unique discourse surrounding of the national air quality standards, different countries typically have own premises on how to calculate (i.e. formula and ranges) and categorize (i.e. colors and health advices) the AQI.

A separate AQI is computed from the concentration of each air pollutant over a specified averaging period. The pollutant with the *worst* index value determines the color of the consolidated total AQI. In Germany, three pollutants are considered, namely NO_2, PM10, and O_3 [5]. In the United States, on the other hand, five pollutants are considered, namely NO_2, SO_2, CO, PM (both PM2.5 and PM10), and O_3 [11]. The AQI is grouped into ranges, where each is assigned a specific description, color, and health advices based on epidemiological research. Above-defined AQI thresholds, agencies typically warn the public of the pollution and invoke emergency plans to abate the health threat.

As an example, Table 12.1 shows the AQI for outdoor PM2.5 and PM10 with the corresponding six categories in the United States (US) [11]. The AQI is a piecewise linear function of the pollutant concentration (C):

$$\text{AQI}(C) = \left(\frac{\text{AQI}_{\text{high}} - \text{AQI}_{\text{low}}}{C_{\text{high}} - C_{\text{low}}} \right) (C - C_{\text{low}}) + \text{AQI}_{\text{low}}$$

An AQI value of 50 or below represents good air quality, while an AQI value over 300 represents hazardous air quality. The value of 100 corresponds to an ambient air concentration that equals the level of the short-term NAAQS for protection of public health [8]. When the AQI values exceed 100, air quality is unhealthy – first for certain sensitive groups of people and then for the general public.

Table 12.1 AQI for outdoor PM2.5 and PM10 in the United States.

Category	AQI range	24-hour-average PM2.5 ($\mu g/m^3$)	24-hour-average PM10 ($\mu g/m^3$)	Health advice
Good	0–50	0–12.0	0–54	None
Moderate	51–100	12.1–35.4	55–154	Unusually sensitive people should consider reducing prolonged or heavy exertion
Unhealthy for sensitive groups	101–150	35.5–55.4	155–254	People with heart or lung disease, older adults, children, and people of lower socioeconomic status should reduce prolonged or heavy exertion
Unhealthy	151–200	55.5–150.4	255–354	People with heart or lung disease, older adults, children, and people of lower socioeconomic status should avoid prolonged or heavy exertion; everyone else should reduce prolonged or heavy exertion
Very unhealthy	201–300	150.5–250.4	355–424	People with heart or lung disease, older adults, children, and people of lower socioeconomic status should avoid all physical activity outdoors. Everyone else should avoid prolonged or heavy exertion
Hazardous	301–500	250.5–500.4	425–604	Everyone should avoid all physical activity outdoors; people with heart or lung disease, older adults, children, and people of lower socioeconomic status should remain indoors and keep activity levels low

Source: Based on U.S. Environmental Protection Agency [11].

12.1.4 Air Monitoring Network

Unless available in relatively high, yet harmful, concentrations, air-borne pollutants are often imperceptible to the human senses. Some air pollutants are poisonous already at low concentrations, while others require an accumulation of exposure over a longer time until effects manifest. Not only the chemical composition of the air is constantly changing, but also air is moving around the globe, crossing nations, and oceans. Therefore, large differences in ultra-localized air pollution levels are present and can be transported across different locations around the globe [12]. Accordingly, active and continuous air quality monitoring at various locations is necessary.

In order to guarantee clean air or, wherever and whenever necessary, to improve air quality, agencies typically operate countrywide networks of air monitoring stations that track different types of air pollutants. These networks that report, for example hourly- and daily-averaged concentration levels are operated by tribal, state, or local governmental/private agencies. Deciding upon the exact locations of the monitoring stations depend on the purpose. The intention may be to monitor busy roads in city centers or to determine background pollution levels in rural regions away from urban areas and emission sources. Typical sources include industrial, motor vehicle, and incinerator emissions. Moreover, the intention may also be to regularly measure at locations of particular concern, such as schools, hospitals, volcanoes, and forests.

What is the collected data used for? [13]

- Assess the extent of pollution
- Provide continuous and timely air pollution data to the general public
- Derive forecasts, trends, warnings, and recommendations to the public, particularly to vulnerable groups
- Present data over geographical locations to determine local and regional conditions
- Support implementation of air quality goals or standards
- Evaluate the effectiveness of air quality improvement measures

The data are typically uploaded online in databases [14]. Apps, such as AirVisual, MyAmbience, IQ Air, Luftqualität, and Air Matters illustrate accordingly the data in helpful, lean, and easy-to-follow visualizations on smartphones. Furthermore, using the local stations-reported data along with other data sources, such as weather and traffic, the startup Breezometer derive high-resolution air quality information at the user's close vicinity.

The number of monitoring stations is growing, which increases our knowledge and understanding of air quality. The EEA brings together air quality measurements from more than 7500 monitoring stations across whole Europe [3]. In the German state of Baden-Württemberg (BW) alone, there are over 80 air monitoring stations [15].

Referring to Figure 12.2, the locally prevailing pollutant load consists to a significant extent of a regional background load [16, 17]. The reasons for this are environmental influences such as weather conditions, pollution from the energy industry, but also from agriculture. In addition, there is urban background pollution from urban sources such as construction sites, households, and small consumers. In addition to the background pollution, additional pollution occurs locally. Often, these are due to high-traffic volumes. Especially in combination with poor air exchange, this can lead to so-called "hotspots" with particularly highly polluted air. The area "Am Neckartor" in Stuttgart (capital city of state BW) is regarded, for example, as a hotspot.

Figure 12.3 shows the air quality monitoring station at Stuttgart Am Neckartor [15]. As depicted in Figure 12.3, the station is typically a water-protected container comprising expensive equipment that regularly samples the air through inlets positioned at a height of about >1 m above the ground. To gain insight about the context, there is a high volume of traffic at this location with three lanes in each direction. Around 510 people live along this section of the road. On the other side of the road, there is a garden with a dense stand of trees parallel to the road. This particular station, for example, is able to measure NO_2, PM2.5, and PM10. The data can be easily fetched online on the official website of the BW State Institute for the Environment, Survey, and Nature Conservation [15].

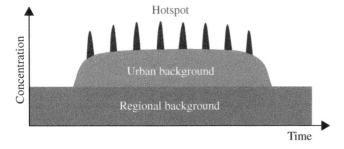

Figure 12.2 Schematic representation of background and additional pollution. Source: Modified from Müller and Warth [16].

Figure 12.3 Air quality monitoring station (EU Nr. DEBW118) at Stuttgart Am Neckartor. Source: Heike Robakowski/LUBW.

Figure 12.4 shows PM10 data measured at this station [18]. The daily emission limit value for the protection of human health is $50\,\mu g/m^3$ with 35 permitted exceedances in the calendar year. As depicted in the graphs, data of two measurement methods are available:

- Optically measured one-hour-averaged PM10 for rapid information of the public
- Gravimetric 24-hour-averaged PM10 for legally compliant determination of the days of exceedance

The "provisional values" for PM10 are collected using the optically measured one-hour-averaged method to provide the public with timely information. In this method, the PM concentration is determined by optical scattered light measurement. The air is sucked in through an integrated fan, and depending on the ambient temperature, it is automatically preheated to remove humidity effects. Using integrated LED light source and photodiodes, particles are detected by optically scattered light. Knowing the geometrical dimensions of the sampling system and by measuring the air flow velocity, the PM concentration in $\mu g/m^3$ can be derived accordingly. Examples of such optical PM measurement equipment are FDS15/FDS18 from Dr. Födisch and Fidas 200s/Frog from Palas.

Decisive for the assessment of air quality in BW, however, are the results based on the gravimetric reference method. Here, the PM is deposited on a filter over a period of 24 hours. The filters are collected by an operator, then weighed and assessed manually in the laboratory to derive the concentration in $\mu g/m^3$. For this reason, the gravimetric measurement results are available after about one month. As shown in Figure 12.4b, the different measurement principles (i.e. optical vs. gravimetric) can lead to – commonly accepted – deviations in the measurement results.

12.1.5 Hand-Held Devices

In Section 12.1.4, official air quality monitoring stations have been described. They are very accurate and optimal for regulatory purposes, yet bulky, expensive, restricted to specific outdoor locations, and provide averaged environmental data only. Furthermore, changing the environmental

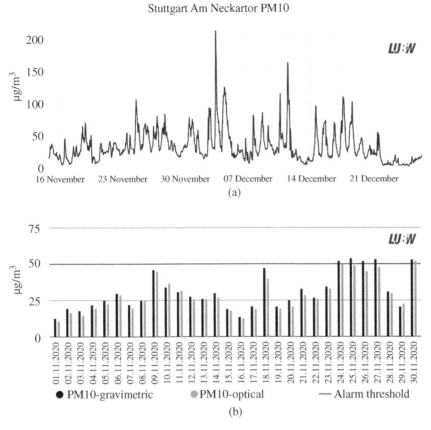

Figure 12.4 PM10 values at Stuttgart Am Neckartor: (a) hourly-averaged PM10 values measured by optical equipment; (b) comparison between 24-hour-averaged PM10 values measured by gravimetric and optical equipment. Source: Data and images are from Landesanstalt für Umwelt Baden-Württemberg [18].

conditions around these stations, for example by avoiding nearby traffic lights and congestion, can significantly alter the localized readings to better, yet impractical, reported data [19].

On the other hand, having compact, portable, real-time, and personalized air quality monitoring devices would enable individuals to track their performance at any location (i.e. indoors and outdoors) and at any time, preferably all the time. By that means, users would gain deeper insight into their personal exposure to hazardous pollutants, and accordingly, improve their habits and health.

Indeed, advanced semiconductor technologies have enabled us to considerably reduce the size of air-quality sensors. Figure 12.5 shows two examples of commercial, hand-held air-quality monitors. Such devices integrate sensors that are able to measure numerous gaseous or particle pollutants, including PM, CO, NO_2, and VOC. Rather than taking you laboriously through each sensing technology, Sections 12.2 and 12.3 focus on miniaturized VOC and PM sensors as two prominent examples. The low cost and compact sizes of these sensors enable integration in endless consumer electronic devices, such as air purifiers, smart thermostats, and other IoT (Internet of Things) applications.

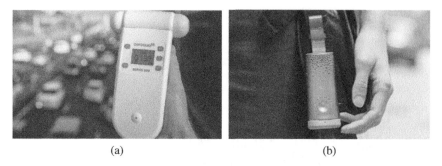

(a)   (b)

Figure 12.5 Hand-held air-quality monitors: (a) Aeroqual Series 500, (b) Plume Labs Flow 2. Source: (a) Aeroqual; (b) Plume Labs.

12.2 Particulate Matter (PM) Sensing

12.2.1 Particulate Matter (PM)

PM, known also as atmospheric aerosol particles, are microscopic particles of solid or liquid matter that are suspended in the air. Such particles, which are emitted, for example from construction sites and wildfires, adversely affect health, including respiratory diseases and increased risk of cancer [6]. In fact, PM levels can vary from 5000 to 10,000 particles/cm^3 in outdoor air and can increase to 300,000 particles/cm^3 or even 1,000,000 particles/cm^3 on streets with high traffic volumes [20, 21]. Read the last sentence again; the numbers are quite high.

PM includes particles with different properties, including shape, reflectivity, size, and composition; it is most commonly though divided into subcategories based on the particle size information [22]. Figure 12.6 illustrates the particle size (i.e. diameter) ranges of common pollutant sources. Fine particles (sometimes referred to as fine dust) of a diameter of 10 μm or less are designated as PM10, while the more dangerous ultrafine particles of a diameter of 2.5 μm or less are designated as PM2.5.

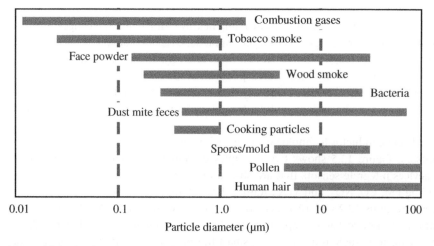

Figure 12.6 Particle size ranges of common pollutant sources. Source: Data and image from Lattanzio [22] and John Wiley and Sons, Best Particles Guide to Residential Construction, 2006.

Figure 12.7 A spherical particle vs. a nonspherical particle. Source: Modified from Horiba Instruments Inc. [23].

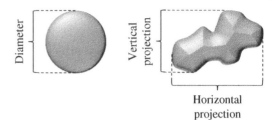

Particles generated during the combustion process are classified as primary particles, which are directly emitted into the atmosphere. These are typically composed of fine particles, with diameter less than 2.5 μm. On the other hand, particles generated by mechanical or chemical reactions in the atmosphere are coarse (diameter > 2.5 μm) and are classified as secondary particles. These secondary particles are from many sources because they can be formed by the accumulation of several other particles [20].

Obviously, if we take a look at the atmospheric aerosol particles under the microscope, we are going to see some 2D projection of the particles. A spherical particle can be easily described using a single number, e.g. the diameter. What about the abundant nonspherical particles? As shown in Figure 12.7, a nonspherical particle can be described using multiple parameters [23]. Yet for simplicity, PM measurement techniques typically make the convenient assumption that every particle is a sphere. In other words, the device determines the diameter of a spherical-equivalent particle that could produce the reported measured value, e.g. scattered light. Unless critical for the target use case, this assumption is broadly accepted.

If hypothetically a single grain of sand has indeed a cylindrical shape (with base diameter d_c, height h_c), it can be modeled as a sphere (with diameter d_s) of the same volume (or mass) [24]:

$$\pi h_c \left(\frac{d_c}{2} \right)^2 = \frac{1}{6} \pi d_s^3$$

If the particle has $d_c = 20\,\mu m$ and $h_c = 100\,\mu m$, it would appear as a sphere with $d_s = 39\,\mu m$.

PM is quantified as the ambient "mass concentration" of particles in μg/m^3. This is due to historical reasons as the most traditional, and the most accurate method to measure PM is the gravimetric-based sensors. These sensors draw preheated dry air at a known flow rate for 24-hours. PM is sorted based on their size and sampled on a preweighted filter. The filter is then weighted postexposure to pollutants in the laboratory to determine the total accumulated PM mass in μg. The average mass concentration in μg/m^3 over 24-hours is then derived simply by dividing the PM mass by the total air volume that passed through the filter. Accordingly, this quantification approach has prevailed until today, and for comparability reasons, other sensing mechanisms as presented later in the chapter strive to comply with it.

12.2.2 Sensing Mechanisms

Measuring PM can be classified, but not limited, to five main methods, namely gravimetric (manual), microbalance (semiautomatic), radiation (semiautomatic), electrical (automatic), and optical (automatic) mechanisms [20, 25]. An overview of the characteristics, advantages, and disadvantages of each mechanism is given herein.

- **Gravimetric mechanism**: The most accurate PM measurements are obtained from instruments that use this gravimetric (sometimes referred to as weighing) mechanism. PM is determined by weighing before and after the air sampling period, typically 24 hours. It is a costly mechanism as the weighing process has to be done manually in the laboratory under standardized and controlled conditions. This mechanism has the additional advantage that the collected PM can be further analyzed chemically. Owing to its accuracy, this mechanism is widely adopted by regulatory institutions. E-FRM-DC from MetOne is an example of a gravimetric PM monitoring equipment.

- **Microbalance mechanism**: This mechanism is based on an oscillatory microbalance element, which alters its resonance frequency depending on the sampled and accumulated PM over its surface. Tapered Element Oscillation Microbalance (TEOM) is the most commonly used instrument based on this approach. TEOM uses a filter which mounted on the end of an oscillating hollow tapered quartz tube. The element is periodically cycled to return to its natural frequency, yet this approach is semiautomatic because the filter needs still to be changed nearly every 30 days. TEOM 1405-F from Thermo Fischer is an example of an instrument that utilizes this sensing mechanism.

- **Radiation mechanism**: This mechanism is based on the absorption of β radiation by PM. The PM sensors that use this mechanism are referred to as Beta Attenuation Monitors (BAM). Such devices collect PM on a filter media (or tape for continuous monitoring) and measure the attenuation of radiated β particles (e.g. Carbon-14) due to the solid matter of PM by integrated Geiger Mueller or photodiode detectors [26]. The filter media need to be changed from time to time, making this mechanism semiautomatic. TEOM 5028i from Thermo Fischer is an example of an instrument that utilizes this sensing mechanism.

- **Electrical mechanism**: This mechanism relies on multilayer electrodes with an initially infinite electrical resistance [27]. During the sensor operation, particles are collected onto the inter-digital electrodes and form conductive paths between the electrode fingers, leading to a drop of the electrical resistance. This principle is considered to be automatic because a controlled regeneration of the detecting electrodes is possible by heating the sensing element at an elevated temperature to burn off the deposited particles.

- **Optical mechanism**: This mechanism enables fully automatic and real-time monitoring of PM. It is based on light scattering, absorption, or extinction caused by the sampled PM. Optical Particle Counters (OPCs) are the most commonly used instrument of this type. OPCs use a light source to illuminate particles at a given angle and a photodetector to measure their properties. Fidas 200 from Palas, EDM180 from Grimm and DustTrak/SidePak from TSI are examples of instruments that employ optical PM sensing mechanism.

The EPA in United States classifies the PM measurement equipment, which are used to monitor the atmospheric air quality for purposes of determining compliance with the NAAQS, into two main categories, namely Federal Reference Methods (FRM) and Federal Equivalent Methods (FEM). Owing to the strict accuracy requirements, gravimetric-based PM monitoring stations are typically classified as FRM, while microbalance- and optical-based PM measurement equipment are classified as FEM. In Germany, FRM-like devices with 24-hour average PM values are used for regulatory purposes, while FEM-like devices with 1-hour average PM values are used to promptly inform the public about the environmental conditions.

There are attempts in research to miniaturize microbalance-based PM sensors with the use of microfabricated, air-microfluidic, and electromechanical (MEMS) structures [28]. Nevertheless, optical-based PM sensors are gaining a wider prominence in the industry and research. As described in more details in Sections 12.2.3 and 12.2.5, this is owed to their offered various virtues for use in low-cost, compact, fast, maintenance-free, and personal air quality monitors.

12.2.3 Optical Particle Counter (OPC)

An OPC is an instrument that detects the counts and sizes of particles [29]. These particles may be in the form of aerosol, dry powders, or liquid suspensions. Therefore, OPCs are widely used in contamination and atmospheric analysis of liquid-borne or air-borne particles. The development of OPCs started in the middle of the twentieth century and accelerated in the 1960s after the invention of lasers [30]. Today, OPCs find prominent use, for example in water, cleanroom, and environmental monitoring. The latter is the focus of this section.

Figure 12.8 sketches a typical OPC system, consisting of a light source, a photodetector, and a particle flow channel. Particles are introduced into the sensitive region by means of fan, pressure, vacuum, sheath, or natural flow [29]. In some systems, particles are separated and aligned by hydrodynamic focusing. An internal feedback loop is present to ensure a constant and stable airflow speed through the sensor's channel, where the light beam, either by an incandescent lamp or by a laser source, is focused to a small sensing volume. Knowing the exact dimensions of the system together with the airflow speed, the volume of the sampled air can be easily calculated. This is helpful to derive, as described in Section 12.2.4, the particle concentration/mass distribution in $\mu g/m^3$. By means of light scattering and/or light extinction/obscuration, particles of different sizes (about 50 nm to 1 mm) passing through the light source are detected by the photodetector.

An alternative, yet similar, method to scattering-based detection of particles is by using direct imaging [31]. This vision-based approach uses a high-definition, high-magnification camera to record the passage of particles that are illuminated by halogen light. The two-dimensional images are analyzed and processed by means of sophisticated software algorithms to obtain the properties of the particles being measurement, including size, color, shape, etc.

Back to the scattering-based detection method. As the dependence of the scattered light intensity on the refractive index of the particles is less pronounced around 90° of scattering angle (α, see Figure 12.8), the photodetector is typically placed perpendicular to the airflow [30]. As it is important to avoid coincidence errors resulting from many particles in the sensing volume, the light source is focused to a small region and the instrument specifies a maximum detectable particle concentration.

The passage of particles into the small sensitive region gives rise to a detectable pulse signals, the magnitude of which depends on the dimension and optical properties of the particles [29]. The recorded signals are counted for each intensity class, then processed, sorted, and tabulated by means of Mie-Theory into standardized particle size bins in order to construct a histogram of particle number size distribution (Figure 12.9).

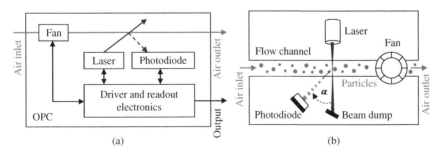

(a)   (b)

Figure 12.8   Optical particle counter (OPC) system: (a) block diagram, (b) schematic. Source: Data from Lattanzio [22].

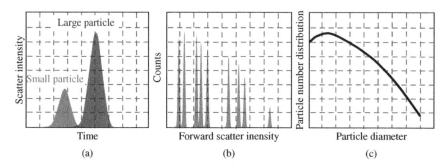

Figure 12.9 Sketch of typical OPC signals: (a) detected intensity signal for small and large particles, (b) particle counts vs. detected forward scatter intensity, (c) particle number distribution.

A first-order approximation, the particle diameter can be estimated by multiplying the pulse width (time) and the average airflow (rate) during the pulse. In practice, however, the scattered light is a function of the illuminating wavelength, intensity, optics, particle, and medium properties. Therefore, OPCs are carefully designed and constructed to incorporate all these attributes into the particles count and size estimation. In principle, the correspondence between the recorded pulse height and particle size is resolved by comparing the data to a standard calibration curve, which is obtained from a set of uniform particles of known diameter in a controlled environment [30].

The estimation of the particle number size distribution is imperfect because accounting for particle density, refractive index, particle velocity, relative humidity, and other meteorological parameters is complicated. For example, assuming a constant mass density, like in many low-cost PM sensors, will create an error when measuring particles of different types, for example "heavy" house dust vs. "light" combustion particles [22]. Some commercial devices use meteorological sensors and dehumidification devices, which can be as complex as a Peltier-heater or as simple as a resistor, to reduce or to eliminate the impact of relative humidity on the readings. Other novel approaches, instead, suggest utilizing multiangle light scattering [32]. Furthermore, the particle density can be calculated from the motion equations in the case of accelerating particles, while the particle velocity and refractive index can be measured using laser Doppler and dual-wavelength methods, respectively [30].

12.2.4 Particle Size Distributions

The particle size distribution can be represented as a number-, volume-, or mass-size distribution. The earlier (i.e. particle number-size distribution) was presented so far in Section 12.2.3. To understand the difference, consider the following scenario [33]: three particles with a diameter of $1\,\mu m$, three particles with a diameter of $2\,\mu m$, and three with a diameter of $3\,\mu m$ are present in the air. As shown in Figure 12.10, generating a number-size distribution for these particles will generate a uniform distribution, where each particle size accounts for one-third of the total. Plotting the volume-size distribution, on the other hand, results in 75% and <3% of the total volume coming from the 3 and $1\,\mu m$ particles, respectively. Noting that the mass-size distribution is nothing, but the multiplication of the volume-size distribution by the volumetric mass density ρ of the particles.

PM10 and PM2.5 are derived from the mass-size distribution and are defined to be the mass concentrations in $\mu g/m^3$ for particles with diameters of ≤ 10 and $\leq 2.5\,\mu m$, respectively. Unlike the gravimetric PM sensors, OPCs output particle number-size distribution and not concentrations in $\mu g/m^3$. Assuming spherical particles with a volume of $(1/6) \cdot \pi d^3 n(d)$, where d is the particle

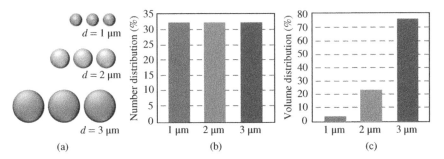

Figure 12.10 Particle size distributions: (a) particles with diameter d, (b) number distribution, (c) volume distribution. Source: Adapted from Horiba Instruments Inc. [33].

diameter and $n(d)$ is the particle number-size distribution (in $\mu m^{-1} m^{-3}$ because it is normalized to the sampled air volume, see Figure 12.9) described in Section 12.2.3, PMx concentration in $\mu g/m^3$ can be calculated as follows [34]:

$$\text{PMx} = \frac{C_F \pi \rho}{6} \int_0^x d^3 n(d) \, dd$$

where C_F is a correction factor with a default value of 1.0.

12.2.5 Miniaturized Optical PM Sensing

Although gravimetric-based PM sensors are long established as the most accurate way of determining mass concentration, they have some practical limitations for use in consumer devices [22]. This is because gravimetric-based PM sensors are bulky, stationary, located outdoors, expensive, and slow. They do not generate personalized information and only measure air quality in their immediate vicinity that is averaged-out over 24 hours, and thus lack information for monitoring the fluctuations in local PM levels and for tracking the rapidly changing environment around us [35].

Optical-based PM sensors, on the other hand, have dramatically lower operational cost and can be miniaturized, thereby offering the opportunity to measure both indoors and outdoors as well as to improve both the spatial and temporal resolution of PM data. In order to enable more and different use cases, it is quite impressive to see how engineers have been able to reduce the dimensions of optical-based PM sensors from the size of a home printer (e.g. Fidas-200) to the size of a matchbox (e.g. SPS-30, PMS-5003, and GP2Y1010AU0F) and even recently down to the size of a match head (e.g. Bosch Sensortec's PM sensing technology [35]). Figure 12.11 shows the evolution.

Following the same principle described in Section 12.2.3, all devices shown in Figure 12.11 rely on a built-in fan to sample air through a flow channel. The limitation, however, of such sensors is the fan noise, the sheer physical size, and the required cumbersome maintenance against cloggage, making them impractical for use in flat portable devices, such as smartphones. On the other hand, a new optical PM sensing technology announced by Bosch Sensortec in [35] promise a significantly smaller solution (i.e. just ~0.2% of the volume of the state of the art) by getting rid of the cumbersome fan, inlet, and flow channel. Instead, the sensor uses eye-safe lasers to detect, in real time, particles that are naturally and freely moving in ambient air. This ultracompact design would enable seamless integration into nearly all consumer devices, such as mobile- and IoT-devices, and allows users to safeguard their health against environmental setbacks by monitoring their personal exposure to PM.

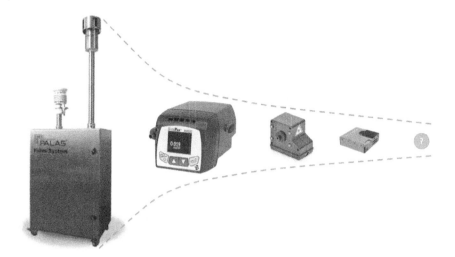

Figure 12.11 Evolution of optical-based PM sensors (size is not to scale). From left to right: Fidas 200 from Palas, SidePak from TSI, OPC-N2 from Alphasense and SPS-30 from Sensirion. This wide spectrum of devices enables to cover a broad range of different use cases.

12.3 Volatile Organic Compounds (VOCs) Sensing

12.3.1 Volatile Organic Compounds (VOCs)

VOCs are organic (carbon-containing) chemicals that are present mostly as gases at room temperature [36, 37]. These organic compounds are called volatile because they evaporate easily at ordinary room temperature. They are found in various man-made and naturally occurring solids and liquids, some of which are harmful to health and can cause environmental damage. Although natural biological VOC emissions tend to be larger overall, man-made sources are of greater concern in urban areas. The severity, however, depends on the type of organic compound, its concentration and the duration of exposure. Among the adverse health effects of VOCs are, but not limited to, sensory irritation, allergies, asthma, cancer, and damage to internal organs.

Examples of dangerous VOCs include the following [37]:

- **Acetone**: It is a colorless, highly flammable liquid chemical, which is considered to be one of the most famous of organic solvents. Besides, the natural sources of acetone, such as volcanoes and forest fires, it is industrially used to make plastic, fibers, drugs, and chemicals and is also found in exhaust of cars and tobacco smoke. It can irritate the eyes, throat, lungs, and nose. It may also cause a faster pulse, headaches and dizziness, nausea and vomiting, a shorter woman's menstrual cycle, and the possibility of passing out [38].
- **Benzene**: It is a colorless or light-yellow liquid, which evaporates into the air very quickly. Benzene is part of crude oil, gasoline, and cigarette smoke. Furthermore, it is industrially used to make resins, nylon, and synthetic fibers. It may cause drowsiness, dizziness, confusion, headaches, and unconsciousness. Additionally, it can decrease red blood cells when exposed to it for a long period of time [39].
- **Toluene**: It is a colorless liquid but has a distinctive smell. Besides, its use in the production of plastics, nylon, and benzene, Toluene is used in making paints, paint thinners, fingernail polish, lacquers, adhesives, and rubber. It can affect the nervous system in a number of ways. In the most

serious of incidences, it can affect the kidneys and liver, and can cause retardation in unborn babies [40].

- **Formaldehyde**: It is a colorless and flammable chemical with a strong smell. As it is used in pressed-wood, formaldehyde is present in building materials, and household products. It may cause burning sensations in the eyes, nose, and throat. In addition, it can result in skin irritation, coughing, wheezing, and watery eyes. Furthermore, it has also been linked by some studies to myeloid leukemia [41].

VOCs can be categorized in several ways, such as by structure (e.g. straight-chained, branched, or ring), by the type of chemical bonds (e.g. alkanes, alkenes, alkynes, saturated, or unsaturated), by the function of specific parts of the molecules (e.g. aldehydes, ketones, or alcohols), or by specific element included (e.g. hydrogen or chlorine) [36]. Furthermore, they can be classified by the ease of which they can be emitted into air. For example, WHO classifies organic pollutants into the following [42, 43]:

- Very volatile organic compounds (VVOCs)
- Volatile organic compounds (VOCs)
- Semivolatile organic compounds (SVOCs)

Intuitively, the higher the volatility (lower the boiling point), the more likely the compound will be emitted from a product or surface (e.g. furnishings and building materials) into the air. In other words, VVOCs are hardly present in materials or on surfaces due to their very high volatility, while SVOCs are hardly present in the air due to their low volatility. Table 12.2 lists concrete examples.

VOCs are present indoors and outdoors with different concentrations [44]. Typical indoor VOC sources include paint, cleaning supplies, personal care products, gas stoves, ordinary heating systems, air fresheners, furnishing, glues, permanent markers, and printing equipment [36, 45]. Some indoor VOCs can be at high concentrations. For example, in some homes ethanol concentrations are above $1000\,\mu g/m^3$, while 1,4-dichlorobenzene, α-pinene, and D-limonene concentrations are around $100\,\mu g/m^3$ [46]. On the other hand, typical outdoor VOC sources include emissions from oil and gas industry, smoking, burning, solvent usage, transportation, and building materials [47, 48]. D'Souza and Batterman have made a thorough population-based study in [49], presenting distributions of personal VOC exposures.

VOC concentrations are typically measured and reported in parts per billion (ppb), parts per million (ppm), or micrograms per cubic meter ($\mu g/m^3$) [36]. If the concentration is 1 ppb, it means that for every billion molecules of air, there is one molecule of the VOC. If the concentration is $1\,\mu g/m^3$, on the other hand, it means that for every cubic meter volume of air there is $1\,\mu g$ of mass (weight) of the VOC.

Table 12.2 Classification of organic compounds according to WHO [42, 43].

Classification	Abbreviation	Boiling point ranges (°C)	Examples
Very volatile organic compounds	VVOCs	<0 to 50–100	Propane, butane, methyl chloride
Volatile organic compounds	VOCs	50–100 to 240–260	Formaldehyde, D-limonene, toluene, acetone, ethanol (ethyl alcohol), 2-propanol (isopropyl alcohol), hexanal
Semivolatile organic compounds	SVOCs	240–260 to 380–400	Pesticides (DDT, chlordane, plasticizers [phthalates], fire retardants [PCB, PBB])

Measuring the concentration of VOCs is important in order to increase user's awareness and to compare against standard regulatory limits, such as the threshold limit value (TLV), which is the limit value a worker can be exposed to for a working lifetime without adverse health effects [50]. Particularly for air quality devices, where a simple indication for the user is easier to understand and react upon, the concentration of VOCs is typically reported as the total concentration (i.e. sum) of multiple airborne VOCs that are present simultaneously in the air. This is sometimes being referred to in literature as TVOCs. Since technologies differ in their sensitivity to individual VOCs, one has to be careful in interpreting the data and in comparing different VOC sensors.

12.3.2 Sensing Mechanisms

The so-called "golden-standard mechanism" for measuring VOC is the EPA Method TO-15 [51, 52]. In this approach, discrete air sample is first collected in a passivated, evacuated steel canister and then measured in the lab using gas chromatography coupled to mass spectrometry (GC/MS). Besides being expensive, time-consuming, and requiring skilled operators, this mechanism is not amenable to reactive species that cannot be readily captured and transported.

Accordingly, engineers have been developing various other field-deployable approaches to make compact and portable VOC sensors that can better cover special and temporal resolutions with a relatively low price tag. If such technologies fulfill their promise, applications such as personal air-quality and forest network monitoring would be feasible. The following are among the most prominent mechanisms for sensing VOCs [53]:

- **Ionization sensors**: Photo-ionization detectors (PID) use high-energy photons typically in the ultraviolet (UV) range (e.g. xenon lamp) to excite and ionize the gas molecules. Owing to the typical photon energy range (about 10 eV), only gases with low ionization (i.e. not oxygen, not nitrogen) can be ionized. The resulting ions produce an electric current that is proportional to the signal output of the detector. Note, however, that VOCs such as benzene, toluene, and xylene cannot be detected by this mechanism because of their higher ionization potential than that of the lamp. Examples of commercial PID-based VOC sensors are PID-AH from Alphasense LTD and VOC-Traq from Baseline-Mocon.
- **Amperometric sensors**: These electrochemical sensors consist of electrodes and an electrolytes (e.g. solid, gel-like, liquid, or gas). The detection of VOCs relies on a chemical redox reaction at the electrode and the transport of charges throughout the electrolyte. In other words, the VOCs diffuse to the sensor's electrode, and thereby, a direct electron transfer takes place which produces a measurable electric current that is proportional to the gas concentration. Examples of commercial amperometric-based VOC sensors are ETO-A1 from Alphasense LTD and 7ETO CiTiceL from City Technology.
- **Resistive sensors**: Metal oxide (MOX)-based sensors consist of electrodes and a MOX (e.g. SnO_2 and WO_3), the electrical property of which changes depending on the exposed VOC gases. The resistance between the electrodes can be accordingly measured to detect the VOC gases in the air. Note that MOX-based sensors are not only sensitive to VOCs but also to inorganic harmful gases, such as NO, NO_2, and CO. The small size and low cost of MOX-based sensors make them attractive for many applications. Examples of commercial MOX-based VOC sensors are TGS 2201 from Figaro, iAQ-100 from AppliedSensor, VM from Aeroqual, SGP40 from Sensirion, and BME680 from Bosch Sensortec.
- **Micro-gas chromatograph (μGC) sensors**: These sensors rely on the principles of gas chromatography but with miniaturized integration [54]. In such sensors, a mixture of sample VOC

vapors are injected at the input of the column by a valve and are swept through it by inert carrier gas. The column separates the mixture and outputs a series of VOC vapors separated by regions. A specific property (e.g. thermal conductivity) of these vapors is then measured to derive the gas concentration. Examples of commercial μGC-based VOC sensors are Frog-4000 from Defiant Technologies and Model-312 from PID Analyzers.

Unlike the other sensing mechanisms, MOX-based VOC sensors offer the combined virtues of being low cost, small in size, and sensitive also to inorganic harmful gases. They are able to supply near or real-time air pollution measurements by relatively simple and regenerative means. Accordingly, the system breakdown and the operating principle of MOX-based VOC sensors are the focus of Section 12.3.3.

12.3.3 MOX-Based Sensors

The basic elements of a MOX-based gas sensor are electrodes, a MOX, and a heater (see Figure 12.12). The sensing mechanism is based on changes in the conductivity of the active MOX-layer associated with adsorption and desorption of gas molecules on the MOX surface. The MOX reacts to almost all reducing and oxidizing gases, making the sensor sensitive not only to complex aromas, such as VOCs, but also to harmful trace gases such as carbon monoxide (CO), nitric oxides (NO_x), ammonia (NH_3), and sulfurous gases (H_2S, SO_x) [56].

The chemical reaction occurring at the MOX layer is referred to as reduction–oxidation (REDOX) reaction. Consider, for example an indoor MOX gas sensor which is sensitive to carbon monoxide (CO). The corresponding REDOX reactions can be described in a simplified form as follows [57, 58]:

- **Adsorption**: Reduction of an oxidizing gas: $0.5O_2 + e^- \rightleftharpoons O^-$
- **Desorption**: Oxidation of a reducing gas: $CO + O^- \rightleftharpoons CO_2 + e^-$

How to interpret these reactions? Abundant oxygen molecules in the air get adsorbed to the MOX surface, trapping electrons. When the sensor is then exposed to the target gas, which is CO in this example, these oxygen molecules react and are accordingly removed, allowing the electrons to flow in the MOX. This results in a reduction of the electrical resistance that is measured by the sensor.

Among the harmful gases which are detected by MOX-based sensing technology are the following:

- **Reducing gases (mostly available indoors)**: Carbon monoxide (CO), benzene (C_6H_6), acetone (C_3H_6O), ethanol (C_2H_6O), formaldehyde (CH_2O), naphthalene ($C_{10}H_8$), hydrogen sulfide (H_2S), and methane (CH_4)
- **Oxidizing gases (mostly available outdoors)**: Sulfur oxides (SO_x), nitric oxides (NO_x), and ozone (O_3)

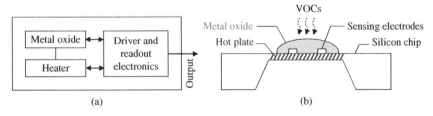

Figure 12.12 MOX-based gas sensor: (a) block diagram and (b) schematic cross-section. Source: Adapted from Sensirion [55].

Depending on the MOX material and the target gas, elevated operating temperatures of around 300–900 °C are necessary to enable the desired REDOX reactions. This is the main reason behind the need of a well-controlled heater element (see Figure 12.12).

Examples of commonly used MOX-materials are tin oxide (SnO_2) [59–63], tungsten oxide (WO_3) [64, 65], zinc oxide (ZnO) [66, 67], and titanium oxide (TiO_2) [68]. The MOX layer can be deposited on the MEMS-structure and electrodes using various fabrication methods, including sputtering, evaporation, and printing [56]. Besides the MOX-material, MOX-structure, -deposition and -doping are carefully selected to enhance the reliability of the sensor and to optimize for the target analyte(s). For example, one of the main manufacturing objects is to deposit a porous MOX structure to ensure high sensitivity by means of favorable surface-to-volume ratio. This is because the larger the MOX surface area exposed for a REDOX chemical reaction, the higher the sensor's sensitivity.

Thanks to innovative semiconductor manufacturing methods, miniaturized MOX-based gas sensors can be produced in large volumes and at low cost. Monolithic cointegration of the MOX-sensor-element together with the digital/analog driver and read-out electronics on a single chip is nowadays feasible [55, 60, 69]. To improve accuracy, to give the user a complete environmental picture, and to ensure a stable operation independent of the surrounding meteorological conditions, the MOX-based gas sensor BME680 from Bosch Sensortec, as one prominent example, also integrates temperature, pressure, and relative humidity sensors in a single component [70].

The power consumption of MOX-based gas sensors depends strongly on their construction and structure. For example, sensors on silicon-bulk substrates require a power of about 1 W at 400 °C [56]. Using the so-called "micro-hotplates," which are thin MEMS membranes as illustrated in Figure 12.12, thermal decoupling of the active sensor element from the housing is guaranteed and a low power consumption (i.e. in the range of 15 mW at 400 °C) is achieved [56]. This opens the possibility to use MOX-based sensors in battery-powered applications. Furthermore, such micro-hotplates, which are typically produced by wet-chemical etching, allows operation with rapidly adjustable temperature levels. Thus, the active temperature can be set and controlled in mere milliseconds. This is beneficial to improve the selectivity of MOX-based sensors by using a so-called "temperature-modulation operating scheme."

Selectivity of MOX-based sensors can be obtained using a wide variety of approaches, including (i) the use of filters or chromatographic columns to discriminate between gases based on their molecular size, (ii) the use of catalysts and promoters or more specific surface additives, (iii) the use of temperature modulation either by setting a selected fixed temperature or by programming various temperature steps [71]. The latter is probably the most cost-effective solution as it is based "only" on software (i.e. doesn't add costs to the bill of material). So how temperature-modulation operating scheme works? The micro-hotplate is pulsed to a specific temperature or cycled between two or more temperatures. The measured raw data is then extracted during these temperature steps and/or transitions to uniquely identify a target analyte in the air [63, 65, 71–74]. For example, Urasinska-Wojcik and Gardner have demonstrated in [72] that using such temperature-modulation operating scheme between 200 and 350 °C, H_2S impurity can be uniquely identified in a harsh environment and using only few seconds of the signal response.

Owing to this huge diversity and miniaturized integration, there are sensors which comprise an array of different MOX-based sensitive elements, either with different MOX-materials, operating scheme, or both [55, 56, 65, 69, 75]. Such devices are sometimes referred to as "electronic noses" or "e-noses" because of their capability to detect odors (i.e. mixture of different gases). Each element in the sensor array responds to a single/number of specific gas molecules. Using smart algorithms and machine learning, fingerprints for gas compositions can thus be uniquely identified. The corresponding use cases are endless and only limited by one's ingenuity. For example, monitoring cleanliness of public places, classification of bad breath, and detection of spoiled food are going to be a viable reality [35].

Humans use a family of more than 400 olfactory receptors to detect odors [76]. Can a MOX-based e-nose get close? Not quite yet. That's why working on advancing environmental sensing technology is interesting.

References

1 National Geographic. (2021) Atmosphere. [Online]. Available: https://www.nationalgeographic .org/encyclopedia/atmosphere-RL/

2 European Environment Agency. (2021) Trends in Atmospheric Concentrations of CO_2 (ppm), CH_4 (ppb) and N_2O (ppb) Between 1800 and 2017. [Online]. Available: https://www.eea.europa .eu/data-and-maps/daviz/atmospheric-concentration-of-carbon-dioxide-5#tab-chart_5_filters= %7B%22rowFilters%22%3A%7B%7D%3B%22columnFilters%22%3A%7B%22pre_config_polutant %22%3A%5B%22CO2%20(ppm)%22%5D%7D%7D

3 J. McGlade (ed.), *Every Breath We Take: Improving Air Quality in Europe*. Luxembourg, Luxembourg: Publications Office of the European Union, 2013.

4 World Health Organization. (2021) Air Pollution. [Online]. Available: https://www.who.int/ health-topics/air-pollution#tab=tab_1

5 Umweltbundesamt. (2021) Air Pollution. [Online]. Available: https://www.umweltbundesamt .de/en

6 United States Environmental Protection Agency. (2021) Air Pollution. [Online]. Available: https://www.epa.gov

7 World Health Organization, *WHO Guidelines for Indoor Air Quality: Selected Pollutants*. Copenhagen, Denmark: WHO, 2010.

8 United States Environmental Protection Agency. (2021) NAAQS Table. [Online]. Available: https://www.epa.gov/criteria-air-pollutants/naaqs-table

9 A. King. (2021) Linking Air Pollution to Higher Coronavirus Death Rates. [Online]. Available: https://www.hsph.harvard.edu/biostatistics/2020/04/linking-air-pollution-to-higher-coronavirus-death-rates/

10 C. Nagel, W. Spangl, and I. Buxbaum, *Sampling Points for Air Quality: Representativeness and Comparability of Measurement in Accordance with Directive 2008/50/EC on Ambient Air Quality and Cleaner Air for Europe*, European Union, 2019.

11 U.S. Environmental Protection Agency, *Technical Assistance Document for the Reporting of Daily Air Quality — The Air Quality Index (AQI)*. North Carolina, USA: EPA, 2018.

12 E. von Schneidemesser, K. Steinmar, E. C. Weatherhead, B. Bonn, H. Gerwig, and J. Quedenau, "Air pollution at human scales in an urban environment: impact of local environment and vehicles on particle number concentrations," *Science of the Total Environment*, vol. 688, pp. 691–700, 2019, https://www.sciencedirect.com/science/article/pii/S0048969719328827.

13 United States Environmental Protection Agency. (2021) Managing Air Quality – Ambient Air Monitoring. [Online]. Available: https://www.epa.gov/air-quality-management-process/ managing-air-quality-ambient-air-monitoring

14 United States Environmental Protection Agency. (2021) Air Data: Air Quality Data Collected at Outdoor Monitors Across the US. [Online]. Available: https://www.epa.gov/outdoor-air-quality-data

15 Landesanstalt für Umwelt Baden-Württemberg. (2021) Messstelleninformation. [Online]. Available: https://www.lubw.baden-wuerttemberg.de/luft/messstelleninformation

16 T. Müller and T. Warth, *Wirksamkeit von Filtersäulen im Bereich Stuttgart Am Neckartor: Abschlussbericht*. Whitepaper, Ludwigsburg, Germany: MANN+HUMMEL GmbH, 2020.

17 P. Lenschow, H.-J. Abraham, K. Kutzner, M. Lutz, J.-D. Preuß, and W. Reichenbächer, "Some ideas about the source of PM10," *Atmospheric Environment*, vol. 35, 1, pp. S23–S33, 2001.

18 Landesanstalt für Umwelt Baden-Württemberg. (2021) Feinstaub PM10: vorläufige Tagesmittelwerte. [Online]. Available: https://www.lubw.baden-wuerttemberg.de/luft/feinstaub-stuttgart?stationId=DEBW118

19 Extra 3 NDR. (2020) Realer Irrsinn: Schummel-Luftmessung in Reutlingen. [Online]. Available: https://youtu.be/AUhbpDsFLxs

20 S. S. Amaral, J. An De Carvalho, Jr., M. A. M. Costa, and C. Pinheiro, "An overview of particulate matter measurement instruments," *Atmosphere*, vol. 6, no. 9, pp. 1327–1345, 2015.

21 M. J. Utell and M. W. Frampton, "Acute health effects of ambient air pollution: the ultrafine particle hypothesis," *Journal of Aerosol Medicine*, vol. 13, no. 4, pp. 355–359, 2000.

22 L. Lattanzio. (2021) Particulate Matter Sensing for Air Quality Measurements, Sensirion: The Sensor Company. [Online]. Available: https://www.sensirion.com/en/about-us/newsroom/sensirion-specialist-articles/particulate-matter-sensing-for-air-quality-measurements/

23 Horiba Instruments Inc. (2019) Particle Size Essentials Guidebook, Technical Note. [Online]. Available: https://www.horiba.com/en_en/en-en/products/by-segment/scientific/particle-characterization/particle-guidebook/

24 Malvern Instruments Worldwide. (2014) Basic Principles of Particle Size Analysis, Application Note. [Online]. Available: https://www.atascientific.com.au/wp-content/uploads/2017/02/AN020710-Basic-Principles-Particle-Size-Analysis.pdf

25 J. Whalley and S. Zandi. (2016) Particulate Matter Sampling Techniques and Data Modeling Methods. Open Access Peer-Reviewed Chapter. [Online]. Available: https://www.intechopen.com/books/air-quality-measurement-and-modeling/particulate-matter-sampling-techniques-and-data-modelling-methods

26 Thermo Fischer Scientific. (2021) Beta Attenuation Technology for Particulate Matter Measurement. [Online]. Available: https://www.thermofisher.com/de/de/home/industrial/environmental/environmental-learning-center/air-quality-analysis-information/beta-attenuation-technology-particulate-matter-measurement.html

27 T. Ochs, H. Schittenhelm, A. Genssle, and B. Kamp, "Particulate matter sensor for on board diagnostics (OBD) of diesel particulate filters (DPF)," *SAE International Journal of Fuels and Lubricants*, vol. 3, no. 1, pp. 61–69, 2010.

28 I. Paprotny, F. Doering, P. A. Solomon, R. M. White, and L. A. Gundel, "Microfabricated air-microfluidic sensor for personal monitoring of airborne particulate matter: design, fabrication and experimental results," *Sensors and Actuators A: Physical*, vol. 201, pp. 506–516, 2013, https://www.sciencedirect.com/science/article/abs/pii/S0924424712007637.

29 R. Xu, *Particle Characterization: Light Scattering Methods*. Dordrecht, Netherlands: Springer, 2002.

30 A. Czitrovszky, Environmental applications of solid-state lasers, *Handbook of Solid-State Lasers*. Woodhead Publishing Series in Electronic and Optical Materials, pp. 616–646, 2013, https://www.sciencedirect.com/science/article/pii/B9780857092724500221?via%3Dihub.

31 C. Igathinathane, S. Melin, S. Sokhansanj, X. Bi, C. Lim, L. O. Pordesimo, and E. Columbus, "Machine vision based particle size and size distribution determination of airborne dust particles of wood and bark pellets," *Powder Technology*, vol. 196, no. 2, pp. 202–212, 2009.

32 W. Shao, H. Zhang and H. Zhou, "Fine particle sensor based on multi-angle light scattering and data fusion," *Sensors (Basel)*, 17(5), 2017.

33 Horiba Instruments Inc. (2017) Particle Size Result Interpretation Number vs. Volume Distributions, Technical Note. [Online]. Available: https://www.horiba.com/en_en/static-light-scattering-sls-particle-size-result-interpretation/

34 Y.-H. Cheng and Y.-L. Lin, "Measurement of particle mass concentrations and size distributions in an underground station," *Aerosol and Air Quality Research*, vol. 10, no. 1, pp. 22–29, 2009.

35 A. Herrmann and R. Fix. (2020) Air Quality Measurement Gets Personal, Bosch White Paper. [Online]. Available: https://www.bosch-sensortec.com/media/boschsensortec/downloads/white_papers/white-paper_air-quality-measurement_june-2020-p2.pdf

36 Berkeley Lab. (2021) Introduction to VOCs and Health. [Online]. Available: https://iaqscience.lbl.gov/voc-intro

37 A. Lafond, Foobot. (2021) The Unit of Measure of Volatile Organic Compounds (VOCs). [Online]. Available: https://foobot.io/guides/volatile-organic-compounds-unit-of-measure.php

38 Delaware Health and Social Services: Division of Public Health. (2015) Acetone. [Online]. Available: https://www.dhss.delaware.gov/dph/files/acetonefaq.pdf

39 Centers for Disease Control and Prevention: Emergency Preparedness and Response. (2018) Facts About Benzene. [Online]. Available: https://emergency.cdc.gov/agent/benzene/basics/facts.asp

40 Agency for Toxic Substances & Disease Registry. (2015) Public Health Statement for Toluene. [Online]. Available: https://www.atsdr.cdc.gov/phs/phs.asp?id=159&tid=29

41 National Center Institute. (2011) Formaldehyde and Cancer Risk. [Online]. Available: https://www.cancer.gov/about-cancer/causes-prevention/risk/substances/formaldehyde/formaldehyde-fact-sheet

42 United States Environmental Protection Agency. (2021) Technical Overview of Volatile Organic Compounds. [Online]. Available: https://www.epa.gov/indoor-air-quality-iaq/technical-overview-volatile-organic-compounds

43 World Health Organization, "Indoor air quality: organic pollutants," in *Report on a WHO Meeting*, Berlin, 23–27 August 1987. EURO Reports and Studies 111. Copenhagen: World Health Organization Regional Office for Europe, 1989.

44 G. De Gennaro, G. Favela, A. Marzocca, A. Mazzone, and M. Tutino, "Indoor and outdoor monitoring of volatile organic compounds in school buildings: indicators based on health risk assessment to single out critical issues," *International Journal of Environmental Research and Public Health*, vol. 10, no. 12, pp. 6273–6291, 2013.

45 W. G. Tucker, Chapter 31: Volatile organic compounds. In: J. D. Spengler, J. M. Samet, and J. F. McCarthy (ed.), *Indoor Air Quality Handbook*. New York City: McGraw Hill, pp. 31.1–31.20, 2000.

46 J. M. Logue, T. E. McKone, M. H. Sherman, and B. C. Singer, "Hazard assessment of chemical air contaminants measured in residences," *Indoor Air*, vol. 21, no. 2, pp. 92–109, 2012.

47 Aeroqual. (2021) VOC Sensors & Monitors. [Online]. Available: https://www.aeroqual.com/voc-sensors-monitors

48 S. M. Charles, C. Jia, S. A. Batterman, and C. Godwin, "VOC and particulate emissions from commercial cigarettes: analysis of 2,5-DMF as an ETS tracer," *Environmental Science and Technology*, vol. 42, no. 4, pp. 1324–1331, 2008.

49 C. Jia, J. D'Souza, and S. Batterman, "Distributions of personal VOC exposures: a population-based analysis," *Environment International*, vol. 34, no. 7, pp. 922–931, 2008.

50 Wikipedia. (2021) Threshold Limit Value. [Online]. Available: https://en.wikipedia.org/wiki/Threshold_limit_value

51 W. A. McClenny and M. W. Holdren, *Compendium Method TO-15: Determination of Volatile Organic Compounds (VOCs) in Air Collected in Specially-Prepared Canisters and Analyzed by Gas Chromatography/Mass Spectrometry (GC/MS)*. Ohio, USA: EPA, 1999.

52 Los Gatos Research. (2021) VOC Analyzer, Application Note. [Online]. Available: https://www.lgrinc.com/documents/App%20Note%20-%20ABB%20VOC%20Analyzer.pdf

53 L. Spinelle, M. Gerboles, G. Kok, S. Persijn, and T. Sauerwald, "Review of portable and low-cost sensors for the ambient air monitoring of benzene and other volatile organic compounds," *Sensors (Basel)*, vol. 17, no. 7, p. 1520, 2017.

54 S. C. Terry, J. H. Jerman, and J. B. Angell, "A gas chromatographic air analyzer fabricated on a silicon wafer," *IEEE Transactions on Electron Devices*, vol. 26, no. 12, pp. 1880–1886, 1979.

55 Sensirion. (2021) VOC Sensor SGP40. [Online]. Available: https://www.sensirion.com/en/environmental-sensors/gas-sensors/sgp40/

56 M.-L. Bauersfeld. (2021) Semi-Conductor Gas Sensors, White Paper. [Online]. Available: https://www.ipm.fraunhofer.de/content/dam/ipm/en/PDFs/product-information/GP/ISS/semiconductor-gas-sensors.pdf

57 A. M. Collier-Oxandale, J. Thorson, H. Halliday, J. Milford, and M. Hannigan, "Understanding the ability of low-cost MOx sensors to quantify ambient VOCs," *Atmospheric Measurement Techniques*, vol. 12(3), pp. 1441–1460, 2019.

58 P. Shankar and J. B. B. Rayappan, "Gas sensing mechanism of metal oxides: the role of ambient atmosphere, type of semiconductor and gases – a review," *Science Letters Journal*, vol. 4, p. 126, 2015, http://www.cognizure.com/scilett.aspx?p=200638572.

59 C.-L. Dai and M.-C. Liu, Nanoparticle SnO_2 gas sensor with circuit and micro heater on chip fabricated using CMOS-MEMS technique. In: *IEEE International Conference on Nano/Micro Engineered and Molecular Systems*, pp. 16–19, 2007, https://ieeexplore.ieee.org/document/4160480.

60 M. Graf, D. Barrettino, M. Zimmermann, A. Hierlemann, H. Baltes, S. Hahn, N. Barsan, and U. Weimar, "CMOS monolithic metal-oxide sensor system comprising a microhotplate and associated circuitry," *IEEE Sensors Journal*, vol. 4, no. 1, pp. 9–16, 2004.

61 I. Elmi, S. Zampolli, E. Cozzani, M. Passini, G. Pinzochero, G. C. Cardinali, and M. Severi, "Ultra low power MOX sensors with ppm-level VOC detection capabilities," *IEEE Sensors*, pp. 28–31, 2007, https://ieeexplore.ieee.org/document/4388363.

62 L. Z. Yan, H. F. Hawari, and G. W. Djaswadi, Highly sensitive SnO_2-reduced graphene oxide hybrid composites for room temperature acetone sensor. In: *IEEE International Colloquium on Signal Processing & Its Applications (CSPA)*, pp. 8–9, 2019, https://ieeexplore.ieee.org/document/8695987.

63 M. Leidinger, T. Sauerwald, W. Reimringer, G. Ventura, and A. Schütze, "Selective detection of hazardous VOCs for indoor air quality applications using a virtual gas sensor array," *Journal of Sensors and Sensor Systems*, vol. 3, no. 2, pp. 253–263, 2014.

64 M. Stankova, X. Vilanova, J. Calderer, E. Llobet, P. Ivanov, I. Gràcia, C. Cané, and X. Correig, "Detection of SO_2 and H_2S in CO_2 stream by means of WO_3-based micro-hotplate sensors," *Sensors and Actuators B: Chemical*, vol. 102, no. 2, pp. 219–225, 2004.

65 M. Penza, G. Casino, and F. Tortorella, "Gas recognition by activated WO_3 thin-film sensors array," *Sensors and Actuators B: Chemical*, vol. 81, no. 1, pp. 115–121, 2001.

66 J. Gut, J. Zhang, M. Zhou, D. Ju, H. Xu, and B. Cao, "High-performance gas sensor based on ZnO nanowire functionalized by Au nanoparticles," *Sensors and Actuators B: Chemical*, vol. 199, pp. 339–345, 2014, https://www.sciencedirect.com/science/article/abs/pii/S0925400514004109.

67 S.-W. Choi and S. S. Kim, "Room temperature CO sensing of selectively grown networked ZnO nanowire by Pd nanodot functionalization," *Sensors and Actuators B: Chemical*, vol. 168, pp. 8–13, 2012.

68 W. Maziarz, A. Kusior, and A. Trenczek-Zajac, "Nanostructured TiO_2-based gas sensors with enhanced sensitivity to reducing gases," *Beilstein Journal of Nanotechnology*, vol. 7, pp. 1718–1726, 2016, https://www.beilstein-journals.org/bjnano/articles/7/164.

69 D. Rüffel, F. Höhne, and J. Bühler, "New digital metal-oxide (MOx) sensor platform," *Sensors*, vol. 18, no. 4, p. 1052, 2018.

70 Bosch Sensortec. (2021) Gas Sensor BME680. [Online]. Available: https://www.bosch-sensortec.com/products/environmental-sensors/gas-sensors-bme680/

71 A. P. Lee and B. J. Reedy, "Temperature modulation in semiconductor gas sensing," *Sensors and Actuators B: Chemical*, vol. 60, no. 1, pp. 35–42, 1999.

72 B. Urasinska-Wojcik and J. W. Gardner, "Identification of H_2S impurity in hydrogen using temperature modulated metal oxide resistive sensors with a novel signal processing technique," *IEEE Sensors Letters*, vol. 1, no. 4, 2017, https://ieeexplore.ieee.org/document/7942142.

73 Z. Wen and L. Tian-Mo, "Gas-sensing properties of SnO_2-TiO_2-based sensor for volatile organic compound gas and its sensing mechanism," *Physica B: Condensed Matter*, vol. 405, no. 5, pp. 1345–1348, 2010.

74 Y. Sing, T. A. Vincent, M. Cole, and J. W. Gardner, "Real-time thermal modulation of high bandwidth MOX gas sensors for mobile robot applications," *Sensors (Basel)*, vol. 19, no. 5, p. 1180, 2019.

75 J. Palacín, D. Martínez, E. Lotet, T. Pallejà, J. Burgués, J. Fonollosa, A. Pardo, and S. Marco, "Application of an array of metal-oxide semiconductor gas sensors in an assistant personal robot for early gas leak detection," *Sensors*, vol. 19, no. 9, p. 1857, 2019.

76 C. Trimmer, A. Keller, N. R. Murphy, L. L. Snyder, J. R. Willer, M. H. Nagai, N. Katsanis, L. B. Vosshall, H. Matsunami, and J. D. Mainland, "Genetic variation across the human olfactory receptor repertoire alters odor perception," *Proceedings of the National Academy of Sciences of the United States of America*, vol. 116, no. 19, pp. 9475–9480, 2019.

13

Insulated Gate Bipolar Transistors (IGBTs)

Thomas Laska

Infineon Technologies AG, Germany

13.1 Introduction

As a high-voltage power switch, the insulated-gate bipolar transistor (IGBT) is a combination of vertical silicon (Si) power metal-oxide-semiconductor field-effect-transistor (MOSFET) and bipolar transistor. A typical cross section of an IGBT is shown in Figure 13.1 with a MOS structure on the front side, a thick n^--doped drift zone and a thin p-doped layer on the back.

Basic research on such devices was carried out in the late 1970s and early 1980s [1]. In this context, the work of several inventors was pivotal, e.g. B. J. Baliga [2], J. D. Plummer [3], or H. W. Becke [4], to mention but a few.

Subsequent development activities then led in the mid-1980s to the first commercially available IGBT products. From that time onwards, the IGBT has been used as a key component in nearly all medium to high-power electronic systems ranging from energy generation and distribution to energy consumption, such as renewables (wind, solar), high-voltage direct-current (HVDC) transmission, industrial AC drive motion control, and home appliances (air conditioners, refrigerators, and washing machines). Other major applications include hybrid and fully electric cars as well as traction (trams, metros, urban railways, heavy locomotives, and high-speed trains).

In most of the abovementioned applications, electrical energy is converted to mechanical energy, or vice versa. One example is depicted in Figure 13.2. "Grid 1" is rectified here to a DC link voltage. The inverter generates a three-phase current/voltage with variable frequency and amplitude for "Grid 2" thus enabling an energy-efficient electric motor operation.

In the depicted inverter, the IGBT acts as a switch, and the diode as a freewheeling path (or valve) typically in a three-phase configuration of six IGBTs with inverse-connected freewheeling diodes (FWDs). Common switching frequencies for the IGBT and FWD inside the inverter are in the range of a few to 15 kHz, and typical DC link voltages are between a few hundred volt (e.g. for air conditioners, washing machines, and hybrid vehicles) up to several kV (e.g. for trains). Key requirements for the switch include the following:

- blocking voltage capability as required from the DC link, plus the additional inductive voltage overshoot spikes (resulting in 600 V rated blocking voltage for a DC link of 300–400 V or 1200 V for a DC link of 500–800 V),
- low ON state losses and low switching losses to achieve highest energy efficiency,

Advances in Semiconductor Technologies: Selected Topics Beyond Conventional CMOS, First Edition. Edited by An Chen.
© 2023 The Institute of Electrical and Electronics Engineers, Inc. Published 2023 by John Wiley & Sons, Inc.

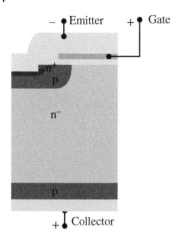

Figure 13.1 Cross section of an insulated-gate bipolar transistor (IGBT).

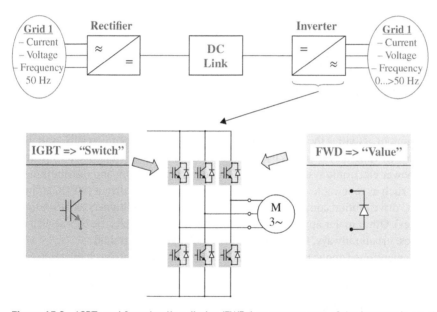

Figure 13.2 IGBTs and freewheeling diodes (FWDs) as components of the inverter in an electrical drive.

- soft switching behavior to ensure low electro-magnetic interference (EMI) noise,
- robustness against electrical malfunctions such as short-circuit and/or overcurrent,
- robustness against cosmic ray particle occurrences.

During the last three decades, many evolutionary, and revolutionary, innovations have improved the electrical performance of the IGBT, especially in terms of reduced ON state voltage and switching losses, and increased maximum junction temperature. Thus, overall power density and power efficiency could be increased tremendously beginning from the first-IGBT generation in the second half of the 1980s up to the advanced seventh-generation IGBTs that are nowadays available. Despite increased competition from wide-bandgap semiconductors, ongoing Si IGBT development and improvement activities are still in progress [5].

13.2   State-of-the-Art IGBT Technology

13.2.1   Structural Basics with Respect to Blocking, ON State and Switching

The required blocking capability of the IGBT is achieved by the inherent vertical p–n junction of the p-body region of the MOS on the upper silicon surface, and the thick n⁻-doped layer underneath it, illustrated in Figure 13.3a, which is called the base or drift zone. If a reverse bias V is applied to this p–n junction, a depletion layer and a high electrical field is formed. The blocking voltage that can be obtained depends on the thickness and the doping concentration of the n⁻-doped layer (e.g. a nominal voltage V_{nom} of 600 V requires an n⁻-doped Si layer of about $1 \times 14\,cm^{-3}$ doping and 60 μm thickness; 1200 V requires about $5 \times 13\,cm^{-3}$ and 120 μm). In Figure 13.3b, the ON state condition with a positive gate voltage is illustrated. The MOS channel on the front side enables an electron current from the emitter to the collector. From the p-layer on the collector side, also called the back emitter, holes are injected into the n⁻ base providing a high-density, electron–hole plasma. This plasma has a factor of about 100 times higher carrier concentration (holes and electrons) as compared to a unipolar device of the same blocking voltage class (such as a super junction Si MOSFET). As a result, the on-resistance or the ON state voltage V_{CEsat} of the IGBT is much lower. However, there is a trade-off in terms of switching losses. They are higher in the case of an IGBT as compared to a unipolar device, as the high number of carriers must be cleared out when switching off, and built up when switching on.

In summary, the IGBT as a combined MOS/bipolar device has clear advantages in terms of losses during ON state, but disadvantages in switching losses when compared to a unipolar MOS device at a given voltage class. Therefore, the IGBT is the device of choice in power applications with switching frequencies up to several kHz, as the losses during ON state are more numerous than the switching losses. However, for applications with switching frequencies of about 100 kHz and higher, in which switching losses are dominant, the unipolar device is more favorable.

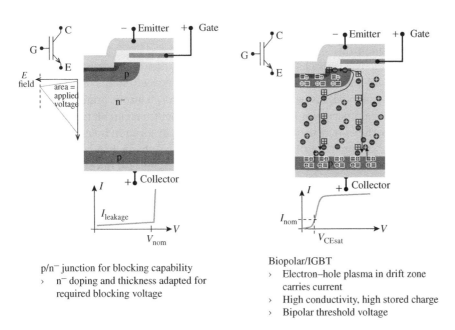

p/n⁻ junction for blocking capability
> n⁻ doping and thickness adapted for
   required blocking voltage

Biopolar/IGBT
> Electron–hole plasma in drift zone
   carries current
> High conductivity, high stored charge
> Bipolar threshold voltage

Figure 13.3   IGBT cross section in the case of blocking (a) and conducting (b).

13.2.2 Cell and Vertical Design

Some key elements in IGBT devices have a big impact on overall performance. These include the MOS transistor cell on the front side of the device (usually a trench cell in modern IGBTs) and the vertical structure consisting of a base or drift region, the back emitter, and (in modern IGBTs) an intermediate layer called buffer or field-stop layer (Figure 13.4).

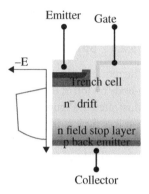

Figure 13.4 State-of-the-art IGBT with a trench cell and vertical structure including a field-stop layer.

Compared to a planar cell concept (used for the very first IGBTs), the advantage of a trench technology, applied widely in various configurations [6–8], is the effect of an enhanced hole and electron concentration in the upper region of the chip during the ON state [9, 10].

The cell size can be reduced to submicron size [11] with additionally reduced V_{CEsat}. The carrier plasma profile, however, must be carefully balanced by the proper selection of gate trenches, source trenches, and dummy trenches in order to stay within the range of reasonable turn-off losses.

In addition to the cell, the vertical structure is also important for vertical bipolar power devices. For a given blocking voltage, the thinner the structure, the lower the ON state *and* the lower the switching losses. Originally, a rather thick "non-punch-through" vertical structure was used, but this was later changed to a state-of-the-art vertical construction with an additional n-doped field-stop layer between the n⁻ drift zone and back emitter. By implementing this field-stop layer, a trapezoid-shaped field distribution can be obtained (in the case of blocking), which enables the thickness of the drift zone to be reduced for a given blocking voltage. Favorable also is the use of a low-efficiency p back emitter that adjusts the carrier plasma to a suitable level without the need for any carrier lifetime reduction [12]. This vertical field-stop concept was first introduced for 1200 V devices [13] up to 6.5 kV, and later for 600 V and even 400 V devices. In the latter cases, ultrathin chips of only 60 μm or even 40 μm thickness are required, which is quite challenging. This is a topic to be discussed in the next chapter.

The vertical structure of the IGBT also has an impact on a key feature of the power device, its cosmic ray robustness. The OFF state (when a high voltage is applied between emitter and collector) is a critical condition in case of incidents with high-energy protons or neutrons originating from cosmic radiation. As a statistically occurring event, these protons or neutrons can hit the chip in its space charge region. Such particles may then create a very high-density electron hole plasma that turns into a rapidly moving "streamer" resulting in a highly conductive channel and a shortening of the electrodes of the device, which may finally end up destroying the chip [14, 15]. Such a cosmic ray induced failure rate is strongly dependent on the applied voltage and can be influenced by the special shape and height of the electrical field. Hence, a vertical design with a low-doped base and additional field-stop layer could end up with a lower failure rate than a nonpunch-through design having a slightly higher doped base zone (at the same applied V_{CE} voltage). Also, the transistor cell on top of the chip may influence the electrical field at the p–n junction, although as a rule this contributes only slightly to the cosmic ray robustness.

Overall, the IGBT supplier should be concerned about the required low cosmic "fit" (failure in time) rate at a specified V_{CE} voltage and has to allow for this low fit rate by using a proper chip design.

13.2.3 Wafer Technology

The n$^-$-doped base or drift zone of the IGBT needs a very precisely doped and very pure silicon crystal, usually achieved by using float-zone (FZ) grown silicon material. However, FZ silicon is available only up to 200 mm wafer diameter. For targeting a cost-effective, high-volume production with a 300 mm IGBT technology, there was a need to define and use special magnetic Czochralski material for the IGBTs, which has since become possible [16].

An additional challenge for this modern IGBT technology was how to combine a large diameter with the need for an ultrathin wafer (e.g. about 60 µm for 600 V; Figure 13.5). To enable such a combination, leading-edge doping processes had to be developed and implemented at the end of the wafer process. These processes included high-energy hydrogen implants [17], and melting-mode laser annealing on the wafer back, which keep the wafer sufficiently cool on the top side [18].

13.2.4 Reverse-Blocking and Reverse-Conducting IGBTs

In the context of state-of-the-art IGBTs, the integration of additional functions within the chip is a topic worth mentioning. One example is the reverse-blocking IGBT [19] in Figure 13.6 not only with some advantages for special applications like the matrix converter, but also with some drawbacks in terms of process complexity and vertical optimization restrictions compared to a standard field-stop IGBT.

Another promising integration candidate is the reverse-conducting IGBT or RC IGBT with the implementation of the power diode function within the IGBT.

Initial RC IGBT approaches were well suited for special applications like inductive heating and cooking with relatively weak diode requirements [20]. However, in the meantime, the RC IGBT concept has been optimized to include typical consumer and general-purpose drives applications in the 600–1200 V range taking the needs for diode hard commutation into consideration [21, 22]. The challenge now is mainly to integrate technologically very different optimized features (lifetime killing of the diode, high p-dose in the IGBT cell) within one single die. However, meanwhile, even traction and HVDC applications in the high-voltage arena of 3.3 and 6.5 kV have been discussed [23, 24].

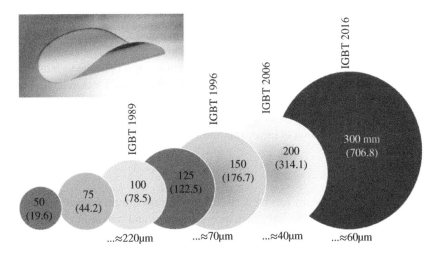

Figure 13.5 Si IGBTs on their path to larger wafer diameters and thinner wafers, example from Infineon.

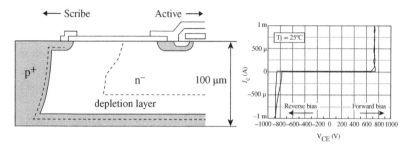

Figure 13.6 600 V reverse-blocking IGBT cross section and blocking characteristics. Source: Takei et al. [19], © 2001 IEEE.

What competition will be like between the standard concept of separated IGBTs and diodes, and integrated solutions such as RB IGBT and RC IGBT is yet to be seen. Depending on inverter topologies, cooling conditions, and gate driving efforts, one or the other concept may be favorable.

13.2.5 Increasing Maximum Junction Temperature

Evolution from the very first IGBTs to current, state-of-the-art IGBTs has also occurred with respect to the maximum specified virtual junction temperature (T_{vjmax}) of the dies within the package. Starting with a T_{vjmax} of 125 °C, today 150 °C as maximum temperature seems to be standard for IGBTs in conventional silicone gel modules. For discrete IGBTs in molded packages, even 175 °C is specified. Measures to achieve success with these increased junction temperatures are coming from both the chip side (reduced hot leakage currents, high intrinsic overcurrent, and short-circuit robustness at higher temperatures) as well as from the packaging side (material for packages and interconnect technologies of the chips in the package).

13.2.6 Assembly and Interconnect Technology

Besides low-power IGBTs in discrete packages, several IGBT and FWD chips are mounted in a multichip module in the case of medium- and high-power applications. The chips are placed on an isolated substrate (DCB, or direct copper-bonded substrate), which is in most cases mounted on a copper base plate and covered by a silicone gel [1]. The back of the chip (collector) is soldered, and the front contacts (emitter and gate) are connected via ultrasonic aluminum bonding to outside terminals (Figure 13.7). However, there are also more sophisticated interconnect technologies in place nowadays such as sintering instead of soldering, or copper bonding instead of aluminum bonding, with better thermal performance and advantages in long-term reliability [25, 26]. This will be of importance for the full exploitation of next-generation IGBTs.

13.2.7 Power Density Increase

Power density in a power electronic system can be derived from the maximum output current or power in a special inverter with given DC link, switching frequency, and maximum junction temperature. The higher the power density, the more compact the power electronic system can be designed, resulting in smaller volume and weight. Two impressive examples of power density increase at the IGBT module level ranging over roughly 20 years are shown in Figure 13.8, one with a reduced module footprint at a given output current, and one with an increased output current at

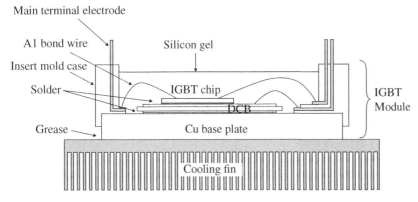

Figure 13.7 Cross section of typical IGBT module on a cooling fin. Source: Iwamuro et al. [1], © 2017 IEEE.

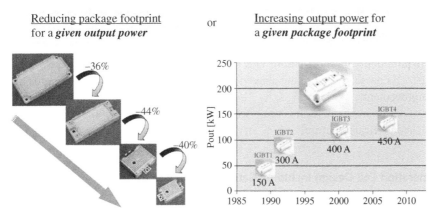

Figure 13.8 Module footprint reduction of a 35 A/1200 V six-pack module (left, vendor A) or output power increase for a given module footprint (62 mm Dual) over 4 IGBT generations (right, vendor B). Source: [5] T. Laska et al., 2019, IEEE.

a given module footprint. Both examples were enabled by several IGBT generation improvement steps in cell (e.g. change from planar transistor cell to trench cell) and vertical technology (e.g. change from nonpunch through to field-stop structure) as well as increased junction temperature, and this trend in increasing power density for Si IGBTs is not yet over.

13.3 Future Prospect of IGBT

13.3.1 Application Requirement Aspects

Since the first IGBT prototypes appeared some 30 years ago up until today with industrialized, mass-market seventh-generation IGBTs, questions have arisen about the further development and perspective of these devices, especially with the new wide-band successor (WBG) candidates in place, mainly SiC ($\geq$600 V) and GaN ($\leq$1200 V).

The closer the Si IGBTs are driven to their limits, the more important it is to clarify what properties are needed from an application point of view, and which ones are only "nice to have" and even, which ones should be eliminated. Also to be considered are the new requirements that may develop. Key examples include the required short-circuit robustness in terms of duration time and voltage slope (dV/dt) restrictions at turn-on and turn-off as well as the current slopes (dI/dt).

These aspects not only define the further path of Si IGBT development, but must also be clarified for WBG devices. In particular, the WBG advantages of very low switching losses are only valid as long as they are allowed to switch very fast. This might be in contradiction with today's requirements from many drives applications in terms of $dV/dt < 10\,\text{kV/}\mu\text{s}$, mainly not only due to motor deterioration prevention but also due to EMI issues. Allowing higher dV/dt values would enable a reduction of switching losses not only for WBG devices but also for Si IGBTs. However, especially the correlation of these dV/dt values, as well as dI/dt values, to a real EMI spectrum within the application is not always fully clear, and depends on specific parasitic elements. These elements differ according to the application and may also vary in terms of specific customer solutions used for the same application. With respect to the dV/dt restrictions preventing electric motor deterioration, new materials in the motor might help to allow higher dV/dt values in the future. Reducing today's short-circuit requirements, usually 10 to 5 μs or even below, would present the opportunity to increase the channel width-to-length ratio allowing both SiC MOSFETS as well as Si IGBTs to reduce ON state losses.

Even if today's requirements in short circuit and dV/dt remain the same, there are ways to improve the Si IGBT. Of course, the more the abovementioned requirements change, the greater the potential for improvement.

To discuss the next development steps, it is again worth taking a closer look at the IGBT device elements of the cell and vertical structure.

13.3.2 Next Generation Cell Design Including Gate Driving Schemes

Inspired by Nakagawa's study on the theoretical limit of the IGBT ON state voltage [27], many research activities deal with investigations on cells with deep sub-μm mesas between adjacent deep trenches [28], as shown in Figure 13.9.

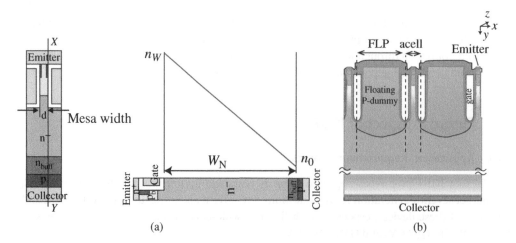

(a) (b)

Figure 13.9 Examples for investigations on IGBTs with sub μm mesas (a: theoretical study with 20–40 nm mesa width). Source: Nakagawa [27], ©2006 IEEE. (b: theoretical and experimental proof of sub μm mesa concept). Source: Based on Sumitomo et al. [28].

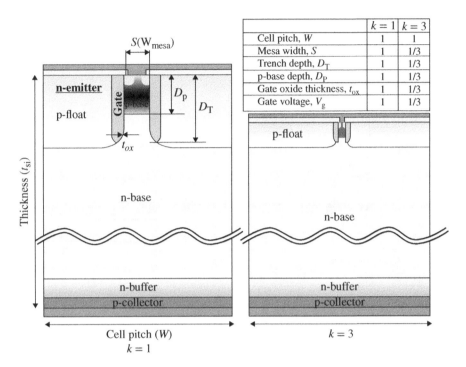

	$k=1$	$k=3$
Cell pitch, W	1	1
Mesa width, S	1	1/3
Trench depth, D_T	1	1/3
p-base depth, D_P	1	1/3
Gate oxide thickness, t_{ox}	1	1/3
Gate voltage, V_g	1	1/3

Figure 13.10 Scaled IGBT approach with a scaling factor of 3. Source: Saraya et al. [30], © 2018 IEEE.

Therefore, an ON state voltage of about 1 V for a 1200 V device seems feasible. However, drawbacks such as high switching losses due to large quantities of carrier plasma as well as an insufficient short-circuit robustness still have to be solved.

Another trend is to use thinner gate oxide with lower gate voltage (e.g. 5 V instead of the established 15 V), either in combination with very small mesas [29] or in a "scaled" cell approach as in Figure 13.10, resulting in – besides lower ON state voltage – lower gate-driving efforts [30]. However, apart from such benefits, potential drawbacks such as the enhanced tendency toward parasitic turn-on also have to be taken into account, especially for high-power IGBTs, and countermeasures have yet to be defined.

If even more sophisticated gate driving schemes are acceptable (perhaps for high-power applications), a so-called "dual gate" IGBT is an attractive candidate [31]. With this device concept, both a low ON state voltage V_{CEsat} and low turn-off losses E_{off} can be achieved by decreasing the carrier plasma from a very high level down to a lower level immediately before turn-off by using a second control gate as shown in Figure 13.11.

Also, an RC IGBT can improve its overall performance by a more complex gate-driving scheme (Figure 13.12). Here two gates are not necessary. One gate can be used not only in IGBT mode, but also in diode mode to modulate the carrier plasma for low forward voltage as well as low recovery losses at commutation [24, 32].

With this gate-driving measure, the high diode recovery losses within a conventional RC IGBT can be reduced significantly, und thus the attainable power density can be increased.

The important IGBT trade-off between sufficient short-circuit robustness and ON state voltage can be improved by the use of IGBT chips with a lower intrinsic short-circuit robustness having the benefit of lower ON state voltage combined with additional measures. By current sensing that

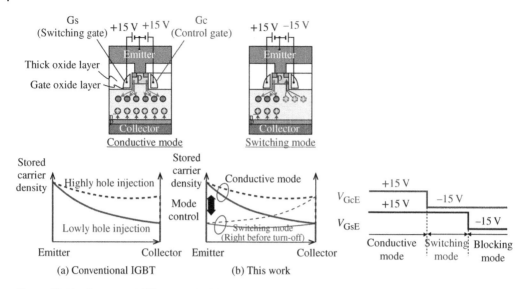

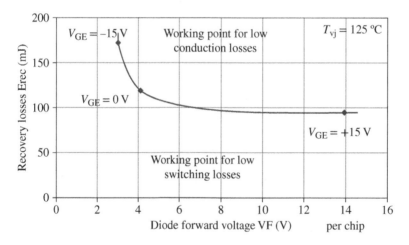

Figure 13.11 Dual gate IGBT proposal with control and switching gate, resulting carrier plasma in comparison to a standard IGBT and respective gate-driving scheme. Source: Miyoshi et al. [31].

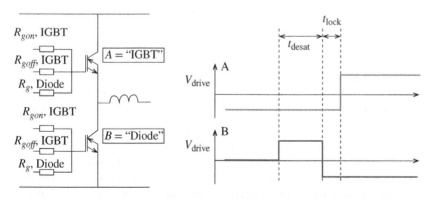

Figure 13.12 E_{rec}/V_F "working" points of a gate-controlled diode within an RC IGBT and respective gate-driving scheme of upper and lower device within a half-bridge configuration. Source: Werber et al. [24].

is integrated on the IGBT chip, a dedicated gate driver can reduce the gate voltage in a short circuit within the very first μs during the short-circuit pulse [33]. This method of short-circuit detection and its respective reaction is not new for dedicated intelligent power modules (IPM), but may gain increasing importance also for common IGBT modules, as it enables additional performance improvement because of the lower ON state voltage. Especially when using cell concepts with very narrow mesas, and critical intrinsic short-circuit performance, a short-circuit detection method that is fast enough might become even more important.

13.3.3   Next Generation Vertical Structure Concepts

Besides innovation in the cell concepts (case by case supported by proper gate-driving schemes), also the vertical IGBT structure can be improved.

1200 V IGBTs with a thickness of about 110 μm, and 600–750 V IGBTs with thicknesses of 60–70 μm are common today. Theoretically, for a 1200 V blocking capability, chip thicknesses of 85–90 μm should be feasible (100 μm already demonstrated [34]), for 600 V about 45–50 μm. Figure 13.13 shows the continuous path of chip thickness reduction, also called vertical chip shrink, from the past until today and outlook into the future. The outcome will again be lower ON state and switching losses. As a rule of thumb, a 10% chip thickness reduction can result in a 10% switching loss reduction and reduced ON state voltage.

However, the thinner the IGBT chip, the more critical such requirements as thermal short circuit, cosmic ray robustness, and switching softness will become. The softness topic can be mitigated by low inductive module and inverter setups also by more sophisticated field-stop engineering as well as the right choice of back emitter efficiency. Also the three-dimensional relation between switching and ON state loss tradeoff on the one hand, and the tradeoff between switching losses and surge voltage on the other hand, as shown in Figure 13.14, can be taken into account.

It becomes clear that an – at a first glance – unfavorable, highly efficient back emitter results in the best compromise in terms of overvoltage and overall losses, at least as long as high parasitic stray inductances occur.

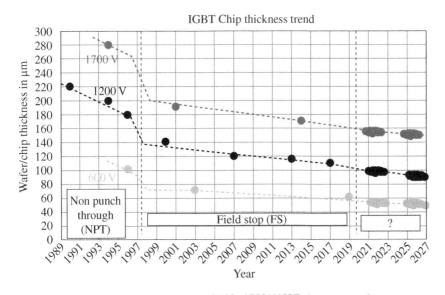

Figure 13.13   Vertical chip shrink path of 600–1700 V IGBTs from past to future.

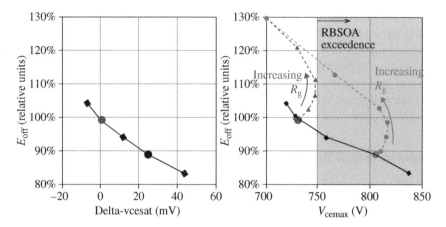

Figure 13.14 Simulated three-dimensional tradeoff of the p-emitter variation: Turn-off energy E_{off} vs. ON state voltage V_{CEsat} (left) and E_{off} vs. maximum collector emitter voltage V_{CEmax} (right): a lower p back emitter (dashed line with circles) provides lower E_{off} than the strong p emitter (dashed line with triangles) at the cost of much higher surge voltage. If the product of the stray inductance and collector current $L_{stray}{}^*I_C$ in the application is too high, the device needs to be slowed down by increasing the gate resistor R_g. In this case, the losses will increase dramatically without benefiting from the lower V_{CEsat}. Source: Wolter et al. [11], © 2015 IEEE.

The second critical topic when reducing the chip thickness is how to keep the cosmic ray robustness at a sufficiently high level. There is still a chance to compensate by appropriate low drift zone doping.

The third critical aspect of thinner chips, as mentioned before, is thermal short-circuit robustness. A sufficiently high level of thermal short-circuit capability despite reduced chip thickness can be achieved by implementing a thick metal layer on top of the chip, which will be shown in the next sub-chapter.

13.3.4 Next Level of Thermal Management and Interconnect Technique Innovation

The increasing junction temperature is an important aspect of IGBT technology. This temperature will most likely reach at least 200 °C. It has been shown that the Si IGBT is in principle able to operate up to 200 °C without reservations under nominal conditions and even under overload conditions (reverse-bias safe operation area, short-circuit safe operation area) [35, 36]. It is perhaps worth mentioning that a higher junction temperature is not a value in itself, but can contribute to the next steps in power density increase. This is illustrated in Figure 13.15 showing the calculated inverter output current as a function of the switching frequency for different allowed junction temperatures. Thus, the step from, e.g. 175–200 °C would once again allow 10–15% higher output for a given IGBT technology.

Of course, the periphery around the chip (package with plastic material, terminals, and the area outside the package) must also be able to sustain higher temperature levels than so far. In detail, the mold compound of a discrete IGBT or the silicone gel performance within a standard IGBT module will be a challenge at increased temperatures; however, there are some initial solutions with gels up to 200 °C [37].

Other serious issues to be solved are the increased temperature swings that stress the chip-to-baseplate or chip-to-lead frame connection as well as wedge bonding on the chip front side, which results in a reduced number of power and temperature cycles until the end of life,

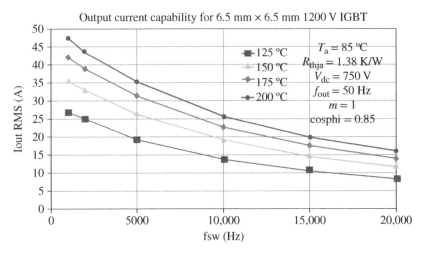

Figure 13.15 Inverter current (I_{out} RMS) as a function of switching frequency (f_{sw}) for different junction temperatures from 125 °C up to 200 °C. Source: Schlapbach et al. [35].

as long as no countermeasures are addressed. However, advanced, interconnect technologies are now available for these challenges, which have greatly increased cycling capabilities that may also be feasible for use up to even 200 °C. These include a sintering interconnect technology on the back and front side of the chip [25, 38, 39] or, alternatively, the combination of front-side, copper wedge bonding on copper chip surfaces, and back sintering, called ".XT" technology.

A very advantageous second effect of such a technology with thick copper chip-front metalliza-tion and back sintering or back diffusion soldering is shown in Figure 13.16 (both of these back interconnect techniques have in common a thin interconnecting layer in the range of 10 μm instead of a soft solder layer of typically 100 μm). Improved short-circuit robustness was proven in both the simulation test and in the experiment. The thick metal on the front of the chip has a positive effect. A diffusion solder layer on the back also has a positive effect, but the combination of both represents

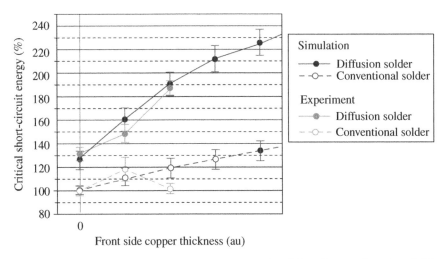

Figure 13.16 Increased critical short-circuit energy of a 1200 V IGBT as a function of different interconnect technologies and Cu chip metal thickness. Source: Hille et al. [40].

more enhanced robustness. Combining both measures results in an increase of the short-circuit energy-destruction level by about a factor of 2 for a reasonable chip-front metal thickness [40].

An increase in the destruction limit of short-circuit energy can counteract in principle the reduced short-circuit destruction energy of the next-generation IGBT chips, which is caused by vertical and area shrink.

It is expected that such combined chip/module solutions will support the process of strongly increased power density (driven by cell-concept innovation as well as vertical shrink). Such solutions allow for a higher maximum junction temperature and provide an additional margin for thermal short-circuit energy, the critical robustness parameter.

Of course, these measures in the area of advanced thermal management will also be important for the roll-out of wide bandgap power semiconductors with their potential for even higher temperatures exceeding 200 °C.

13.4 Outlook

After over 30 years, the IGBT has yet to reach the end of its technological roadmap. Ongoing joint development activities combining advanced interconnect technologies with innovations in cell and vertical structure, diode function integration, and advanced gate-driving concepts will lead to further significant gains in power density and efficiency.

However, with respect to economic parameters such as cost and power for the overall power electronic system (including cooling measures and passive components as well as the number of power semiconductors), the gap between Si switch solutions and those solutions using wide bandgap transistors should become smaller as the cost-cutting measures for the latter accelerate. It can be assumed, nonetheless, that there will be a coexistence between Si power devices and wide-bandgap power devices beyond the next decade.

Acknowledgment

The author would like to thank Karen Hanson for editing the English text and providing valuable suggestions for improvement from the first manuscript to the final draft.

References

1 N. Iwamuro and T. Laska, "IGBT history, state-of-the-art, and future prospects," *IEEE Transactions on Electron Devices*, vol. 64, no. 3, pp. 741–752, 2017.

2 B. J. Baliga, "Enhancement and depletion mode vertical channel MOS-gated thyristors," *Electronics Letters*, vol. 15, no. 20, pp. 645–647, 1979.

3 J. D. Plummer, "Monolithic semiconductor switching device," US Patent 4, 199,774.

4 H. W. Becke and C. F. Wheatley, Jr., "Power MOSFET with an anode region," US Patent 4, 364,073.

5 T. Laska, "Progress in Si IGBT technology – as an ongoing competition with WBG power devices," in *2019 IEEE International Electron Devices Meeting (IEDM)*, 7–11 Dec. 2019, pp. 12.2.1–4.

6 T. Laska, F. Pfirsch, F. Hirle, J. Niedermey, C. Schaffer, and T. Schmidt, "1200 V-trench-IGBT study with square short circuit SOA," in *Proceedings of the International Symposium on Power Semiconductor Devices and ICs*, 3–6 June 1998, pp. 433–436.

7 M. Harada, T. Minato, H. Takahashi, H. Nishihara, K. Inoue, and I. Takata, "600 V trench IGBT in comparison with planar IGBT," in *Proceedings of the International Symposium on Power Semiconductor Devices and ICs*, 31 May–2 June 1994, pp. 411–416.

8 H. Takahashi, H. Haruguchi, H. Hagino, and T. Yamada, "Carrier stored trench-gate bipolar transistor – a novel power device for high voltage application," in *Proceedings of International Symposium on Power Semiconductors and ICs*, 23–23 May 1996, pp. 349–352.

9 M. Kitagawa, I. Omura, S. Hasegawa, T. Inoue, and A. Nakagawa, "A 4500 V injection enhanced insulated gate bipolar transistor (IEGT)," in *IEEE International Electron Devices Meeting Technical Digest*, 5–8 Dec. 1993, pp. 679–682.

10 I. Omura, T. Ogura, K. Sugiyama, and H. Ohashi, "Carrier injection enhancement effect of high voltage MOS devices," *Proceedings of International Symposium on Power Semiconductors and ICs*, 26–29 May 1997, pp. 217–220.

11 F. Wolter, W. Roesner, M. Cotorogea, T. Geinzer, M. Seider-Schmidt, and K. -H. Wang, "Multi-dimensional trade-off considerations of the 750 V micro pattern trench IGBT for electric drive train applications," in *Proceedings of International Symposium on Power Semiconductors and ICs*, 10–14 May 2015, pp. 105–108.

12 T. Matsudai, K. Kinoshita, and A. Nakagawa, "New 600 V trench gate PT IGBT concept with very thin wafer and low efficiency p-emitter, having an ON state voltage drop lower than diode," in *Proceedings of International Power Electronics Conference-Tokyo*, 3–7 April 2000, pp. 292–296.

13 T. Laska, M. Miinzer, F. Pfirsch, C. Schaeffer, and T. Schmidt, "The field stop IGBT (FS IGBT) – a new power device concept with a great improvement potential," in *Proceedings of International Symposium on Power Semiconductors and ICs*, 22–25 May 2000, pp. 355–358.

14 H. Kabza, H. -J. Schulze, and Y. Gerstenmaier, "Cosmic radiation as a cause for power device failure and possible countermeasures," in *Proceedings of International Symposium on Power Semiconductors and ICs*, 31 May–2 June 1994, pp. 9–12.

15 F. Pfirsch and G. Soelkner, "Simulation of cosmic ray failures rates using semi-empirical models," in *Proceedings of International Symposium on Power Semiconductors and ICs*, 6–10 June 2010, pp. 125–128.

16 H. J. Schulze, H. Öfner, F. -J. Niedernostheide, J. G. Laven, H. P. Felsl, S. Voss, A. Schwagmann, M. Jelinek, N. Ganagona, A. Susiti, T. Wübben, W. Schustereder, A. Breymesser, and M. Stadtmüller, "Use of 300 mm magnetic Czochralski wafers for the fabrication of IGBTs," in *Proceedings of International Symposium on Power Semiconductors and ICs*, 12–16 June 2016, pp. 355–358.

17 F. -J. Niedernostheide, H. -J. Schulze, H. P. Felsl, F. Hille, J. G. Laven, M. Pfaffenlehner, C. Schäffer, H. Schulze, and W. Schustereder, "Tailoring of field-stop layers in power devices by hydrogen-related donor formation," in *Proceedings of International Symposium on Power Semiconductors and ICs*, 12–16 June 2016, pp. 351–354.

18 T. Gutt and H. Schulze, "Deep melt activation using laser thermal annealing for IGBT thin wafer technology", in *Proceedings of International Symposium on Power Semiconductors and ICs*, 6–10 June 2010, pp. 29–32.

19 M. Takei, Y. Harada, and K. Ueno, "600 V-IGBT with reverse blocking capability," in *Proceedings of International Symposium on Power Semiconductors and ICs*, 7 June 2001, pp. 413–416.

20 O. Hellmund, L. Lorenz, and H. Ruthing, "1200 V reverse conducting IGBTs for soft-switching applications," in *China Power Electronics Journal*, Edition 5, p. 20–22, 2005.

21 K. Satoh, T. Iwagami, H. Kawafuji, S. Shirakawa, M. Honsberg, and E. Thai, "A new 3 A/600 V transfer mold IPM with RC (Reverse Conducting) –IGBT," in *Proceedings of PCIM Europe*, May 30–June 1 2006, pp. 73–78.

22 H. Rüthing, F. Hille, F. -J. Niedernostheide, H. -J. Schulze, B. Brunner, "600 V reverse conducting (RC-) IGBT for drives applications in ultra-thin wafer technology," in *Proceedings of International Symposium on Power Semiconductors and ICs*, 27–30 May 2007, pp. 89–92.

23 M. Rahimo, U. Schlapbach, A. Kopta, J. Vobecky, D. Schneider, and A. Baschnagel, "A high current 3300 V module employing reverse conducting IGBTs setting a new benchmark in output power capability," in *Proceedings of International Symposium on Power Semiconductors and ICs*, 18–22 May 2008, pp. 68–71.

24 D. Werber, T. Hunger, M. Wissen, T. Schütze, M. Lassmann, B. Stemmer, V. Komarnitskyy, and F. Pfirsch, "A 1000 A 6.5 kV power module enabled by reverse-conducting trench-IGBT-technology," in *Proceedings of PCIM*, 19–21 May 2015, pp. 351–358.

25 P. Beckedahl, M. Hermann, M. Kind, M. Knebel, J. Nascimento, and A. Wintrich, "Performance comparison of traditional packaging technologies to a novel bond wireless all sintered power module," in *Proceedings of PCIM Europe*, 17–19 May 2011, pp. 247–251.

26 K. Guth, D. Siepe, J. Görlich, H. Torwesten, R. Roth, F. Hille, and F. Umbach, "New assembly and interconnects beyond sintering methods", in *Proceedings of PCIM Europe*, 4–6 May 2010, pp. 232–237.

27 A. Nakagawa, "Theoretical investigation of silicon limit charact. Of IGBTs," in *Proceedings of International Symposium on Power Semiconductors and ICs*, 4–8 June 2006, pp. 5–8.

28 M. Sumitomo, J. Asai, H. Sakane, K. Arakawa, Y. Higuchi and M. Matsui, "Low loss IGBT with partially narrow Mesa structure (PNM-IGBT)," in *Proceedings of International Symposium on Power Semiconductors and ICs*, 3–7 June 2012, pp. 17–20.

29 M. Tanaka and A. Nakagawa, "Novel 3D narrow mesa IGBT suppressing CIBL," in *Proceedings of International Symposium on Power Semiconductors and ICs*, 13–17 May 2017, pp. 124–127.

30 T. Saraya, K. Itou, T. Takakura, M. Fukui, S. Suzuki, K. Takeuchi, M. Tsukuda, Y. Numasawa, K. Satoh, T. Matsudai, W. Saito, K. Kakushima, T. Hoshii, K. Furukawa, M. Watanabe, N. Shigyo, K. Tsutsui, H. Iwai, A. Ogura, S. Nishizawa, I. Omura, H. Ohashi, and T. Hiramoto, "Demonstration of 1200 V scaled IGBTs driven by 5V gate voltage with superiorly low switching loss," in *2018 IEEE International Electron Devices Meeting (IEDM)*, 1–5 Dec. 2018, pp. 189–192.

31 T. Miyoshi, Y. Takeuchi, T. Furukawa, M. Shiraishi, and M. Mori, "Dual side-gate HiGT breaking through the limitation of IGBT loss reduction," in *Proceedings of PCIM*, 16–18 May 2017, pp. 315–322.

32 M. Rahimo, M. Andenna, L. Storasta, C. Corvasce, and A. Kopta, "Demonstration of an enhanced trench Bimode insulated gate transistor ET-BIGT," in *Proceedings of International Symposium on Power Semiconductors and ICs*, 12–16 June 2016, pp. 151–154.

33 M. Otsuki, M. Watanabe, and A. Nishiura, "Trends and opportunities in intelligent power modules (IPM)," in *Proceedings of International Symposium on Power Semiconductors and ICs*, 10–14 May 2015, pp. 317–320.

34 J. Vobecký, M. Rahimo, A. Kopta, and S. Linder, "Exploring the silicon design limits of thin wafer IGBT technology," in *Proceedings of International Symposium on Power Semiconductors and ICs*, 18–22 May 2008, pp. 76–79.

35 U. Schlapbach, M. Rahimo, C. von Arx, A. Mukhitdinov, and S. Linder, "1200 V IGBTs operating at 200 °C? An investigation on the potentials and the design constraints," in *Proceedings of International Symposium on Power Semiconductors and ICs*, 27–30 May 2007, pp. 9–12.

36 T. Laska, M. Münzer, R. Rupp, and H. Rüthing, "Review of power semiconductor switches for hybrid and fuel cell automotive applications," in *Proceedings of Automotive Power Electronics*, Paris, 21–22 June 2006.

37 T. Seldrum, E. Vanlathem, V. Delsuc, H. Enami, "New silicone gel enabling high temperature stability for next generation of power modules," in *Proceedings of PCIM Europe*, 10–12 May 2016, pp. 1017–1020.

38 R. Amro, J. Lutz, J. Rudzki, M. Thoben, and A. Lindemann, "Double-sided low-temperature joining technique for power cycling capability at high temperature," in *European Conference on Power Electronics and Application*, 11–14 Sept. 2005.

39 U. Scheuermann and P. Beckedahl, "The road to the next generation power module – 100% solder free design," in *Proceedings of the CIPS*, 11–13 March 2008.

40 F. Hille, F. Umbach, T. Raker, and R. Roth, "Failure mechanism and improvement potential of IGBT's short circuit operation," in *Proceedings of International Symposium on Power Semiconductors and ICs*, 6–10 June 2010, pp. 33–37.

14

III–V and Wide Bandgap

Mohammed Alomari

Institut für Mikroelektronik Stuttgart, Stuttgart, Germany

14.1 Introduction

We live in an age where semiconductor devices are present in almost every aspect of our lives, from very simple home appliances to the most sophisticated space exploration vehicles, in lightening and power conversion, in communication, in medical applications, and in sensors. So far, silicon-based electronics have filled all of those needs with a stable production platform and steady development of the devices, coupled with increased sophistication in fabrication technology, pushing the limits of silicon to the physical limit of the material. Yet, there was the realization of the need to make a step change in electronics, by introducing a different class of semiconductors, distinguished from silicon either by the wider bandgap (E_g), or by composition. Making such a step change is even more pressing nowadays, with the wide reliance on mobile communication electronics, and power conversion electronics, driven by mass markets such as smart phones, the internet of things, and electric vehicles, with all the required electronic infrastructure, from base stations to charging stations. Moreover, the mobile nature of those applications demands faster, light weight, small form factor components, with very high efficiency, especially in power conversion [1–3]. Although silicon will still dominate those fields, replacing some of the core silicon components with wide bandgap or III–V components will be necessary to overcome the physical limit imposed by the silicon material.

The wide bandgap semiconductors are generally defined as semiconductors having bandgaps above 1.7 eV, such as silicon carbide (SiC), diamond, and gallium nitride (GaN). The III–V semiconductors on the other hand are composed of a combination of group III elements (mostly Al, Ga, and In) with group V elements (mostly N, P, As, and Sb). Among the possible combinations, the most famous are gallium arsenide (GaAs), indium phosphide (InP), and GaN. It should be noted that some of the III–V semiconductors, such as GaN, can be also classified as wideband gap semiconductors. The very first direct implication of using a compound semiconductor is the direct bandgap, allowing the ability to emit light, as in light emitting diodes (LEDs) and solid-state lasers. This is indeed the exclusive domain of III–V semiconductors, where the emitted wavelength can be engineered by engineering the bandgap, which in turn is engineered by changing the III–V ratio or composition. Wavelengths from blue, as first demonstrated in an LED by Nakamura et al. [4], to even ultraviolet ranges [5, 6] could be demonstrated. For more in-depth discussion of the use of III–V materials in optoelectronics, the reader is referred to [7, 8]. The discussion in this chapter will be limited to the use of those semiconductor classes in electronic device applications, focusing

Advances in Semiconductor Technologies: Selected Topics Beyond Conventional CMOS, First Edition. Edited by An Chen.
© 2023 The Institute of Electrical and Electronics Engineers, Inc. Published 2023 by John Wiley & Sons, Inc.

Table 14.1 Material properties of III–V and wide bandgap semiconductors in comparison to silicon.

	Bandgap	Mobility	Saturation velocity	Breakdown strength	Thermal conductivity	Relative dielectric constant	Intrinsic concentration
	E_g (eV)	μ (cm²/V s)	$v_{sat,peak}$ (10⁶ cm/s)	E_{Br} (MV/cm)	K (W/cm K)	ε_r	n_i @300 K (cm⁻³)
Si	1.11	1360	9.7	0.37	1.5	11.7	1.38×10^{10}
GaAs	1.42	8800	6	0.4	0.55	13	2.2×10^{6}
GaN	3.42	2000	27	3.3	1.3	9.5	9.8×10^{-12}
4H-SiC	3.26	1140	22	2.4	3.7	9.7	2.5×10^{-10}
Diamond	5.43	4500	20	10	24	5.7	$<10^{-20}$

The mobility and breakdown strength are for low doping concentrations ($<10^{15}$ cm⁻³) and are for electrons except for the case of diamond (hole channel).
Properties listed are after Baliga [9], Pernegger et al. [10], Shur [8], and Wort and Balmer [11].

on the latest GaN, SiC, and diamond advances. GaN-based devices will be discussed further due to the unique properties of the material system.

The promotion of wide bandgaps and III–V semiconductors as supplementary, or replacement of silicon in some cases, stems from the intrinsic material properties, as listed in Table 14.1. Note that the listed metrics are for an ideal case, which could be true for the very mature silicon technology, but not for the emerging wide bandgap semiconductors. The application fields are mainly the high frequency power amplification, and DC power conversion. The wide bandgap of GaN, SiC, and diamond should allow higher breakdown voltages (V_{Br}), thus, enabling the application of higher supply voltages, resulting in higher output powers [12]. Moreover, the larger bandgap and lower intrinsic carrier concentrations make these materials less susceptible to thermal noise and parasitic leakage currents, thus reducing off-state losses. For very high frequency applications, the product of mobility and saturation velocity in GaAs and InP enables devices up to the THz regime. Indeed, GaAs- and InP-based devices do dominate the very high frequency range. For more details about this established technology, the reader is referred to [7, 8]. However, the relatively low breakdown strength limits GaAs and InP devices to low voltages and consequently low output powers. In addition, the higher thermal conductivities of GaN, SiC, and diamond allow better thermal management for high power applications, where increased device self-heating is coupled with increased output power. A better way to appreciate the importance of the intrinsic material parameters is to evaluate their impact on power efficiency. Taking the example of a switch mode DC power converters, which are used in almost all power conversion systems [13, 14], increasing the switching speed of the core semiconductor power converter device enables the reduction of the size of the accompanying passive components such as inductors, leading to a more compact and light weight system [15]. In addition, the losses in the power conversion semiconductor device determine the overall system efficiency. In general, the power dissipation in a switched mode, causing the internal heating, will limit the maximum on-state current of the device to the maximum acceptable junction temperature. The dissipated power (P_{diss}) is described as:

$$P_{diss} = (V_{on} \cdot I_{on}) + (V_{off} \cdot I_{off}) + P_{switch}$$

where V_{on} is the on-state voltage, I_{on} is the on-state current, V_{off} is the off-state voltage, I_{off} is the off-state (leakage) current, and P_{switch} is the losses during the switching between on-state and off-state. The losses in the on-state scales with desired current operation point, and the on-state voltage drop required to achieve the on-state current. In other words, reducing the on-state resistance (R_{on}) of the device is important in reducing the on-state losses. Here, the higher channel mobility and channel density play a critical role in reducing R_{on}. At the same time, in order to maximize the output power by operating at high off-state voltages, the wide bandgap semiconductors are prime candidates due to their high critical electric field E_{crit}. This enables the increase of the off-state voltage without the need to extend the width of the drift region (or channel length for lateral devices), and therefore, without increase in R_{on}. The losses related to a higher operating voltage are the leakage current losses (I_{off}). The losses induced by switching between the on-state and off-state are minimized by switching the device at higher frequencies. This is determined by a higher channel mobility, and most importantly by a smaller gate capacitance. Those features are available in GaN high electron mobility transistors (GaN HEMTs). The impact of the semiconductor material properties in switched mode operation can be visualized by plotting the relationship between the specific on-resistance ($R_{on,sp}$) and E_{crit} as shown in Figure 14.1, using the Baliga Figure-of-Merit, for the case of SiC and GaN compared to silicon. For more details on the derivation of this relationship for lateral and vertical devices, the reader is referred to Baliga [9].

For Radio Frequency (RF) power amplifiers, the high cutoff frequencies can be achieved due to the high mobilities and high saturation velocities. The power loss in this case is not determined only by R_{on} but also by the operation class of the device. Here the application plays an important role. But in general, the smaller gate and drain capacitances, higher mobility, and higher breakdown voltage also promote the wide bandgaps as prime candidates for high RF power – high frequency devices.

Despite the clear theoretical potential of wide bandgap semiconductors, many challenges have to be met first before considering the wide replacement of silicon power electronics with those materials. Reliability issues related to the maturity of material growth, and the resulting defect density, had to be overcome. In addition, the complexity of integrated circuits using those materials are still far from what could be achieved by silicon and impact the cost-benefit relationship negatively. While a silicon-based power metal-oxide-semiconductor field-effect-transistor (MOSFET) can be readily monolithically integrated with the appropriate gate driver, including complimentary metal-oxide-semiconductor (CMOS)-based control and monitoring electronics, the wide bandgap

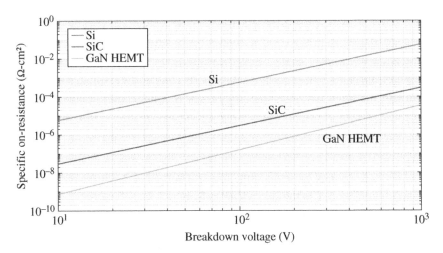

Figure 14.1 Specific on-resistance vs. critical electric field for SiC and GaN compared to Si. Source: Baliga [9]/World Scientific Publishing Company.

power modules still have to rely on heterogeneous integration with the silicon circuit, and exist mostly as single power device dies. But even if used in a single die format, the cost of wide bandgap power devices is far higher than their silicon counterparts, due initially to the higher material (substrate) cost, smaller available wafer diameters, and the complexity of fabricating the devices in the established CMOS compatible cleanroom facilities. In addition, wide bandgap devices will also reach the physical performance limits, set by the high fields which can be applied to the material, and its effect on the other device components such as the gate dielectric or device passivation, in addition to high field effects such as impact ionization. A review of those effects is presented in [16].

This article will highlight the status of those power devices, with highlights on the main challenges and the possible solutions. A more detailed discussion will be given for the case of GaN HEMTs, not only due to the unique material and device structure but also since GaN is so far the only wide bandgap which could be grown on large diameter silicon substrates, thus overcoming a major barrier toward mass market commercialization.

14.2 Diamond Power Devices

By considering only the ideal intrinsic material properties, diamond should be the ultimate material for high frequency and high-power applications. Diamond has the widest bandgap, and the highest breakdown strength, coupled with the highest thermal conductivity found in nature. In addition, diamond is an inert material and is the hardest material we know. This should enable devices capable of handling high power densities, with very high voltages and very high currents. The high thermal conductivity of diamond enables handling the high-power densities without sophisticated heat extraction methods. Here, the device itself is no more than the thermal barrier to the heat sink, as in all other electronics. However, many challenges had to be met first before demonstrating the potential of diamond devices, at only a fraction of the ideally expected performance.

The extreme properties of diamond as an electronic material were measured on naturally occurring diamond of type IIa [17]. This type of diamond has the highest purity but is extremely rare and prohibitively expensive. So, the first step in enabling diamond technology was the development of synthetic single crystalline diamond growth techniques, using high-temperature, high-pressure growth chambers, and chemical vapor deposition (CVD) techniques [18]. Nowadays, electronic grade substrates are commercially available, and are used as radiation detector material (due to the diamond radiation hardness), heat spreaders for high energy optical applications, and are also used to develop diamond electronics. However, the cost and size ($\sim$1.5 mm $\times$ 1.5 mm) of such substrates are still very far from being considered for a mass market and are only suitable for niche applications. Coupled with the small size substrates, the hardness of diamond presents another barrier to wide commercialization, and requires specialized fabrication technology, especially when sub-micrometer structures are required in the device. This limited the fabrication technology of diamond devices to research groups and it is difficult to envision the transfer of fabrication technology to already existing CMOS-based fabs, such as in the case of SiC and GaN. But probably the biggest challenge for diamond devices is the difficulty of doping and dopant activation, in addition to forming ohmic contacts, due to the very wide bandgap.

The typical n-type dopants for diamond are nitrogen and phosphorous (with activation energies of 1.7 and 0.6 eV, respectively), and the p-type dopant is boron with activation energy of 0.37 eV [11]. Therefore, only the p-type doping can be partially activated at room temperature, which restricts the diamond devices to p-type unipolar devices. Moreover, due to impurity scattering, the mobility drops drastically with increased doping concentration (450 cm^2/V s for a doping concentration of

1×10^{19} cm^{-3} [10]), 10 times drop compared to the ideal situation at low doping. The challenges of doping activation and low mobility at high doping concentrations seems to have severely hampered the development of the boron delta-doped diamond Field Effect Transistor (FET), which is one of two concepts perused in diamond FETs. Although devices were demonstrated with breakdown voltage more than 1.3 kV [19] at room temperature, the device needs to be operated at 300 °C to fully activate the channel dopants. More details regarding the delta-doped diamond FET can be found in [20, 21].

The second more successful concept is the hydrogen-terminated diamond field effect transistor (HD-FET). Here, a surface channel is formed when terminating the surface carbon with hydrogen, without the need of doping. This is explained by spontaneous polarization due to the difference in the electronegativity in the carbon–hydrogen dipole, which is very similar to the case of GaN HEMTs as will be described later. Chen et al. [22] presents a comprehensive review of the hydrogen terminated devices along with a timeline of the device development and performance metrics. The limitation of those devices is the very unstable nature of the hydrogen–carbon bond, causing it to dissolve in very short time and at higher device temperature. But with the continuous research and the introduction of more efficient surface passivation, the stability of the hydrogen termination could be somewhat controlled. For RF power amplification, devices were reported to deliver 3.8 W/mm at 1 GHz, and devices delivering 0.7 W/mm at 10 GHz. For DC power applications, HD-FETs could be demonstrated with breakdown voltage up to 500 V [23]. Still, the stability of those devices is measured in weeks, and all those devices rely on the small size substrates.

To achieve a true commercial breakthrough a move must be made toward large area substrates, but this is not possible using single crystalline diamond. Alternatively, research also was conducted on polycrystalline diamond, which could be grown on foreign substrates, such as silicon and SiC. High cutoff frequencies up to 45 GHz have been demonstrated on free standing polycrystalline substrates, using the HD-FET concept [24]. HD-FET polycrystalline diamond grown on SiC substrates showed breakdown voltages up to 500 V [25]. More recently, HD-FET grown on GaN/Si with blocking voltage up to 400 V was demonstrated [26]. This enables not only the integration of diamond and GaN devices, but also greatly enhances the thermal management in GaN devices, since the heat can be removed from the top side of the device. Although in the initial research state, those results indicate the potential for using diamond-based power devices in a more commercial manner, once the technology is mature, the stability of the device is increased, and the ability to integrate the devices with control circuits is feasible. As in the cases of GaN and SiC, the monolithic integration of silicon-based CMOS circuits is not possible due to the fundamentally different substrate used.

14.3 SiC Power Devices

SiC is probably the most mature wide bandgap semiconductor so far. SiC has many crystal-stacking sequences, but the most commonly used is the 4H-SiC stacking, which will be referred to here as simply SiC. Covering all the aspects of the technology and devices development is out of the scope of this chapter. Here will be given a short summary of the history, potential, and challenges of SiC power devices. The reader is referred to [9, 27–29] for more in-depth discussion of the history, current state, and challenges of SiC power devices. The development of SiC power devices is very similar to the development of silicon power devices, with similar device structures, such as the unipolar SiC-MOSFET and the bipolar SiC-insulated gate bipolar transistor (IGBT). In addition, the native oxide of SiC is SiO_2, which enabled to use the extensive knowledge gained from silicon, where MOSFET devices rely on a well-defined high quality gate oxide. The research in

SiC devices started as early as the 1980s and still continues today. What gave SiC an edge over other wide bandgap semiconductor is the commercial introduction of high voltage–high current Schottky diodes, as early as the year 2000 which was capable of voltages above 400 V, and therefore did not compete with existing silicon-based solutions. Currently, SiC junction barrier diodes (SiC JBS) with voltage blocking capabilities between 600 V and 3.3 kV, and currents up to 50 A, are commercially available. SiC is particularly attractive for DC power conversion applications, not only due to the high breakdown strength but also due to the high thermal conductivity, second only to diamond, which facilitates the thermal management of high-power devices. The SiC power devices exist commercially in both unipolar (MOSFET and junction field-effect-transistor [JFET]) and bipolar (IGBT) forms. SiC IGBTs can handle more than 15 kV blocking voltages with currents up to 40 A. But being bipolar devices, they are limited by the switching speed, and more attractive solution for high speed–high power devices is the SiC unipolar devices, which could be capable of switching in the 10s of MHz regime [30], with proper parasitic inductance reduction through advanced packaging, compared to only several hundreds of Hz for IGBTs. This drive toward higher switching speeds is gaining additional traction nowadays with the planned wide electrification of vehicles. The discussion here will be focused on the unipolar SiC devices. Han et al. [28] represents a comprehensive review of SiC IGBTs and their performance compared to silicon IGBTs.

In general, the development and commercialization of SiC devices passed through the typical new semiconductor barriers to market introduction, such as the increase of wafer diameter (150 mm wafers were introduced in 2012) and the drop of wafer cost, in addition to reducing the bulk defect density. For SiC MOSFETs in particular, the performance available now still did not approach the theoretically expected performance. The n-type bulk mobility is still around $950 \, cm^2/V \, s$, and the channel mobility is as low as $100 \, cm^2/V \, s$, which is quite low compared to the theoretical $1400 \, cm^2/V \, s$. Those comparatively low values are due to bulk and interface defects, which indicate that more effort put in improving the material quality is key to push SiC performance to the limit. The improvement in channel mobility is key to reduce the on-resistance of the device and thus reducing size and cost. As in silicon MOSFETs, the channel mobility of SiC MOSFET is strongly related to the quality of the SiC/SiO_2 interface, as discussed in detail in [31]. The initially defective and poor-quality interface causes the devices to suffer from reduced mobility due to interface scattering, in addition to introducing threshold voltage instabilities due to charge trapping in the oxide. This introduces not only a degraded device performance but also reliability issues. Improving the oxide and oxide interface quality is therefore key to increasing the mobility, reducing the on-resistance. Additionally, a thicker oxide could then be used, which enables the gate diode to withstand higher voltage operation ranges. A common method used so far to increase the oxide interface quality is the nitridation of the formed SiO_2. This practice could somehow mitigate the problem, but not completely. Currently, the commercially available SiC devices are actually operated (or rated) to voltages below what the device can handle, in order to avoid the charge trapping issue. Working within the safe operation range, SiC-based MOSFETs and JFETs are indeed commercially available and are used currently in many applications, such as motor drives, power inverters, traction systems, power supplies, and electric vehicles [29], with industrial standards being contentiously reviewed and updated [32]. Commercially available SiC MOSFETS are capable of handling voltages up to 1.8 kV and currents up to 120 A. 3.3 kV SiC JFETs are used in Bullet Trains in Japan [31].

In addition to developing suitable packing material, mostly being ceramic-based in order to handle the high-power densities, one other challenge being addressed is the ability to develop efficient gate drivers out of SiC [27], and to integrate the gate driver and control electronics monolithically, as in the case of system-on-chip for silicon MOSFETs. Since SiC can be doped with both n-type and p-type doping, CMOS like logic circuits can be built [33, 34]. A comprehensive review of the state-of-the-art logic-based circuits in wideband gap semiconductors (SiC and GaN) is presented in [35].

14.4 GaN Power Devices

GaN belongs to the III-Nitride group, along with other binary alloys (AlN and InN) and ternary and quaternary alloys (AlGaN, AlInN, InGaN, etc.), which retains thermodynamically stable hexagonal (Wurtzite) structure. For simplicity, the III-N alloys will be referred to as GaN- or GaN-based devices. Those alloys are widely used in optoelectronic applications due to the ability to engineer the bandgap (and therefore the emitted light wavelength) by changing the alloy composition. Those alloys are also very attractive for electronic applications due to the unique property of obtaining large polarization fields, which promotes the ability to form heterostructures with large carrier densities at the interface as a two-dimensional electron gas (2DEG), with high mobilities. In addition, the ability to engineer the bandgap gives another design degree of freedom, unlike the fixed bandgap of SiC or diamond. This forms the basis of the HEMT, which is common in GaAs technology. However, unlike GaAs, GaN HEMTs do not require extrinsic doping. The high channel density and mobility, together with the wide bandgap and the absence of doping, puts the theoretical limits of GaN HEMTs beyond what can be achieved by SiC as shown in Figure 14.1.

GaN is a polar material. Due to the lattice asymmetry along the *c*-axis and the ratio of the *c*-axis lattice constant to the basal plane lattice constant (*a*-direction), and the difference in the electron affinity between Ga and N, a spontaneous polarization exists in the crystal. In addition, when the basal plane lattice constant is changed due to external strain (without changing the composition), a piezoelectric polarization component is added depending on the value and direction of strain (compressive or tensile). Besides the bandgap, the value of the polarization therefore depends on the lattice constants, which in turn depend on the alloy composition. Detailed analysis regarding the pyroelectric properties of III-N alloys, the related material properties, and the formation of the 2DEG are found in [36–38]. When a III-N alloy (usually called a barrier, typically AlGaN) is grown pseudomorphically on top of another alloy (usually called a buffer, typically GaN), a net polarization difference at the interface is present and induces a fixed, net charge at the interface. For Ga-face alloys, this is a net positive charge when the buffer has a smaller bandgap than the barrier. To maintain charge neutrality, the fixed positive charge is compensated by an electron channel, in the form of a 2DEG. The source of electrons in the channel was identified to be from surface traps [39], which is driven into the quantum well formed by the bandgap discontinuity by the polarization field in the barrier, which scales with the barrier thickness. The channel will be partially depleted by the surface potential, and the degree of depletion depends on the barrier thickness. Therefore, the channel density is dependent on both the composition and thickness of the barrier. For a thick enough barrier, where the effect of surface depletion is marginal, the channel density approaches the polarization discontinuity. Figure 14.2 shows the calculated bandgap for III-Nitrides, and Figure 14.3 shows theoretically expected channel density for different alloys grown on a relaxed GaN buffer, after the analysis presented in [36]. Figure 14.4 shows a cross section of a typical GaN HEMT grown on a foreign substrate. Considering for the moment the top layers (barrier and buffer), it should be noted that the HEMT is already formed and defined by epitaxy, and for the very basic designs, requires only forming the ohmic contacts and the gate contact. A more detailed description of the device processing is in [37, 40], and only specific technological issues will be discussed

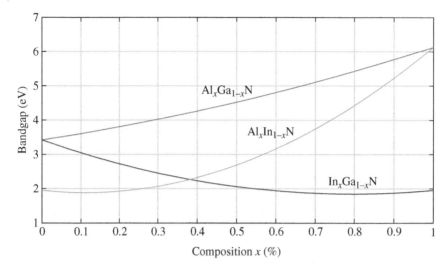

Figure 14.2 Calculated bandgap for III-N alloys as function of the alloy composition (x). Source: Based on Ambacher et al. [36].

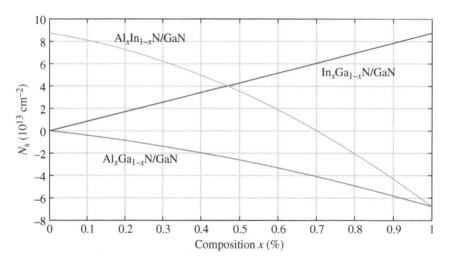

Figure 14.3 Calculated maximum 2DEG channel density for III-N alloy barrier grown on relaxed GaN buffer. Source: Based on Ambacher et al. [36].

here. The GaN HEMT therefore resembles a MOSFET, but with the channel being capacitively coupled to the gate, and the coupling is dominated by the dielectric properties of the barrier.

Ideally, all the degrees of freedom in buffer composition and thickness, barrier composition and thickness, and subsequently the many varieties of device design could be possible. However, this is limited by the fact that not all the alloys could be grown with sufficient quality. Underpinning this fact is that there is no native GaN substrate. GaN does not occur naturally, and has to be epitaxially grown on a substrate, thus limiting the possibilities within what would be achieved due to the thermal and lattice mismatch between the substrate and the grown alloy. Table 14.2 summarizes the main substrates available so far and the typically resulting defect density, due to the thermal and lattice mismatch after the data presented in.

Figure 14.4 Cross section of a typical GaN HEMT grown on a foreign substrate. Dimensions are not to scale.

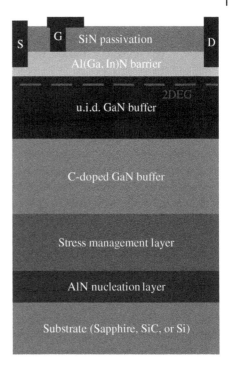

Table 14.2 Main substrates and their properties available for GaN epitaxy, with the typically resulting defect density.

	Sapphire	SiC	Si	GaN
Lattice mismatch (%)	16	3.4	17	0
Thermal expansion coefficient mismatch (%)	25.3	25	53	0
Resulting defect density (cm^{-2})	8×10^8	10^9	10^8	10^4
Substrate isolation (resistivity)	High	High	Medium (HR Si) or conductive (p-doped Si)	High
Maximum available diameter (mm)	100	150	200	76.2

Properties listed are after Meneghini et al. [37] and Quay [40].

Until recently, GaN is grown only on foreign substrates, such as c-plane Sapphire and SiC, and (111) silicon. Commonly used is the metal organic chemical vapor deposition (MOCVD) technique, but also molecular beam epitaxy (MBE) can be used. A review of the growth techniques and conditions is presented in [41]. For all foreign substrates, a nucleation AlN layer is needed to initiate the growth. This is followed by strain and defect management layers, which enables the growth of the GaN buffer, and the subsequent barrier layer. Managing the strain, especially for growth on silicon, where the wafer diameter is the largest and the thermal and lattice mismatches are also the largest, is essential in order to achieve wafers with acceptable bow (typically below 50 μm). The n-type unintentionally doped GaN buffer (u.i.d. GaN) is slightly conductive and limits GaN devices from reaching the theoretically predicted breakdown strength. For RF applications, where the maximum applied voltages are around 50 V, the breakdown is limited by the gate to drain distance, and the slight buffer conductivity is tolerated. This conductivity results from the unintentional incorporation of silicon and oxygen (both present in the growth gases), which are shallow n-type dopants for GaN, with activation energies around 0.2 eV below the conduction band, and therefore activated at room temperature. The resulting doping concentration is in the order of 1×10^{16} cm^{-3}. This concentration is marginal given the high concentration in the channel (1–2×10^{13} cm^{-2}, which for the approximately 5 nm wide quantum well is equivalent to doping densities in the order of 1×10^{21} cm^{-3}). However, in the case of conductive substrates, such as p-type silicon, an insulating buffer is required (usually called compensated buffer), to achieve the required vertical isolation between the channel and the grounded substrate. This layer is not needed in the case of insulating substrates such as sapphire and SiC but is essential in enabling high voltage GaN on silicon. The slightly n-type background doping of the buffer is compensated by doping the buffer with carbon, either from the growth carrier gases or by introducing extrinsic carbon containing gas. Depending on the growth conditions, carbon can act as n-type doping as a donor trap, or as p-type doping as an acceptor trap [42]. Under Ga-rich growth conditions common for MOCVD, carbon acts as an acceptor trap (p-type doping) and will compensate the background n-type doping, until the Fermi level is pinned at ~0.9 eV above the valence band (deep acceptor). After this point, another mechanism for the occupation of carbon in the lattice takes place, and addition of carbon does not yield higher p-type doping, since carbon also starts to occupy the donor state with a similar probability to the acceptor state. The resulting compensated buffer is slightly p-doped (~1×10^9 cm^{-3}) and is sufficiently insulating for high voltage applications.

GaN on GaN is also recently available with sufficient quality. Two main methods are used, namely the ammonothermal method and the Hydride Vapor Phase Epitaxy (HVPE). The ammonothermal method relies on GaN growth from gallium solution under high pressure. The resulting GaN is a true crystal, but the method is limited in the wafer size to several centimeters, and very slow growth rates [43]. The HVPE method [44] on the other hand delivers similar crystal quality with high growth rates. Both methods require very tight control of the epitaxial conditions and very tight control of impurities. Very high quality and low impurity HVPE substrates were demonstrated recently [45].

The choice of substrate is governed by the desired application. For high frequency applications, an insulating substrate is preferred, in order to avoid the losses due to the substrate capacitance. In addition, for high power applications (RF or DC), the thermal conductivity of the substrate determines the heat dissipation efficiency. Although the best expected performance would come from using GaN on GaN, due to the very low defect concentration, lattice and thermal matching, and high isolation, the resulting device cost is higher due to the higher substrate cost and smaller available diameter. In cost effectiveness terms, GaN on silicon has the lead, but again, the resulting device operation limits have to be considered. For example, for GaN high power RF applications, SiC substrates are favored due the higher thermal conductivity, which enables more reliable device operation and simpler thermal and packaging solutions than in the case of silicon substrates, thus enabling the device to operate at higher output power levels for the same die area, with less self-heating effects, and less cooling requirements. Other approaches to manage the device heating are described later.

Despite the relatively high defect density compared to silicon or SiC, where the device would fail to operate with such high defect density, GaN HEMTs proved to be very defect-tolerant. This is due to nature of channel formation being not dependent on the micro-bulk properties, the strong polarization fields, and the high channel mobility in a 2DEG. Indeed, impressive power handling capabilities and very high operation frequencies could be demonstrated. GaN on SiC HEMTs could demonstrate very high output power densities at frequencies above 1 GHz, more than a decade ago [46, 47], and up to 40 W/mm [48], with proper management of the high electric fields using field plates. However, sustained operation at such very high power densities is not possible without proper thermal management, despite the high thermal conductivity of the substrate.

The commercially available GaN on SiC RF devices are rated for power densities around 5 W/mm, which is below what can actually be achieved. Recently, GaN on SiC HEMTs could demonstrate 5 W/mm output power at 40 GHz, with power added efficiency of 43.6% [49]. GaN on SiC could also demonstrate very high cutoff frequencies up to 500 GHz [50, 51]. For GaN on silicon, high RF power capabilities (output power more than 4.2 W/mm and 64% power added efficiency) were demonstrated [52], and very high cutoff frequencies up to 250 GHz [53] were demonstrated, approaching the performance of GaN on SiC. High frequency diodes were also demonstrated using GaN on silicon HEMTs [54].

The same GaN HEMT structure is also used to fabricate high voltage devices for DC power applications. Up to 3000 V blocking voltage could be demonstrated for devices on novel AlN on sapphire substrates [55–57]. High voltages could also be demonstrated on the more commercially viable GaN on silicon, with voltages up to 1200 V and currents up to 30 A for single devices, with conversion efficiencies reaching 99% [52, 58]. Diodes also were demonstrated with blocking voltage up to 3000 V [59]. However, most of the high voltage GaN on silicon development is focused on 650 V operation class, since the higher voltage ranges can be efficiently supplanted by SiC devices, and do not require the high switching frequencies which GaN can deliver.

However, despite the impressive performance of GaN HEMTs, several reliability issues and limitations due to the device nature, were, and are being addressed, especially for the commercially viable GaN on silicon. Those reliability and functionality issues are mainly the charge trapping effects and the resulting degradation in R_{on}, the depletion mode (D-mode) nature of the device, the lack of efficient complementary technology (p-GaN HEMTs and n-GaN HEMTs), and the thermal management issues.

Among several reliability concerns as summarized in [60], charge trapping is the most severe. Charge trapping in GaN HEMTs can occur at the passivation to barrier interface, in the bulk of passivation, or in the bulk of the barrier or buffer, and affects the dynamic response of the device. At high fields (usually in off-state), charges are injected at the interfaces or bulk, and are discharged in the on-state. If the trap life time, composed of the trapping life time, emission life time, and removal of the trapped charge, is slower than the switching frequency of the device, then an increase in R_{on} occurs, accompanied with a shift in the threshold voltage and knee-voltage walk out [61]. This is due to the charge neutrality of the device, where the trapped charges partially deplete the 2DEG channel, in order to preserve the charge neutrality. The surface and passivation traps were actually an issue in the early high frequency devices and the trap speed was the limiting factor in operating the power amplifier at higher frequencies [62]. Those traps are characterized by a relatively short lifetime above 1 MHz. This issue is resolved by proper passivation of the device, including the type of passivation and the way it is deposited. A standard practice in GaN HEMTs is an optimized CVD-based SiN passivation, which largely eliminates this issue. Bulk traps on the other hand have relatively long-life times in the millisecond range and up to weeks. So, for high frequency applications, the bulk traps have minor influence, since the device turns on before the trapping process is complete in the off-state period. But for DC power applications, where the switching frequencies are below 1 MHz, the opposite is true. Here the surface traps (if present) can charge and discharge before the next switching cycle, while the bulk traps remain trapped.

Although dislocations in the material act as trap centers, the dominant species causing the charge trapping is the carbon doping in the buffer [63], and therefore always present in high voltage GaN on silicon and presented a serious impediment in commercializing GaN on silicon for DC power applications. The buffer charge trapping can be mitigated by either properly engineering the buffer conductivity, to allow a faster discharge path as described in [64], or by injecting holes as counter charges using Mg-doped GaN drain contact as described in Tanaka et al. [65]. However, both of those methods are solutions for a certain operation range, and do not address the issue fundamentally, in which case would require the elimination of the carbon doped buffer, while at the same time replacing this layer with another layer capable of blocking high voltages. Such concept is the super lattice buffer (SL-buffer), where the carbon-doped buffer is replaced by alternating III-Nitride layers [66]. This method proved to be effective in suppressing the charge trapping effect to a large degree within typical high voltage operating ranges and frequencies [67], in addition to enabling higher isolation for thinner buffer thicknesses [68], which is a considerable contribution to reducing the GaN device cost by reducing the required growth time, and inducing less wafer stress and bow. However, whether this method does indeed eliminate the charge trapping for all voltage ranges or all operation frequencies is still to be determined. The trend observed so far in GaN HEMTs is that for each application operating boundaries, a highly specialized epitaxial sequence should be developed for that specific application.

Another barrier to commercialization faced by GaN HEMTs is the devices D-mode nature, which introduces a safety concern, especially for high current and high voltage devices. Moreover, in D-mode devices, switching off the channel requires the application of negative gate voltages, which is not compatible with the mainstream gate driver technologies. Several approaches were successfully implemented to convert the device to an enhancement mode type. A circuit-based approach is the GaN HEMT-Si MOSFET cascode approach [69], which is now commercially available. This approach relies on the reliable Si technology to drive the GaN HEMT into depletion and facilitates the interface to other silicon-based control electronics. This method however requires tight matching between the capacitances and leakage currents of the HEMT and the MOSFET, and requires the hetero-integration of both devices. A monolithic approach to obtain an E-mode HEMT is the use of Mg-doped p-GaN layer at the gate contact to deplete the channel [70]. This approach was recently industrially adopted, and enabled the suppression of charging effects using the same p-GaN layer, in addition to turn-on voltages more than 2 V. However, this method limits the switching speed of the devices and introduces reliability issues related to charge storage in the p-GaN layer [71–73]. Recently, it could be shown that p-GaN HEMTs could reach cutoff frequencies in the GHz regime by proper gate structuring [74]. A second monolithic approach is using the FinFET concept to deplete the channel from the side in addition to the barrier top [75–78]. This method avoids introducing additional p-GaN growth steps but requires nano-structuring the gate region with features below 50 nm. A third approach is recessing the gate region, removing the barrier material, and hence removing the formed channel [79]. This method however requires precise etch control and optimization of the resulting side wall etch and the subsequent passivation. All of the abovementioned methods did however yield E-mode devices approaching or exceeding the performance of the traditional D-mode GaN HEMT. A more detailed review of E-mode GaN HEMT concepts is presented in Roccaforte et al. [80].

With the development of E-mode devices, complementary logic on GaN could be demonstrated using E-mode/D-mode complementary devices [81]. However, high-performance logic normally depends on p–n pairs, as usual for silicon CMOS. The limitation in this case for GaN HEMTs is the low mobility of holes ($\sim$100 cm^2/V s) compared to that of the 2DEG, which results in unmatched or over scaled logic circuits. Several attempts however could show successful p–n type logic using bandgap and polarization field engineering [82], or p-doped GaN with n-doped GaN [83, 84]. A detailed review is presented in Bader et al. [35].

With the increased efficiency and power handling capability of GaN HEMTs, in addition to increased integration density, managing the thermal losses is key to enable reliable and safe operation. In SiC devices, managing the heat is facilitated by the high thermal conductivity of the SiC substrate. GaN grown on foreign substrates lacks this advantage, since the nucleation and stress relief layers are very defective and disordered, and present the highest thermal barrier when attempting to extract the heat from the back side. This is somewhat less punishing for GaN on SiC for high-frequency applications, since the resulting defect density is the lowest, coupled with the high thermal conductivity of the substrate. The worst-case scenario is the commercially viable GaN on silicon. Several approaches are perused to effectively enhance the heat extraction from the back side and most of them rely on the removal of the substrate and the nucleation and stress relief layers, and then adding a heat extraction material such as copper or AlN [85]. A more efficient method is the complete Si removal and wafer transfer to highly thermally conductive substrates such as polycrystalline or single crystalline diamond [86, 87], which showed very significant heat extraction efficiency. Ideally, GaN would be grown directly on single crystalline diamond, as was shown in Alomari et al. [88]. However, this requires MBE of the GaN layer on the small size diamond substrates, which is not very attractive in terms of cost efficiency. Alternatively, direct growth of polycrystalline diamond can be applied on the top side of the device [26, 89], therefore utilizing the large area advantage, but requires highly thermally stable contact technology in order to withstand the high growth temperatures of diamond. A more tolerant approach would be the growth of diamond on the back side after the removal of substrate and nucleation layers. With a deeper understanding of the interface and bulk thermal properties of polycrystalline diamond [90], and therefore minimizing the thermal boundary interface resistance, it is possible to approach heat extraction performance similar or exceeding the performance using bonding approaches, even if on single crystalline diamond. This has been demonstrated recently in Tadjer et al. [91], where DC powers up to 56 W/mm could be driven in the device, while maintaining junction temperatures below 180 °C.

The reliability issues of charge trapping and poor heat extraction, stemming from the existence of doped buffer, defective nucleation, and stress relief layers, and the added thermal boundary interface resistance in the substrate transfer or re-growth techniques can be avoided by using bulk GaN substrates. It has been shown in [92] that using GaN bulk substrates enables devices with virtually no trapping effects, compared to GaN on silicon, and enhanced heat extraction. The relatively recent availability of higher quality bulk GaN substrates also enabled the implementation of vertical GaN device concepts, where higher currents can be handled with smaller die area compared to the lateral HEMT, as demonstrated in the case of vertical GaN on silicon [93]. Such devices also include the use of advanced structuring techniques such as FinFETs and P-doped GaN layers [94, 95]. In addition, high performance–high power diodes could be demonstrated using bulk GaN substrates [96]. All of the abovementioned advances in Ga-face HEMTs could theoretically also be applied on nitrogen-face HEMTs, with several thrusts in research in this direction [97, 98].

14.5 Wide Bandgaps for High-Temperature Applications

So far, and due to the very mature and relatively cheap silicon technology, wide bandgaps have to show very clear and cost-effective advantages in order to replace the current silicon devices, which could be the case in high frequency power electronic, and high-power DC converters, but still within a specific application range. High-temperature applications, where electronics are needed to operate in environment temperatures above 300 °C, can be an exclusive field for wide bandgaps.

At such temperatures, silicon fails to operate, which so far limited the available elements for such high temperatures to passive components or vacuum tubes. Examples of such applications are terrestrial applications, such as turbine engine monitoring, oil drilling, exhaust gas monitoring, and geothermal reactors monitoring. In addition, extra-terrestrial applications, such as the exploration of hot planets, Venus as an example, where the temperature is in average 464 °C [99] require the electronics to be able to operate at the elevated temperature and to wirelessly communicate sensor data, usually at frequencies above 1 MHz. Even in terrestrial applications, the high temperatures are usually coupled with vibrations in the kHz range, and to avoid interference with the wirelessly communicated data the device should also be able to operate at higher frequencies. Such requirements can be met by wide bandgaps, due to their physical properties, and therefore enable the population of semiconductors in the high temperature domain. Several demonstrations of the ability of wide bandgaps to operate at very high temperatures indicate the very real applicability of such a concept. Diamond diodes have been demonstrated to operate up to 800 °C [100], and GaN HEMTs up to 1000 °C [101] in vacuum. The high frequency response of those bare-die experiments could not be directly measured due to the lack of any RF probes capable of operating above 300 °C. However, the equivalent circuits extracted from the measured DC characteristics indicate that the GaN HEMT is still capable of switching in the GHz regime at temperatures up to 800 °C at least. Since GaN is also a piezoelectric material, high sensitivity pressure sensors can be fabricated for high-temperature environments [102]. Other sensor types, such as Hall sensors, are also possible [103]. A more thorough progress was shown in the case of SiC in a focused research effort by Glenn Research Centre in the frame of Venus exploration program. SiC JFETs were demonstrated to operate for extended times at 500 °C in the MHz regime [104], and digital circuits were demonstrated to operate up to 1000 °C. Underpinning those extraordinary results is the optimization of the other components, such as packaging and interconnects, to be also high temperature capable [105]. So far, electronic components are optimized and produced with the thermal limitations of silicon in mind, but with wide bandgaps, the semiconductor is not any more the limit. Once those passive components are also optimized for higher operation temperature, it could also be possible to change the perspective regarding power conversion applications at moderate temperatures, by reducing the cooling requirements, and therefore the total cost and weight of the system, saving yet more energy and enabling higher efficiency in using our limited energy resources.

14.6 Conclusion

Owing to the superior electronic properties, wide bandgaps offer superior performance compared to silicon-based devices, in applications requiring high output power handling and switching at higher frequencies, as demonstrated by the review of the state-of-the-art presented here. The increased demand on solid state DC power converters, and on high power–high-frequency amplifiers, promotes wide bandgaps as prime candidates for replacing silicon technology, especially due to the much higher power efficiencies. Wide bandgap power devices are indeed used in many applications, with several niche markets already replacing silicon and GaAs, but cannot yet completely replace silicon electronics due to several commercialization barriers. The complexity of integrated circuits, the high reliability and the low cost of silicon technology is still not matched by wide bandgaps. Therefore, a considerable effort, both on the research and industrial level, to address those limitations is being spent. Meanwhile, the hybrid integration approach of a wide bandgap power module with silicon-based control circuits provides an intermediate approach. Wide bandgaps can also enable completely new markets and applications, not possible using silicon technology, such as high temperature applications. This would also require suitable packaging and interconnect development but is being spearheaded by space exploration activities. Once this new semiconductor frontier is mastered, the inclusion of smart electronic systems will be possible in almost all aspects of our lives, with considerable saving on the energy consumption and environmental load on our planet.

References

1 M. Henke and T.-H. Dietrich, *High Power Inductive Charging System for an Electric Taxi Vehicle*, IEEE, 2017.

2 T. Loher, S. Karaszkiewicz, L. Bottcher, and A. Ostmann, *Compact Power Electronic Modules Realized by PCB Embedding Technology*, IEEE, 2016.

3 P. Sinhuber, W. Rohlfs, and D. U. Sauer, *Conceptional Considerations for Electrification of Public City Buses – Energy Storage System and Charging Stations*, IEEE, 2010.

4 S. Nakamura, T. Mukai, and M. Senoh, "Candela-class high-brightness InGaN/AlGaN double-heterostructure blue-light-emitting diodes," *Applied Physics Letters*, vol. 64, no. 3, pp. 1687–1689, 1994.

5 E. Feltin, A. Castiglia, G. Cosendey, J. F. Carlin, R. Butte, and N. Grandjean, *GaN-Based Laser Diodes Including a Lattice-Matched $Al_{0.83}In_{0.17}N$ Cladding Layer*, OSA, 2009.

6 G. Huber, A. Richter, N-O. Hansen, M. Fechner, and E. Heumann, *GaN Diode Laser Pumped Solid-State Lasers in the Visible and UV Spectral Region*, IEEE, 2009.

7 K. Chang, *GaAs High-Speed Devices: Physics, Technology, and Circuit Applications*, John Wiley & Sons, 1994.

8 M. Shur, *GaAs Devices and Circuits*, Springer US, 1987.

9 B. J. Baliga, *Gallium Nitride and Silicon Carbide Power Devices*, WSPC, 2016.

10 H. Pernegger, S. Roe, P. Weilhammer, V. Eremin, H. Frais-Kölbl, E. Griesmayer, and H. Kagan, "Charge-carrier properties in synthetic single-crystal diamond measured with the transient-current technique," *Journal of Applied Physics*, vol. 97, no. 4, p. 073704, 2005.

11 C. J. H. Wort and R. S. Balmer, "Diamond as an electronic material," *Materials Today*, vol. 11, no. 1, pp. 22–28, 2008.

12 C. Raynaud, D. Tournier, H. Morel, and D. Planson, "Comparison of high voltage and high temperature performances of wide bandgap semiconductors for vertical power devices," *Diamond and Related Materials*, vol. 19, no. 1, pp. 1–6, 2010.

13 B. Brakus, *Switched Mode PWM DC/AC-Inverter for Uninterruptible 50 Hz AC Power Supply*, INTELEC '82 – International Telecommunications Energy Conference, pp. 49–54, 1982.

14 E. Gurpinar, F. Iannuzzo, Y. Yang, A. Castellazzi, and F. Blaabjerg, "Design of low-inductance switching power cell for GaN HEMT based inverter," *IEEE Transactions on Industry Applications*, vol. 54, no. 3, pp. 1592–1601, 2018.

15 M. H. Rashid, *Power Electronics Handbook*, Elsevier, 2007.

16 J. A. Cooper and D. T. Morisette, "Performance limits of vertical unipolar power devices in GaN and 4H-SiC," *IEEE Electron Device Letters*, vol. 41, no. 6, pp. 892–895, 2020.

17 J. E. Field (ed.), *The Properties of Natural and Synthetic Diamond*. London: Academic Press, 1993, ISBN: 0-12-255352-7. *Crystal Research and Technology*, vol. 28, p. 602.

18 M. Schwander and K. Partes, "A review of diamond synthesis by CVD processes," *Diamond and Related Materials*, vol. 20, no. 10, pp. 1287–1301, 2011.

19 H. Umezawa, T. Matsumoto, and S.-I. Shikata, "Diamond metal-semiconductor field-effect transistor with breakdown voltage over 1.5 kV," *IEEE Electron Device Letters*, vol. 35, no. 11, pp. 1112–1114, 2014.

20 R. S. Balmer, I. Friel, S. Hepplestone, J. Isberg, M. J. Uren, M. L. Markham, N. L. Palmer, J. Pilkington, P. Huggett, S. Majdi, and R. Lang, "Transport behavior of holes in boron delta-doped diamond structures," *Journal of Applied Physics*, vol. 113, no. 1, p. 033702, 2013.

21 E. Kohn, M. Adamschik, P. Schmid, A. Denisenko, A. Aleksov, and W. Ebert, "Prospects of diamond devices," *Journal of Physics D: Applied Physics*, vol. 34, no. 8, pp. R77–R85, 2001.

22 Z. Chen, Y. Fu, H. Kawarada, and Y. Xu, "Microwave diamond devices technology: field-effect transistors and modeling," *International Journal of Numerical Modelling: Electronic Networks, Devices and Fields*, vol. 34, no. 8, 2020.

23 H. Kawarada, H. Tsuboi, T. Naruo, T. Yamada, D. Xu, A. Daicho, T. Saito, and A. Hiraiwa, "C–H surface diamond field effect transistors for high temperature (400 °C) and high voltage (500 V) operation," *Applied Physics Letters*, vol. 105, no. 7, p. 013510, 2014.

24 K. Ueda, M. Kasu, Y. Yamauchi, T. Makimoto, M. Schwitters, D. J. Twitchen, G. A. Scarsbrook, and S. E. Coe, "Diamond FET using high-quality polycrystalline diamond with f_T of 45 GHz and f_{max} of 120 GHz," *IEEE Electron Device Letters*, vol. 27, no. 7, pp. 570–572, 2006.

25 M. Syamsul, N. Oi, S. Okubo, T. Kageura, and H. Kawarada, "Heteroepitaxial diamond field-effect transistor for high voltage applications," *IEEE Electron Device Letters*, vol. 39, no. 1, pp. 51–54, 2018.

26 R. Soleimanzadeh, M. Naamoun, R. A. Khadar, R. Van Erp, and E. Matioli, "H-terminated polycrystalline diamond p-channel transistors on GaN-on-silicon," *IEEE Electron Device Letters*, vol. 41, no. 1, pp. 119–122, 2020.

27 L. F. S. Alves, P. Lefranc, P.-O. Jeannin, and B. Sarrazin, *Review on SiC-MOSFET Devices and Associated Gate Drivers*, IEEE, 2018.

28 L. Han, L. Liang, Y. Kang, and Y. Qiu, "A review of SiC IGBT: models, fabrications, characteristics, and applications," *IEEE Transactions on Power Electronics*, vol. 36, no. 2, pp. 2080–2093, 2021.

29 X. She, A. Q. Huang, O. Lucia, and B. Ozpineci, "Review of silicon carbide power devices and their applications," *IEEE Transactions on Industrial Electronics*, vol. 64, no. 10, pp. 8193–8205, 2017.

30 S. Guo, L. Zhang, Y. Lei, X. Li, F. Xue, W. Yu, and A. Q. Huang, *3.38 MHz Operation of 1.2 kV SiC MOSFET with Integrated Ultra-fast Gate Drive*, IEEE, 2015.

31 T. Kimoto and H. Watanabe, "Defect engineering in SiC technology for high-voltage power devices," *Applied Physics Express*, vol. 13, no. 11, p. 120101, 2020.

32 D. A. Gajewski, *Challenges and Peculiarities in Developing New Standards for SiC*, IEEE, 2020.

33 M. Albrecht, T. Erlbacher, A. Bauer, and L. Frey, "Improving 5V digital 4H-SiC CMOS ICs for operating at 400C using PMOS channel implantation," *Materials Science Forum*, vol. 963, no. 7, pp. 827–831, 2019.

34 M. H. Weng, D. T. Clark, S. N. Wright, D. L. Gordon, M. A. Duncan, S. J. Kirkham, M. I. Idris, H. K. Chan, R. A. R. Young, E. P. Ramsay, and N. G. Wright, "Recent advance in high manufacturing readiness level and high temperature CMOS mixed-signal integrated circuits on silicon carbide," *Semiconductor Science and Technology*, vol. 32, no. 4, p. 054003, 2017.

35 S. J. Bader, H. Lee, R. Chaudhuri, S. Huang, A. Hickman, A. Molnar, H. G. Xing, D. Jena, H. W. Then, N. Chowdhury, and T. Palacios, "Prospects for wide bandgap and ultrawide bandgap CMOS devices," *IEEE Transactions on Electron Devices*, vol. 67, no. 10, pp. 4010–4020, 2020.

36 O. Ambacher, J. Majewski, C. Miskys, A. Link, M. Hermann, M. Eickhoff, M. Stutzmann, F. Bernardini, V. Fiorentini, V. Tilak, and B. Schaff, "Pyroelectric properties of Al(In)GaN/GaN hetero- and quantum well structures," *Journal of Physics: Condensed Matter*, vol. 14, no. 3, pp. 3399–3434, 2002.

37 M. Meneghini, G. Meneghesso, and E. Zanoni, *Power GaN Devices: Materials, Applications and Reliability*, 1st Edition. Switzerland: Springer International Publishing, 2016.

38 J. Piprek (ed.), *Nitride Semiconductor Devices: Principles and Simulation*, Wiley, 2007.

39 J. P. Ibbetson, P. T. Fini, K. D. Ness, S. P. DenBaars, J. S. Speck, and U. K. Mishra, "Polarization effects, surface states, and the source of electrons in AlGaN/GaN heterostructure field effect transistors," *Applied Physics Letters*, vol. 77, no. 7, pp. 250–252, 2000.

40 R. Quay, *Gallium Nitride Electronics*, Springer-Verlag GmbH, 2008.

41 G. Meneghesso, M. Meneghini, and E. Zanoni (eds.), *Gallium Nitride-Enabled High Frequency and High Efficiency Power Conversion*, Springer International Publishing, 2018.

42 J. L. Lyons, A. Janotti, and C. G. V. de Walle, "Carbon impurities and the yellow luminescence in GaN," *Applied Physics Letters*, vol. 97, no. 10, p. 152108, 2010.

43 K. Grabianska, R. Kucharski, A. Puchalski, T. Sochacki, and M. Bockowski, "Recent progress in basic ammonothermal GaN crystal growth," *Journal of Crystal Growth*, vol. 547, no. 10, p. 125804, 2020.

44 T. Yoshida, Y. Oshima, K. Watanabe, T. Tsuchiya, and T. Mishima, "Ultrahigh-speed growth of GaN by hydride vapor phase epitaxy," *Physica Status Solidi C Current Topics in Solid State Physics*, vol. 8, no. 4, pp. 2110–2112, 2011.

45 H. Fujikura, T. Konno, T. Yoshida, and F. Horikiri, "Hydride-vapor-phase epitaxial growth of highly pure GaN layers with smooth as-grown surfaces on freestanding GaN substrates," *Japanese Journal of Applied Physics*, vol. 56, no. 7, p. 085503, 2017.

46 J. H. Leach and H. Morkoc, "Status of reliability of GaN-based heterojunction field effect transistors," *Proceedings of the IEEE*, vol. 98, no. 7, pp. 1127–1139, 2010.

47 Y.-F. Wu, A. Saxler, M. Moore, R. P. Smith, S. Sheppard, P. M. Chavarkar, T. Wisleder, U. K. Mishra, and P. Parikh, "30-W/mm GaN HEMTs by field plate optimization," *IEEE Electron Device Letters*, vol. 25, no. 3, pp. 117–119, 2004.

48 Y.-f. Wu, M. Moore, A. Saxler, T. Wisleder, and P. Parikh, *40-W/mm Double Field-Plated GaN HEMTs*, IEEE, 2006.

49 Y. Zhang, S. Huang, K. Wei, S. Zhang, X. Wang, Y. Zheng, G. Liu, X. Chen, Y. Li, and X. Liu, "Millimeter-wave AlGaN/GaN HEMTs with 43.6% power-added-efficiency at 40 GHz fabricated by atomic layer etching gate recess," *IEEE Electron Device Letters*, vol. 41, no. 5, pp. 701–704, 2020.

50 M. Micovic, D. F. Brown, D. Regan, J. Wong, Y. Tang, F. Herrault, D. Santos, S. D. Burnham, J. Tai, E. Prophet, and I. Khalaf, *High Frequency GaN HEMTs for RF MMIC Applications*, IEEE, 2016.

51 J.-S. Moon, J. Wong, B. Grabar, M. Antcliffe, P. Chen, E. Arkun, I. Khalaf, A. Corrion, J. Chappell, N. Venkatesan, and P. Fay, "360 GHz fMAX graded-channel AlGaN/GaN HEMTs for mmW low-noise applications," *IEEE Electron Device Letters*, vol. 41, no. 8, pp. 1173–1176, 2020.

52 T. Ueda, M. Ishida, T. Tanaka, and D. Ueda, "GaN transistors on Si for switching and high-frequency applications," *Japanese Journal of Applied Physics*, vol. 53, no. 9, p. 100214, 2014.

53 L. Li, K. Nomoto, M. Pan, W. Li, A. Hickman, J. Miller, K. Lee, Z. Hu, S. J. Bader, S. M. Lee, and J. C. Hwang, "GaN HEMTs on Si with regrown contacts and cutoff/maximum oscillation frequencies of 250/204 GHz," *IEEE Electron Device Letters*, vol. 41, no. 5, pp. 689–692, 2020.

54 A. Eblabla, X. Li, M. Alathbah, Z. Wu, J. Lees, and K. Elgaid, "Multi-channel AlGaN/GaN lateral Schottky barrier diodes on low-resistivity silicon for sub-THz integrated circuits applications," *IEEE Electron Device Letters*, vol. 40, no. 6, pp. 878–880, 2019.

55 Y. Wu, J. Zhang, S. Zhao, W. Zhang, Y. Zhang, X. Duan, J. Chen, and Y. Hao, "More than 3000 V reverse blocking Schottky-drain AlGaN-channel HEMTs with >230 MW/cm^2 power figure-of-merit," *IEEE Electron Device Letters*, vol. 40, no. 11, pp. 1724–1727, 2019.

56 M. Xiao, Y. Ma, K. Cheng, K. Liu, A. Xie, E. Beam, Y. Cao, and Y. Zhang, "3.3 kV multi-channel AlGaN/GaN Schottky barrier diodes with P-GaN termination," *IEEE Electron Device Letters*, vol. 41, no. 8, pp. 1177–1180, 2020.

57 Y. Zhang, J. Zhang, Z. Liu, S. Xu, K. Cheng, J. Ning, C. Zhang, L. Zhang, P. Ma, H. Zhou, and Y. Hao, "Demonstration of a 2 kV Al$_{0.85}$Ga$_{0.15}$N Schottky barrier diode with improved on-current and ideality factor," *IEEE Electron Device Letters*, vol. 41, no. 3, pp. 457–460, 2020.

58 T. Boles, *GaN-on-Silicon – Present Capabilities and Future Directions*, Author(s), 2018.

59 T. Zhang, J. Zhang, H. Zhou, Y. Wang, and T. Chen, "A >3 kV/2.94 mΩ·cm^2 and low leakage current with low turn-on voltage lateral GaN Schottky barrier diode on silicon substrate with anode engineering technique," *IEEE Electron Device Letters*, vol. 40, no. 10, pp. 1583–1586, 2019.

60 M. Meneghini, A. Barbato, M. Borga, C. De Santi, M. Barbato, S. Stoffels, and M. Zhao, *Power GaN HEMT Degradation: From Time-Dependent Breakdown to Hot-Electron Effects*, IEEE, 2018.

61 G. Meneghesso, M. Meneghini, I. Rossetto, D. Bisi, S. Stoffels, M. Van Hove, S. Decoutere, and E. Zanoni, "Reliability and parasitic issues in GaN-based power HEMTs: a review,". *Semiconductor Science and Technology*, vol. 31, no. 8, p. 093004, 2016.

62 R. Vetury, N. Q. Zhang, S. Keller, and U. K. Mishra, "The impact of surface states on the DC and RF characteristics of AlGaN/GaN HFETs,". *IEEE Transactions on Electron Devices*, vol. 48, no. 3, pp. 560–566, 2001.

63 H. Yacoub, C. Mauder, S. Leone, M. Eickelkamp, D. Fahle, M. Heuken, H. Kalisch, and A. Vescan, "Effect of different carbon doping techniques on the dynamic properties of GaN-on-Si buffers," *IEEE Transactions on Electron Devices*, vol. 64, no. 3, pp. 991–997, 2017.

64 M. J. Uren, S. Karboyan, I. Chatterjee, A. Pooth, P. Moens, A. Banerjee, and M. Kuball, ""Leaky dielectric" model for the suppression of dynamic R_{ON} in carbon-doped AlGaN/GaN HEMTs," *IEEE Transactions on Electron Devices*, vol. 64, no. 7, pp. 2826–2834, 2017.

65 K. Tanaka, T. Morita, H. Umeda, S. Kaneko, M. Kuroda, A. Ikoshi, H. Yamagiwa, H. Okita, M. Hikita, M. Yanagihara, and Y. Uemoto, "Suppression of current collapse by hole injection from drain in a normally-off GaN-based hybrid-drain-embedded gate injection transistor," *Applied Physics Letters*, vol. 107, no. 10, p. 163502, 2015.

66 S. Iwakami, M. Yanagihara, O. Machida, E. Chino, N. Kaneko, H. Goto, and K. Ohtsuka, "AlGaN/GaN heterostructure field-effect transistors (HFETs) on Si substrates for large-current operation," *Japanese Journal of Applied Physics*, vol. 43, no. 6, pp. L831–L833, 2004.

67 L. Heuken, M. Kortemeyer, A. Ottaviani, M. Schröder, M. Alomari, D. Fahle, M. Marx, M. Heuken, H. Kalisch, A. Vescan, and J. N. Burghartz, "Analysis of an AlGaN/AlN super-lattice buffer concept for 650-V low-dispersion and high-reliability GaN HEMTs," *IEEE Transactions on Electron Devices*, vol. 67, no. 3, pp. 1113–1119, 2020.

68 R. Kabouche, I. Abid, R. Püsche, J. Derluyn, S. Degroote, M. Germain, A. Tajalli, M. Meneghini, G. Meneghesso, and F. Medjdoub, "Low on-resistance and low trapping effects in 1200 V superlattice GaN-on-silicon heterostructures," *Physica Status Solidi A Applications and Material Science*, vol. 217, no. 11, p. 1900687, 2019.

69 X. Huang, Z. Liu, Q. Li, and F. C. Lee, *Evaluation and Application of 600V GaN HEMT in Cascode Structure*, IEEE, 2013.

70 Y. Uemoto, M. Hikita, H. Ueno, H. Matsuo, H. Ishida, M. Yanagihara, T. Ueda, T. Tanaka, and D. Ueda, "Gate injection transistor (GIT)-a normally-off AlGaN/GaN power transistor using conductivity modulation," *IEEE Transactions on Electron Devices*, vol. 54, no. 12, pp. 3393–3399, 2007.

71 M. Ge, M. Ruzzarin, D. Chen, H. Lu, X. Yu, J. Zhou, C. De Santi, R. Zhang, Y. Zheng, M. Meneghini, and G. Meneghesso, "Gate reliability of p-GaN gate AlGaN/GaN high electron mobility transistors," *IEEE Electron Device Letters*, vol. 40, no. 3, pp. 379–382, 2019.

72 A. N. Tallarico, S. Stoffels, N. Posthuma, S. Decoutere, E. Sangiorgi, and C. Fiegna, "Threshold voltage instability in GaN HEMTs with p-type gate: Mg doping compensation," *IEEE Electron Device Letters*, vol. 40, no. 4, pp. 518–521, 2019.

73 J. Wei, R. Xie, H. Xu, H. Wang, Y. Wang, M. Hua, K. Zhong, G. Tang, J. He, M. Zhang, and K. J. Chen, "Charge storage mechanism of drain induced dynamic threshold voltage shift in p-GaN gate HEMTs," *IEEE Electron Device Letters*, vol. 40, no. 4, pp. 526–529, 2019.

74 C.-J. Yu, C. W. Hsu, M. C. Wu, W. C. Hsu, C. Y. Chuang, and J. Z. Liu, "Improved DC and RF performance of novel MIS p-GaN-gated HEMTs by gate-all-around structure," *IEEE Electron Device Letters*, vol. 41, no. 5, pp. 673–676, 2020.

75 B. Lu, E. Matioli, and T. Palacios, "Tri-gate normally-off GaN power MISFET," *IEEE Electron Device Letters*, vol. 33, no. 3, pp. 360–362, 2012.

76 L. Nela, G. Kampitsis, J. Ma, and E. Matioli, "Fast-switching tri-anode Schottky barrier diodes for monolithically integrated GaN-on-Si power circuits," *IEEE Electron Device Letters*, vol. 41, no. 1, pp. 99–102, 2020.

77 L. Nela, M. Zhu, J. Ma, and E. Matioli, "High-performance nanowire-based E-mode power GaN MOSHEMTs with large work-function gate metal," *IEEE Electron Device Letters*, vol. 40, no. 3, pp. 439–442, 2019.

78 M. Zhu, J. Ma, L. Nela, C. Erine, and E. Matioli, "High-voltage normally-off recessed tri-gate GaN power MOSFETs with low on-resistance," *IEEE Electron Device Letters*, vol. 40, no. 8, pp. 1289–1292, 2019.

79 J. T. Asubar, S. Kawabata, H. Tokuda, A. Yamamoto, and M. Kuzuhara, "Enhancement-mode AlGaN/GaN MIS-HEMTs with high VTH and high IDmax using recessed-structure with regrown AlGaN barrier," *IEEE Electron Device Letters*, vol. 41, no. 5, pp. 693–696, 2020.

80 F. Roccaforte, G. Greco, P. Fiorenza, and F. Iucolano, "An overview of normally-off GaN-based high electron mobility transistors," *Materials*, vol. 12, no. 5, p. 1599, 2019.

81 M. Zhu and E. Matioli, *Monolithic Integration of GaN-Based NMOS Digital Logic Gate Circuits with E-Mode Power GaN MOSHEMTs*, IEEE, 2018.

82 H. Hahn, B. Reuters, S. Kotzea, G. Lükens, S. Geipel, H. Kalisch, and A. Vescan, *First Monolithic Integration of GaN-Based Enhancement Mode n-Channel and p-Channel Heterostructure Field Effect Transistors*, IEEE, 2014.

83 N. Chowdhury, Q. Xie, M. Yuan, K. Cheng, H. W. Then, and T. Palacios, "Regrowth-free GaN-based complementary logic on a Si substrate," *IEEE Electron Device Letters*, vol. 41, no. 6, pp. 820–823, 2020.

84 K. Fu, H. Fu, X. Huang, T. H. Yang, H. Chen, I. Baranowski, J. Montes, C. Yang, J. Zhou, and Y. Zhao, "Threshold switching and memory behaviors of epitaxially regrown GaN-on-GaN vertical p–n diodes with high temperature stability," *IEEE Electron Device Letters*, vol. 40, no. 3, pp. 375–378, 2019.

85 G. Pavlidis, S. H. Kim, I. Abid, M. Zegaoui, F. Medjdoub, and S. Graham, "The effects of AlN and copper back side deposition on the performance of etched back GaN/Si HEMTs," *IEEE Electron Device Letters*, vol. 40, no. 7, pp. 1060–1063, 2019.

86 T. Gerrer, V. Cimalla, P. Waltereit, S. Müller, F. Benkhelifa, T. Maier, H. Czap, O. Ambacher, and R. Quay, *Transfer of AlGaN/GaN RF-Devices Onto Diamond Substrates via van der Waals Bonding*, IEEE, 2017.

87 T. Ohki, A. Yamada, Y. Minoura, K. Makiyama, J. Kotani, S. Ozaki, M. Sato, N. Okamoto, K. Joshin, and N. Nakamura, "An over 20-W/mm S-band InAlGaN/GaN HEMT with SiC/diamond-bonded heat spreader," *IEEE Electron Device Letters*, vol. 40, no. 2, pp. 287–290, 2019.

88 M. Alomari, A. Dussaigne, D. Martin, N. Grandjean, C. Gaquière, and E. Kohn, "AlGaN/GaN HEMT on (111) single crystalline diamond," *Electronics Letters*, vol. 46, no. 4, p. 299, 2010.

89 M. Alomari, M. Dipalo, S. Rossi, M.-A. Diforte-Poisson, S. Delage, J.-F. Carlin, N. Grandjean, Gaquiere, C., Toth, L., Pecz, B. and Kohn, E., "Diamond overgrown InAlN/GaN HEMT," *Diamond and Related Materials*, vol. 20, no. 4, pp. 604–608, 2011.

90 J. Anaya, S. Rossi, M. Alomari, E. Kohn, L. Toth, B. Pécz, K. D. Hobart, T. J. Anderson, T. I. Feygelson, B. B. Pate, and M. Kuball, "Control of the in-plane thermal conductivity of ultra-thin nanocrystalline diamond films through the grain and grain boundary properties," *Acta Materialia*, vol. 103, no. 1, pp. 141–152, 2016.

91 M. J. Tadjer, T. J. Anderson, M. G. Ancona, P. E. Raad, P. Komarov, T. Bai, J. C. Gallagher, A. D. Koehler, M. S. Goorsky, D. A. Francis, and K. D. Hobart, "GaN-on-diamond HEMT technology with $T_{AVG} = 176°C$ at $P_{DC,max} = 56$ W/mm measured by transient thermoreflectance imaging," *IEEE Electron Device Letters*, vol. 40, no. 6, pp. 881–884, 2019.

92 M. Alshahed, L. Heuken, M. Alomari, I. Cora, L. Toth, B. Pecz, C. Waechter, T. Bergunde, and J. N. Burghartz, "Low-dispersion, high-voltage, low-leakage GaN HEMTs on native GaN substrates," *IEEE Transactions on Electron Devices*, vol. 65, no. 7, pp. 2939–2947, 2018.

93 R. A. Khadar, C. Liu, R. Soleimanzadeh, and E. Matioli, "Fully vertical GaN-on-Si power MOSFETs," *IEEE Electron Device Letters*, vol. 40, no. 3, pp. 443–446, 2019.

94 A. Raj, A. Krishna, N. Hatui, C. Gupta, R. Jang, S. Keller, and U. K. Mishra, "Demonstration of a GaN/AlGaN superlattice-based p-channel FinFET with high ON-current," *IEEE Electron Device Letters*, vol. 41, no. 2, pp. 220–223, 2020.

95 Y. Zhang and T. Palacios, "(Ultra)wide-bandgap vertical power FinFETs," *IEEE Transactions on Electron Devices*, vol. 67, no. 10, pp. 3960–3971, 2020.

96 D. Disney, H. Nie, A. Edwards, D. Bour, H. Shah and I. C. Kizilyalli, *Vertical Power Diodes in Bulk GaN*, IEEE, 2013.

97 S. Rajabi, S. Mandal, B. Ercan, H. Li, M. A. Laurent, S. Keller, and S. Chowdhury, "A demonstration of nitrogen polar gallium nitride current aperture vertical electron transistor," *IEEE Electron Device Letters*, vol. 40, no. 6, pp. 885–888, 2019.

98 P. Shrestha, M. Guidry, B. Romanczyk, N. Hatui, C. Wurm, A. Krishna, S. S. Pasayat, R. R. Karnaty, S. Keller, J. F. Buckwalter, and U. K. Mishra, "High linearity and high gain performance of N-polar GaN MIS-HEMT at 30 GHz," *IEEE Electron Device Letters*, vol. 41, no. 5, pp. 681–684, 2020.

99 T. George, K. A. Son, R. A. Powers, L. Y. Del Castillo, and R. Okojie, *Harsh Environment Microtechnologies for NASA and Terrestrial Applications*, IEEE, 2005

100 A. Vescan, I. Daumiller, P. Gluche, W. Ebert, and E. Kohn, "High temperature, high voltage operation of diamond Schottky diode," *Diamond and Related Materials*, vol. 7, no. 2, pp. 581–584, 1998.

101 D. Maier, M. Alomari, N. Grandjean, J.-F. Carlin, M.-A. Diforte-Poisson, C. Dua, S. Delage, and E. Kohn, "InAlN/GaN HEMTs for operation in the 1000 °C regime: a first experiment," *IEEE Electron Device Letters*, vol. 33, no. 7, pp. 985–987, 2012.

102 C. A. Chapin, R. A. Miller, K. M. Dowling, R. Chen, and D. G. Senesky, "InAlN/GaN high electron mobility micro-pressure sensors for high-temperature environments," *Sensors and Actuators A: Physical*, vol. 263, no. 8, pp. 216–223, 2017.

103 H. S. Alpert, C. A. Chapin, K. M. Dowling, S. R. Benbrook, H. Köck, U. Ausserlechner, and D. G. Senesky, "Sensitivity of 2DEG-based Hall-effect sensors at high temperatures," *Review of Scientific Instruments*, vol. 91, no. 2, p. 025003, 2020.

104 D. J. Spry, P. G. Neudeck, D. Lukco, L. Y. Chen, M. J. Krasowski, N. F. Prokop, C. W. Chang, and G. M. Beheim, "Prolonged 500 °C operation of 100+ transistor silicon carbide integrated circuits," *Materials Science Forum*, vol. 924, no. 6, pp. 949–952, 2018.

105 A. Nasiri, S. S. Ang, T. Cannon, E. V. Porter, K. U. Porter, C. Chapin, R. Chen, and D. G. Senesky, "High-temperature electronics packaging for simulated venus condition," *Journal of Microelectronics and Electronic Packaging*, vol. 17, no. 4, pp. 59–66, 2020.

15

SiC MOSFETs

Peter Friedrichs

IFAG IPC T, Infineon Technologies AG, Neubiberg, Germany

15.1 Introduction to Silicon Carbide for Power Semiconductors

Silicon carbide (SiC) belongs to the so-called wide-bandgap (WBG) semiconductors. A comparison of Si vs. SiC material properties is shown in Figure 15.1.

A WBG usually results in a significantly higher internal breakdown field. Compared to silicon, SiC has a higher breakdown-field value of about 10 times. Thus, active layers of high-voltage devices can be made much thinner, and doped higher, when compared to silicon power switch designs. This effect eventually shifts the transition between fast and unipolar devices such as metal oxide semiconductor field effect transistors (MOSFETs), Schottky barrier diodes (SBDs) or junction field effect transistors (JFETs) and bipolar components like IGBTs to much higher voltages. While with silicon, the transition takes place at around 600 V, SiC components can be implemented in unipolar configurations at several kV.

15.2 SiC Schottky Barrier Diodes

The voltage range for fast and unipolar Schottky diodes as well as field-effect based SiC switches (MOSFET, JFET) can be extended to over 1000 V. This is possible because of inherent properties of the SiC material. The low leakage current in high-voltage Schottky diodes is possible due to the metal semiconductor barrier, which is about two times higher than in Si Schottky diodes. Compared to silicon-based, high-voltage diodes which are nearly exclusively based on pin-structures, the SiC SBD comes along with a basically zero-reverse recovery, as shown in Figure 15.2.

SiC-based SBDs entered the market in 2001, at that time mainly in combination with super-junction silicon MOSFETs. These were used to enable the highly efficient constant current mode (CCM) mode power factor correction (PFC) stage (see Figure 15.3) in power supplies, which can be found in most switch-mode power supplies in the output power range between 100 W and 2 kW.

A specific feature of this PFC subsystem is that the switching frequency is quite high (50–150 kHz at a voltage level of about 400 V), and therefore, a diode is needed that has low switching losses. All the reverse current generated by the diode (D1) will go straight through the switch (P1) and generate huge losses there. Having here a zero-reverse current device like the SiC Schottky diode will allow customers to shrink the transistor significantly, and to save at least part of the

Advances in Semiconductor Technologies: Selected Topics Beyond Conventional CMOS, First Edition. Edited by An Chen.
© 2023 The Institute of Electrical and Electronics Engineers, Inc. Published 2023 by John Wiley & Sons, Inc.

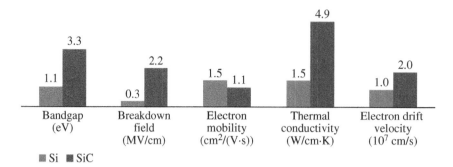

Figure 15.1 Comparison between key physical parameters for silicon and silicon carbide.

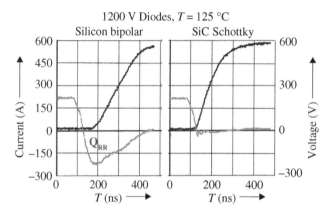

Figure 15.2 Comparison of the turn off behavior between silicon pin diodes and SiC Schottky barrier diodes.

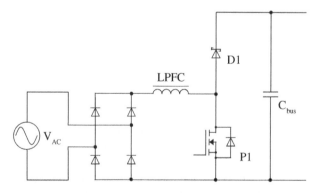

Figure 15.3 Schematic of a typical Boost PFC stage using an SiC (D1) and Si super-junction MOSFET (P1).

increased price required for a SiC diode, as compared to the cost of a conventional, ultrafast Si pin diode. In addition, the complete subsystem can be shrunk significantly by using smaller devices. Circuits, such as snubber circuits, can thus be partly eliminated due to the ideal dI/dt and dV/dt characteristics of the SiC diode.

In this chapter, we explain the benefits more in detail, and also show that in conjunction with compensation or super-junction MOSFETs like CoolMOS™, customers will be able to produce a new generation of power supplies with significantly improved power handling density.

Solar power conversion also started to benefit from the SiC diode technology, basically driven by the same value proposition as originally with PFC. Today, SiC diodes with blocking voltages from 600 V up to 1700 V are established design elements in the library of power system designers.

The major benefit of SiC-based Schottky diodes is the lack of reverse recovery charge. Yet the diodes offer only comparatively low tolerance in terms of surge current stress, which regularly occurs during power-on or after the line cycle drop outs. This is due to the unipolar characteristic of these diodes, resulting in a significant temperature coefficient of its resistivity. An improvement of the surge current capability was therefore mandatory to enable the ultimate success in the above-mentioned use cases.

For this purpose, Infineon launched in 2006 a merged pn Schottky diode (MPS) as sketched in Figure 15.4. Such a structure is well-known for lowering the leakage current of the diodes via field shielding of the Schottky interface. But in the SiC study, the p-areas were optimized as efficient emitter structures. During normal operation, the diode behaves like a normal Schottky diode, whereas during surge current conditions, the diode behaves like a pn diode as shown in the right part of Figure 15.4.

This first-generation of MPS diodes showed an impressive performance. A common measure to quantify the surge current capability is the i^2t value. This value represents the maximum $\int i^2 dt$ of a 10 ms long sine half-wave current pulse that the device can withstand without destruction. For the 8 A rated device, i^2t values of about 50 A^2s with peak currents of about 3700 A/cm^2 were achieved which is roughly 10 times the rated current. For comparison, the destructive current density of a standard Schottky diode is about 1650 A/cm^2. As an additional benefit, these devices show a stable avalanche breakdown. The breakdown voltage increases with the temperature. This stabilizes the avalanche breakdown and enables avalanche operation also for paralleled devices. Typical 8 A devices, for example, withstood a single "shot" of avalanche energy of 12.5 mJ at an avalanche time length of 9.5 μs, a repetitive avalanche energy of 0.3 mJ, and a switching frequency of 100 kHz. The latter was evaluated in a 1000-hour stress test [1].

Meanwhile further generations of diodes have continued the success story of SiC MPS diodes, characterized by innovations like diffusion solder assembly (third generation), thin-wafer technology (fifth generation), and reduced barrier heights for improved *I–V* characteristics (sixth generation).

For voltages exceeding 3 kV blocking capability, it becomes more difficult, even with SiC, to achieve an attractive cost-performance rate for SiC Schottky diodes. The required drift zone design

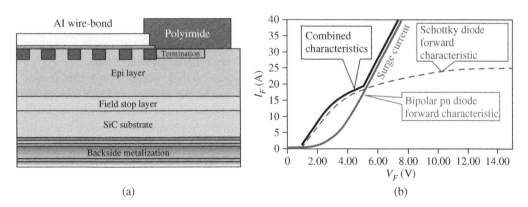

(a) (b)

Figure 15.4 (a) Structure of an MPS diode, the dark gray areas indicate p-doped regions, (b) *I–V* forward behavior of an MPS diode.

would result in a high differential resistance and thus, a large area would be required to reach a low forward-voltage drop. The MPS design is not able to compensate for this deficiency. Thus, one would rather use a pure SiC-based pn diode for high blocking voltages. This is mainly because a high pulse-current handling capability (i^2t) is needed for many target applications requiring such a high voltage diode, which can be achieved with pn devices only. However, those devices would exhibit a high knee voltage of >3 V due to the larger bandgap, again leading to a negative impact on the conduction losses. Thus, the smartest solution would be the use of internal body diodes in SiC MOSFETs, as they can combine low conduction losses, enabled by synchronous rectification, and a high i^2t performance due to the pn junction, which will be explained in the following chapters.

15.3 SiC Transistors

With regard to active semiconductor components for voltages exceeding 300 V, SiC transistors, similar to SiC diodes, are about to become an attractive alternative to today's established IGBT technologies, and even super-junction silicon MOSFETs in industrial power electronics. The dedicated material properties of SiC enable the design of minority carrier-free unipolar devices like MOSFETs instead of the charge-modulated IGBT devices at high blocking voltages. The loss restrictions of IGBTs are caused by the dynamics of minority carriers. In MOSFETs, those minority carriers are eliminated. As an example, extremely high dv/dt slopes in the range above 100 kV/µs have been measured for SiC MOSFETs. Initially, the superior dynamic performance of SiC-based transistors compared to IGBTs in the area of 1200 V and higher was seen as the most important advantage. But there are more fundamental differences between the IGBT and the unipolar SiC switch, which will increasingly attract attention. First of all one can benefit from the linear, threshold-free I–V curve of the output characteristic, which enables a significant conduction-loss reduction mainly under partial load conditions. Secondly, unipolar structures allow integration of a nearly recovery-free body diode with the option of synchronous rectification. Based on this property, the device shows knee-voltage-free conduction behavior even in the third quadrant by synchronous rectification. Thus, compared to many IGBT-based solutions where an external body diode is required, the number of necessary components is reduced by half. This leads to a significant reduction in the required power module footprint and reduced temperature swing within the chips during typical switch mode operation. The body diode performance is also one of the key differentiators compared to the silicon-based super-junction MOSFET technology. Devices like the CoolMOS from Infineon have only limited options in case freewheeling operation is needed. Furthermore, compared to super-junction MOSFETs, the SiC MOSFETs typically have quite a low charge stored in the output capacitance (Q_{oss}), which makes them the perfect candidate for use in modern, ultra-low-loss bridge topologies in power supplies like totem pole.

Nevertheless, compared to the success journey of SiC diodes, it took much longer for SiC transistors to become commercially available. MOSFETs seemed to be the concept of choice considering the success of this technology in silicon, and the unique SiC ability to form a native silicon dioxide like silicon. The major obstacle, however, was the poor inversion-channel conductivity due to the defect signature at the 4H-SiC SiO$_2$ interface. Early samples ended up with a channel mobility of much less than 1% of the bulk mobility for electrons, compared to silicon MOSFETs, where up to 90% can be reached. Thus, the advantage of SiC having a low drift-zone resistivity was devaluated by a practically nonconductive channel. To overcome this dilemma, transistors initially based on different device concepts such as the bipolar junction transistor (BJT) [2], the normally-off vertical JFET [3], the normally-on, quasi-vertical JFET [4], and several exotic MOSFET derivatives [5–7] have been developed and also partially commercialized.

A first revolution in this field was the introduction of NO-annealed MOSFET concepts. This technology led to a channel mobility of about 20 times higher than that of standard process technologies. The door was now open for the development of cost-efficient and attractive SiC MOSFETs. Initially, the devices were based on DMOS, such as lateral-channel concepts. However, since 2017, vertical-channel, trench-based MOSFETs have also become available. Nearly all players in the field have now adopted this concept on the roadmap for upcoming MOSFET generations. This move can be characterized as a second revolutionary step, since the channel mobility in SiC trench MOSFETs is many factors higher than in DMOS-like devices. In addition to the outstanding performance offered by SiC MOSFETs, a reliability level has to be provided by the suppliers mainly for the sensitive gate oxide, as users know from silicon power transistors. More details regarding the various MOSFET concepts will be explained in the next chapters.

15.4 SiC Power MOSFETs

15.4.1 Possible Cell Concepts

A common and straightforward way to run a power MOSFET is to use a planar gate and a quasi-vertical structure, as shown in Figure 15.5 [4].

As SiC devices allow roughly 10 times higher electric fields than their Si counterparts, the electric field in the gate oxide has to be limited in order to maintain the required reliability of the device. The electric field strength has to be reduced or limited by the appropriate device design measures. The inherent JFET formed by the p-wells of the structure can be used as a design element and is, therefore, not a purely parasitic element as in silicon counterparts. However, in general, lateral channel structures limit the downward scalability of the cell size. Thus, the area-specific on-resistance that can be achieved is limited as well. Another disadvantage arises from the specific interface trap structure of MOS elements on the silicon face of 4H-SiC, which is linked to a low channel mobility, to be discussed in the following chapters.

Figure 15.6 shows another design mode using a double trench structure [6, 7]. The implementation of a trench gate at a 90° inclined crystal surface results in an enhanced channel width and exhibits a higher channel mobility. However, as the channel is oriented along different sidewalls

Figure 15.5 SiC MOSFET with planar gate structure. Source: Based on Agarwal and Ryu [5].

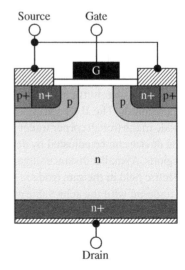

that correspond to different crystal planes, key data, such as threshold voltage e.g. between the channel on the right side and the left side of the trench, will be different from each other. The second (passive) trench forms a deep p-well ensuring a limitation of the electric field at the SiO$_2$/SiC interface, and can provide a sufficient area for body diode operation in bipolar mode. This is in contrast to silicon trench concepts, in which the field usually does not need to be limited to the same extent as required in SiC MOS devices. The second trench increases the cell pitch, and as such, limits the channel width compared to known silicon-based trench devices. However, compared to the discussed planar concepts, a factor of 2 can still be achieved.

Another trench concept introduced by Infineon [8] is depicted in Figure 15.7. The one-sided channel is oriented precisely along the so-called a-plane <11$\overline{2}$0>, which has been proven to deliver the highest channel mobility and lowest interface trap density among the different planes with 90° orientation toward the silicon face of 4H-SiC. The SiO$_2$/SiC interface is again shielded by deep p-wells, formed in each trench by sacrificing the second channel path. This design allows a much narrower placement of the deep regions, and thus, improves the shielding impact. It also no longer requires a second trench, and thus, compared to the concept with the channel on both sides, there is no disadvantage regarding cell density.

The deep p-type regions also act as emitters for the body diode, which can be used for freewheeling operation. More aspects about the body diode, its operation modes, and reliability challenges will be discussed later.

The single-side channel concept enables a very dense cell design, and results in a low area-specific on-resistance that is about half the value of typical DMOS cells, assuming comparable gate-oxide stress conditions for the on-state (see next chapters). Thanks to the channel properties (high channel mobility), the device can be easily turned on by commonly known gate–source bias levels, e.g. of $V_{GS(on)} = +15$ V. Furthermore, the cell construction inherently has a ratio of the so-called Miller charge Q_{GD} to the gate–source charge Q_{GS} that is able to suppress parasitic re-turn-on effects in bridge topologies, which typically increase the switching losses. Q_{GS} is relatively large since a huge part of the trench contributes to it, i.e. the n$^+$-type source region and all p-type areas. Therefore, a well-controlled dynamic behavior exhibiting low switching losses can be achieved [9]. The cell design also offers the implementation of a short-circuit capability according to the needs of several target applications. The JFET part between p-emitter regions not only protects the oxide in the trench corner under reverse bias but also the saturation current of the device can be adjusted by designing the distance and depth of the p-type regions. A smaller distance supports both a lower saturation current and lower electric field in the gate oxide of the trench corner. However, as usual there is a trade-off with the achievable overall on-state resistance. Thus, in the initial design phase, the requirements regarding short circuit and cumulated electric field stress need to be carefully analyzed and translated into a feasible design.

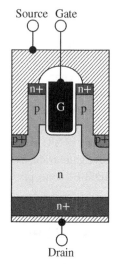

Figure 15.6
A trench SiC MOSFET with double trench structure. Source: Based on Nakumara et al. [6].

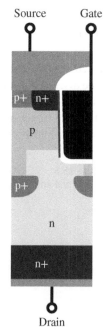

Figure 15.7
Proposed trench SiC MOSFET with asymmetric channel.

15.4.2 SiC MOS Channel Challenges

SiC in power devices can reveal many advantages compared to silicon. Thanks to its property as a WBG material, mainly if the design and resistance of active (bulk) layers are considered. All the same, there are some essential differences to silicon leading to a certain challenge when fabricating MOS-based active devices on the 4H-SiC polytype, which is the commonly used silicon-carbide version for power devices.

SiC MOSFETs built on 4H-SiC typically show a very low inversion channel electron mobility at the SiO_2/SiC interface. The background is still under investigation, but many studies claim that carbon-related interface defects result from thermal oxidation; other reports cannot identify any carbon at the interface. However, traps are present in a high density, and energetically close to the conduction band. Electron scattering at such charged point defects at the interface and charge trapping effects drop the channel mobility in 4H-SiC MOSFETs to the range of 5–70 cm^2/Vs, which is typically only a fraction of the bulk mobility of approximately 400 cm^2/Vs (for doping levels in the range of the channel doping) [10]. Further defects with energy levels across the full bandgap of SiC can have a large spread of time constants for trapping or emitting electrons. This situation can result in a hysteresis of the threshold voltage [11]. Trapping and de-trapping of charged carriers in near interface states (NIT) may even end with in more complex threshold voltage variations, being determined by the gate voltage profile in the application and the temperature [12]. Fortunately, a large amount of the observed parameter shifts are fully reversible under nominal operating conditions. However, advanced gate oxide processes are required to address the more permanent drift effects, e.g. of the threshold voltage.

Another challenge to be solved in modern SiC devices is a typical 4° off-axis tilt of commercially available SiC substrates in order to enable stable growth of epitaxial layers, which later form on the active region of the device. Due to the off-axis orientation, the wafer surface does not perfectly align with the so-called (0001) *c*-plane. A step-like roughness can be identified, as depicted in Figure 15.8. The inclined surface represents a challenge to both the planar type devices and the trench concepts. For example, a traditional vertical etch step to form a trench consequently will have side walls with different roughness, performance, and reliability. The channel mobility in trench-based MOSFETs strongly depends on the chosen crystal plane and difference of close to 2 between worst and best channel mobility has been worked out, as shown in Figure 15.9 [13]. The same crystal-orientation dependence can be found for the threshold voltage. These findings need to be considered when defining the layout and the cell concept of trench-based SiC devices.

The introduction of nitric oxide annealing can reduce the interface state density [14]. However, the process conditions need to be adapted to the individual 4H-SiC crystal face in order to achieve a stable and effective interface state passivation process. Improved passivation power usually leads to

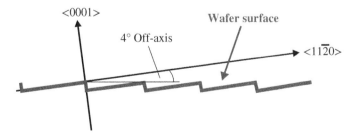

Figure 15.8 Schematic illustration of the 4° off axis cut of 4H-SiC wafers.

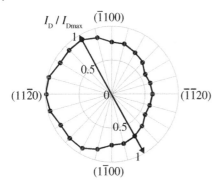

Figure 15.9 Relative channel mobility for various trench planes of 4H-SiC on-axis substrate. Source: Based on Yano et al. [13].

an increase in channel mobility due to the suppression of Coulomb scattering and charge trapping. Furthermore, threshold voltage drift effects will decrease as well.

15.4.3 Typical MOSFET Device Characteristics – Static Behavior, Switching Performance, and Body Diode Aspects

Infineon's first generation CoolSiC™ MOSFET have some typical device characteristics that will be explained in the following paragraph. As a reference, the part with the sales number IMZ120R045M1 (single chip in TO247-4 package) will be used. Details about the underlying device technology can be found in [8].

The analyzed 1200 V rated trench MOSFET is designed for standard gate-bias levels of $V_{GS(off)} = -5\,V \ldots 0\,V$ for off-state and a recommended $V_{GS(on)}$ of +15 V for on-state (+20 V_{max} taking into account transients). The output characteristics for gate voltages beyond threshold are depicted in Figure 15.10, covering the temperature range up to 175 °C.

The typical value for $R_{DS(on)}$ is 45 mΩ, measured at $V_{GS(on)} = +15\,V$, a nominal drain current of $I_D = 20\,A$ at room temperature. With increasing temperature, all parts of the device (individual resistors) making up the total on-resistance, will change its value. The temperature coefficients T_k of the most important contributors (channel, JFET region, drift zone), however, are different. The drift zone and the JFET region show the typical positive coefficient as expected from the device physics of unipolar layers. MOS channels in silicon would also show a positive T_k, however, due to the large defect density described above, SiC MOS channels tend to show a negative T_k. This is due

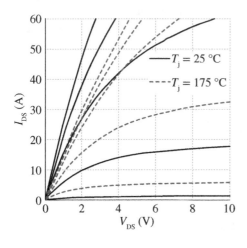

Figure 15.10 Output characteristics showing the drain current I_{DS} as a function of drain–source voltage V_{DS} for gate–source voltage $V_{GS(on)} = 17, 15, 13, 10, 7\,V$ at 25 °C (solid) and 175 °C (dotted), typ. values, gate voltages sorted from top to bottom.

to an accelerated trapping and de-trapping process at higher temperatures, which ends up in a virtually higher channel conductivity. The more defects there are, the more pronounced the negative T_k will be. Thus, since the total T_k of the MOSFET is the combination of all major components, one can say that the more positive the temperature coefficient, the less defects will impact the device performance in the channel. On the other hand, devices with a small or near-zero T_k typically have large channel defect densities. It becomes even more complex if the temperature behavior for different gate bias levels is analyzed. The smaller the T_k for the recommended $V_{GS(on)}$, the higher the danger that the T_k will become negative for a gate bias below this value (e.g. 15 V instead of 20 V) and thus, the device will be exposed to thermal runaway. Figure 15.11 sketches the behavior of the analyzed 1200 V 45 mΩ part across the specified temperature range of $-40\,°C$ up to $175\,°C$. It has its minimum at room temperature and increases from 45 mΩ to about 72 mΩ typically at the maximum allowed junction temperature (at $I_{DS} = 20$ A and $V_{GS(on)} = 15$ V). For the threshold voltage $V_{GS(th)}$ a typical value of 4.5 V (at $I_{DS} = 10$ mA, $V_{GS} = V_{DS}$, $T = 25\,°C$) was chosen. $V_{GS(th)}$ shows the expected negative T_k and decreases with a slope of 6 mV/K to a typical value of 3.75 V at $175\,°C$.

The transconductance behavior is depicted in Figure 15.12, here at $V_{DS} = 20$ V and for room temperature as well as for the maximum rated junction temperature.

Figure 15.11 Typical temperature dependence of $R_{DS(on)}$ and $V_{GS(th)}$: solid curve: $V_{GS} = 15$ V, $I_{DS} = 20$ A, dashed curve: $V_{GS} = 15$ V, $I_{DS} = 40$ A, light gray curve: $V_{GS(th)}$ (at $V_{GS} = V_{DS}$, $I_{DS} = 10$ mA).

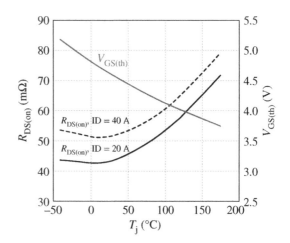

Figure 15.12 Typical transfer characteristics measured in pulse mode at $V_{DS} = 20$ V at $25\,°C$ (solid, black) and $175\,°C$ (dashed).

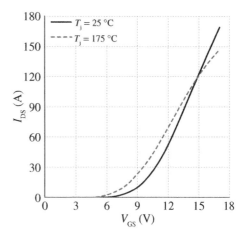

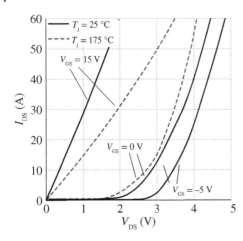

Figure 15.13 Typical 3rd quadrant characteristics at 25 °C (solid) and 175 °C (dashed), V_{GS} = +15 V, 0 V, and −5 V.

For trench MOSFETs in SiC, the transfer curves show not only a parallel shifted toward lower V_{GS} with increasing temperature. Figure 15.12 clearly reveals a crossing point at V_{GS} = 15 V. The temperature dependence decreases with increasing gate source voltage.

The static characteristic of the body diode (third quadrant operation) is shown in Figure 15.13. At a V_{GS} of −5 V, the pure body diode *I–V* curve without a partial MOS channel impact is shown. For a V_{GS} of around zero volt, some impact from the channel is already visible, indicated by a reduced source–drain voltage V_{SD}. However, very low V_{SD} and linear characteristics are found as soon as the channel is turned on, as in the on-state. An on-resistance of only 33 mΩ at 25 °C and 57 mΩ at 175 °C, respectively, can be extracted. These numbers are lower than obtained in forward mode. This is due to the now negative feedback of the drain bias in the JFET region (smaller space charge region and thus, a wider JFET part). Thus, in order to achieve lowest losses in reverse conduction mode, synchronous rectification is mandatory, which is typically done by turning on the channel after a certain dead time.

Typical device capacitance vs. drain-source voltage is shown in Figure 15.14. Both C_{rss} and Q_{GD} are comparably small due to the p-well design mentioned earlier, which helps to reduce parasitic "re-turn-on" effects in bridge topologies.

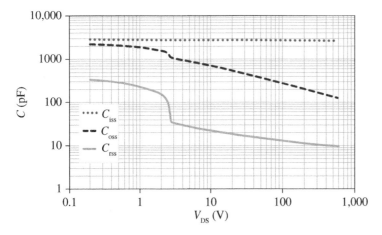

Figure 15.14 Small signal capacities C_{iss}, C_{oss}, C_{rss} as function of V_{DS} measured at 1 MHz, V_{GS} = 0 V, internal gate resistor is typical 4 Ω.

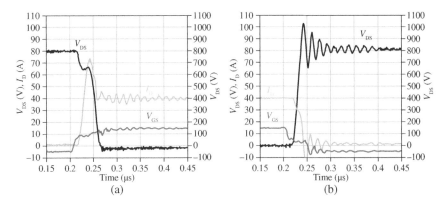

Figure 15.15 Turn-on and turn off waveform in a half-bridge using the internal body diode, for the analyzed 45 mΩ MOSFET in a 3-pin discrete housing ($V_{DS} = 800\,V$, $I_D = 40\,A$, $R_G = 6.0\,\Omega$, $T_j = 175\,°C$, $V_{GS} = -5\,V/+15\,V$).

Figure 15.15 displays typical turn-on and turn-off waveforms using the internal diode as the upper element of a half-bridge leg as freewheeling element. In contrast to IGBTs, no tail current effects take place, which is related to the lack of minority carriers during forward conduction of MOSFETs. The switching performance is predominantly defined by the capacitance of the transistor and thus, nearly independent of the junction temperature. In bridge topologies, however, where the internal body diode is used, the turn-on process can be impacted by the residual Q_{rr} of the internal MOSFET body diode. Since the structure is pn-based, a small amount of minority carriers is injected, and need to be removed by recombination. Even though the actual values are very small compared to typical recovery charges present in corresponding silicon devices, they have to be considered in the total loss balance.

In order to extract the impact turn-on and turn-off energies, the devices have been analyzed in two bridge configurations – one using the internal body diode as freewheeling element and one using a Schottky diode at this position. Figure 15.16 shows the resulting dependencies. Turn-on energies E_{on} are in any mode the dominating part of the total switching losses and show lower

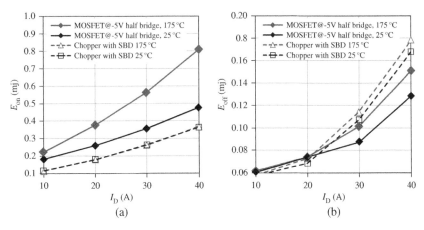

Figure 15.16 Turn-on and turn-off energy as a function of drain current of the analyzed 45 mΩ MOSFET in a 4-pin discrete housing pin ($V_{DS} = 800\,V$, $R_g = 2.2\,\Omega$, $V_{GS} = -5\,V/+15\,V$) for different upper switch elements in a half bridge.

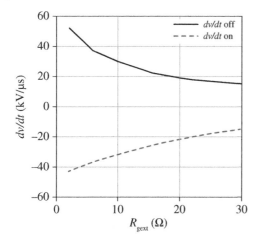

Figure 15.17 Max. dv/dt vs. external gate resistor R_{Gext} ($V_{DS} = 800$ V, $I_D = 20$ A, $T = 175$ °C).

values in configuration with an SBD for the above-mentioned reasons. The SBD completely eliminates the impact of both re-"turn-on" effects as well as Q_{rr} related losses as also indicated by the temperature dependence.

Figure 15.17 shows the controllability of the switching speed at turn-on and turn-off by $R_{g(ext)}$. This feature enables a wide utilization of the technology since even applications like drives which require reduced dv/dt values can be addressed.

A further reduction of the switching losses in a bridge configuration with two active devices can be achieved by implementing a Kelvin source contact in order to de-couple the gate control from the load current path, eliminating feedback loops via the source inductance. An easy way to do this is to use a TO-247 4-pin or to replace through-hole packages with SMD devices like the TO-263-7 in the future. Figure 15.18 shows the impact of this measure on the total losses.

In some cases such as train propulsion, freewheeling diodes are exposed to very high pulse currents, i.e. a failure of a device in an adjacent phase leg, or subsequently after a DC-link short-circuit occurrence.

The stress time can take up to a few dozen ms. In order to characterize power devices with respect to this behavior, usually an i^2t specification can be found in datasheets. Modern applications in which benefits provided by SiC should be utilized as much as possible show typically smaller system inductance levels compared to traditional IGBT solutions. Thus, the pulse current requirements

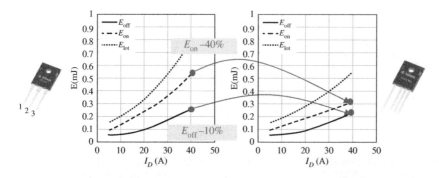

Figure 15.18 Impact of a Kelvin source connection on the total switching losses in a SiC MOSFET package.

will be more challenging than presently for silicon-based setups [15]. Since the chip size is also smaller, the i^2t potential of SiC MOSFETs needs to be specially designed.

The practical limit is usually a thermal one with borders being defined by the melting point of the front-side metal, as shown in Figure 15.19. Thus, in a nutshell, one has to control the temperature rise in the pulse-current phase, which can be done by reducing the forward voltage drop. In most of the currently available SiC MOSFETs, the body diode forward voltage drop at nominal current is quite high, much higher than the 3 V knee voltage of the pn-junction. While at rated currents, opening the parallel current path via the channel reduces the V_F to low values, this measure is not effective at very high current pulses. This can be seen from the measurements conducted on the 1200 V 45 mΩ SiC MOSFETs (Figure 15.20).

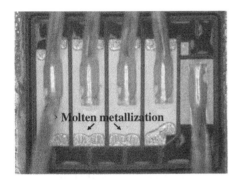

Figure 15.19 Surge-current destruction behavior of a 1200 V SiC MOSFET, surge current applied to body diode (3rd quadrant) with $V_{GS} = -9$ V.

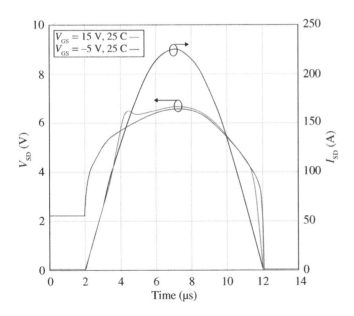

Figure 15.20 Surge-current behavior of a Trench based 45 mΩ 1200 V power MOSFET, last pulse before destruction, 10 ms sine half wave, $V_{GS} = 15$ V (open channel), and $V_{GS} = -5$ V (closed channel) $T_{start} = 25\,°C$.

Top view

Stacking fault

Cross section

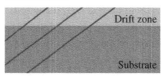

Drift zone

Substrate

Figure 15.21 Top view and cross-section of stacking faults in a SiC device.

In conclusion, if a high i^2t value for a MOSFET body diode is required, both the injection efficiency of the emitter and a good ohmic contact to the p-regions are needed for reducing the thermal stress on the device. Furthermore, the cell design has to be done in a way that the pn-junction knee voltage is exceeded.

A further aspect related to the use of internal body diodes in SiC MOSFETs is the potential shift of important parameters like $R_{DS(on)}$ caused by bipolar degradation effects in SiC. The effects can be triggered in any type of SiC device when pn junctions are biased into forward mode, e.g. the body diode of a MOSFET when conducting current. There is a common understanding that the root cause are basal plane dislocations (BPDs) on the SiC substrate. In bipolar mode, the energy released by the recombination-injected carriers fosters the evolution of stacking faults originating from the BPDs [16]. Such stacking faults continuously grow through the active layer toward the surface. Regions affected by the typically triangular or bar-shaped features, as shown in Figure 15.21 on the left side, are basically nonconductive and thus, the active area of a chip becomes smaller.

A careful analysis of the effect leads to the following conclusions:

- The effect takes place only if BPDs are present and recombination processes occur at the BPDs. For the devices without BPDs (or BPDs not subjected to recombination events), there will be no bipolar degradation effects.
- If the effect occurs at all, it saturates. Once the grown crystal defect (SF) reaches the surface of the device, the process stops, and the related parameter shift saturates at a level corresponding to the density of the respective defects in the region with recombination events. Depending on operating conditions, e.g. the current through the PN junction and the junction temperature, the time from initial state to saturation could take from several minutes to hours of accumulated bipolar operation time.

Numerous studies revealed that bipolar degradation only impacts the active area of SiC devices. In MOSFETs, the $R_{DS(on)}$ as well as the forward voltage drop of the body diode increase at a given current. Other parameters like breakdown voltage, dynamic behavior, or oxide reliability are not impacted.

It should be noted here that concepts based on Schottky diodes as internal body diodes, which are discussed here to help prevent bipolar degradation effects [17], will not be able to handle pulse currents such as the i^2t performance discussed above. Thus, Schottky-based diodes have some limitations for use in e.g. traction applications. In addition, they would require additional space in the cell that automatically reduces the channel width.

15.4.4 Gate-Oxide Reliability Aspects

One of the major concerns of today's SiC MOSFETs is their stability, mainly when compared to the well-established, silicon-based MOS technology. The reasons for this are manifold. On the one hand, every new technology is challenged with respect to its maturity. In addition, numerous studies have revealed that crystal defects being present in SiC have a much higher density compared to silicon and thus, impact the oxide lifetime. Even the first commercial devices available showed peculiarities under gate-oxide stress conditions [18], in particular in the form of high failure rates as a result of high gate-oxide operating fields of around 4 MV/cm and higher [19]. The gate operating field $E_{ox,on}$ is roughly defined by the formula

$$E_{ox,on} \approx \frac{V_{GS(on)}}{t_{ox}} \tag{15.1}$$

with t_{ox} being the nominal gate oxide thickness. Finally, the higher operating fields in SiC devices raise some concerns with respect to the impact on the gate-oxide stability of MOS devices.

The impact of material defects on oxide reliability in SiC MOS devices, in the form of extrinsic defects, is described in detail in [20], and complemented by a numerical model.

Two models are discussed, one claiming an intrinsic oxide weakness, the other extrinsic defects. Most studies tend to favor the defect-related model of gate-oxide failures in SiC MOS structures. For example, tests stressing a large number of capacitor-like vehicles have shown that the intrinsic (and not the defect-impacted) oxide behavior of SiO$_2$ on SiC is nearly the same as for SiO$_2$ on Si [21].

In the extrinsic defect model, early failures of gate oxide are caused by material imperfections (e.g. substrate defects, oxide defects, etc.) decades before a calculated end of life [22, 23]. Thus, oxide stability can be influenced or enhanced by using less-defective substrates, processing perfect oxide layers oxides, and suppressing the formation of new material defects during device processing. Since it can be said that the SiC MOS structure has similar intrinsic material characteristics compared to a Si MOS device, the procedures known from the Si device technology can be transferred to SiC. The major difference today is a higher extrinsic defect density of the thin gate-oxide layer at the end of processing.

One additional difference between Si-based MOS structures and their counterparts in SiC needs some further discussion. Based on the band-structure comparison shown in Figure 15.22, it is clear that the injection barriers for electrons, and especially for holes, are much lower in the case of SiC compared to silicon.

For a given applied electric field across the oxide, the injection barriers result in higher Fowler–Nordheim tunneling currents [24]. This would be relevant in case tunneling currents are responsible for the wear-out of gate oxides, however, experimental work does not support this theory. Tunneling impact can usually be ruled out for devices with an oxide layer having 30 nm or more of nominal thickness, which is typical for power devices. A gate-tunneling current is negligible at nominal use conditions ($E_{ox,on} < 3$–5 MV/cm). In this field-driven regime, the well-known linear E-model with its voltage and temperature acceleration can explain oxide-failure mechanisms [25].

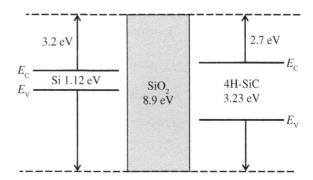

Figure 15.22 Illustration of the band offsets between Si/SiO$_2$ and 4H-SiC/SiO$_2$.

More relevant is this topic in nonstandard operating modes like avalanche. Under such conditions, a large number of holes and electrons are generated within the semiconductor under high field conditions [26]. An improper device design might cause an injection of holes into the oxide under avalanche conditions, which can severely affect stability.

In the case of purely thermal (non-deposited) oxide layers on SiC, the discussion is ongoing whether the carbon atoms have any influence on the integrity of the oxide. Based on the oxidation theory, carbon should first be turned into CO and later into CO_2 during oxidation. This is then expelled in gaseous form, and should not be found therefore in the bulk of the oxide layer. This has meanwhile been confirmed by SIMS or HR TEM experiments [27].

Consequently, the intrinsic long-term behavior of the SiO_2 on SiC and Si seem to be similar. Therefore, MOSFETs with a given area and oxide thickness will have the same stability for a given oxide field, regardless of whether they are based on silicon or SiC, as long as the layer is free of defect-related impurities.

In order to analyze the failure rates of gate oxides, typically end-of-life stress tests under accelerated conditions are conducted, and the time-to-failure is monitored and analyzed via Weibull statistics [28]. If such an approach is applied, identical intrinsic-oxide properties of SiO_2 on SiC and on Si indicate that the intrinsic parts of Si and SiC MOSFETs with the same oxide thickness and area coincide (Figure 15.23).

However, it is well-known that SiC still contains a much higher density of crystal defects compared to silicon. Forming an interface between the semiconductor and the oxide for processing a MOS device, those failures can lead to distortions of the insulating SiO_2. Irrespective of the actual nature of such distortions, they can be described by the local thinning model. In the oxide-thinning approach, an irregular region can be handled like a spot with a defect-free, but thinner oxide, which naturally fails at much lower voltages compared to the remaining layer, but at the same electric field [29] based on Eq. (15.1). If there are no extrinsics in place, t_{ox} equals the bulk oxide thickness, and the failure time of such a device coincides with the intrinsic line in a Weibull plot.

Thus, the extrinsic part of a Weibull plot (e.g. the line with the flat slope in Figure 15.23) represents nothing other than devices with a smaller electrical oxide thickness $t_{ox(ex)}$. The slope of the intrinsic branch is typically distinguished from the extrinsic one by its stepper slope (defined by remaining statistical variations in the oxide thickness e.g.), at least for an already mature technology with the majority of devices being free of extrinsics. Provided a test shows a graph with a more or less constant slope, it can be concluded that nearly all components are in some way affected by extrinsics.

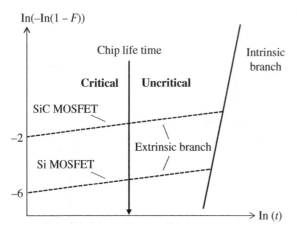

Figure 15.23 Schematic representation of the extrinsic and intrinsic Weibull distribution for SiC MOSFETs and Si MOSFETs having the same oxide thickness and area. Source: Beier-Moebius and Lutz [19].

An extrinsic defect with a thinner electrical thickness $t_{ox(ex)}$ will have a shorter lifetime τ_{ex} which can be estimated using the linear E-model [24]:

$$\tau_{ex} = \tau_{intr} \exp\left(\gamma \left[\frac{V_{GS(on)}}{t_{ox}} - \frac{V_{GS(on)}}{t_{ox(ex)}}\right]\right) = \tau_{intr} \exp(\gamma[E_{ox(on)} - E_{ox(on,ex)}]) \tag{15.2}$$

The voltage acceleration of the linear E-model is quantified by the parameter γ in Eq. (15.2). The remaining parameters can be derived from the previous definitions, being valid for a defect-free, intrinsic oxide (intr) or extrinsic (ex) device. The factor γ is an important input parameter for the extrapolation from wear-out experiments on lifetime at a fixed bias under real operating conditions. It should be assessed for each technology individually. In the case of the trench-based 1200 V SiC MOSFET technology, the assessment was done during the development process.

A larger number of DUTs (two groups of 1000 devices each) were used for this purpose. A constant gate voltage in three steps (starting at 25 V respectively 30 V, with an increase of 5 V after 100 days) were applied. The experiment results (failures) are depicted by the symbols in Figure 15.24. The observed failure rates of 2.9% and 6.5% (group 1 and 2 respectively) can be fit using the linear E-Model [25]. Based on the data, an acceleration factor of $\gamma = 3.9$ cm/MV was extracted; the resulting calculated failure rate is shown by the solid lines in the graph. If a failure rate is calculated using this model, assuming 15 V at the gate for 20 years and 150 °C, one FIT per die can be gained, which is in line with the reliability data for IGBTs e.g. in modern industrial applications [30].

Defect densities in today's state-of-the-art SiC material are between 0.1 and 1 defects/cm^2. Thus, for devices with 10 mm^2 oxide, more than 90% of the components are free of extrinsic defects. This appears to be a fairly good number, however, it would translate into higher field-failure rates of 3–4 orders of magnitudes as compared with Si, as indicated in Figure 15.23. This clearly does not meet the requirements of today's target applications with targets in the ppm or even ppb range.

To derive the appropriate measures to deal with the problem, and to assure an adequate field reliability of MOS-based SiC devices, we first need to understand that not every extrinsic defect is critical with respect to the failure rate in the field. A critical extrinsic defect is an imperfection

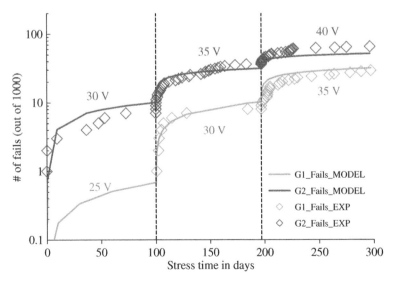

Figure 15.24 Constant gate bias stress test results at 150 °C. Solid lines represent the prediction by a linear E model.

which either fails or leads to parameter degradation during the operating lifetime of the device. Assuming for instance a relatively thick nominal oxide and an extrinsic defect with an electrically active thickness $t_{ox(ex)}$ of about 5% below the bulk oxide thickness, such a defect will be irrelevant if only the target operating life is considered. Thus, one might assume that the problem can be solved by increasing the nominal thickness. However, it was shown in [31] that this measure alone does not lower the failure rate sufficiently; it would also have a negative impact on the performance of the device, mainly $R_{DS(on)}$.

The most appropriate way to decrease the number of critical defects and, therefore, the failure rate in the field, is to apply screening procedures to devices before they are shipped to the customer [27]. Ideally, a gate stress pattern is applied, which is characterized by bias levels and a corresponding duration. The requirements for the screen pattern are clearly identifiable. The pattern should reveal critical extrinsic defects either by destroying an affected device or leading to a measurable parameter change while clean devices must not be affected at all (lifetime is only partially reduced). Thus, the bulk-gate oxide should have a sufficient thickness to provide this headroom. In essence, the field failure risk at the customer side with its negative implications is translated into a yield risk at the supplier side with a manageable burden, as long as the related defect density and yield loss are in the range as described above.

In the Weibull plots in Figure 15.25, the failure rate curves are shown for screened and unscreened populations. The linear slope of the Weibull distribution is, over time, a direct consequence of the constant failure hazard after screening. An aging process is imposed, which corresponds to twice or more the lifespan of the chip. This is possible within a reasonably short time, only if voltages are used that are much higher than the recommended bias for turn-on $V_{GS(on)}$. Nevertheless, to meet the requirement to keep devices intact that have no critical defects, the maximum applicable bias is limited [32]; the gate oxide thickness of a SiC MOSFET cannot be made thinner as for a silicon-based counterpart with comparable use voltage and gate-oxide lifetime requirements.

15.4.5 Short-Circuit Aspects and Avalanche Ruggedness of SiC MOSFETs

SiC MOSFETs and IGBTs (as the component that SiC MOSFETs are intended to replace) differ substantially when it comes to behavior in the event of a short circuit. For typical industrial-use cases, an IGBT can withstand up to 10 μs during a short circuit. If datasheets for SiC MOSFETs are screened with regard to this feature, usually no specifications are given at all, and if they are, the maximum withstand ratings are only a few μs (e.g. up to 3 μs for the trench-based technology

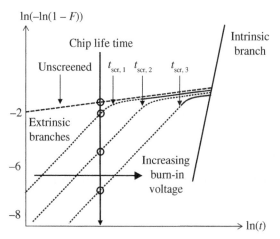

Figure 15.25 Sketch of the SiC MOS extrinsic and intrinsic Weibull distributions for unscreened and screened hardware when increasing the bias of a screen pattern. Source: Beier-Moebius and Lutz [19].

discussed earlier). Keeping in mind that today's target use cases for SiC MOSFETs are fine without short-circuit ruggedness, or even with the few microseconds mentioned above, this is not seen as critical for now. However, once further applications like motor drive inverters or train propulsion are addressed, this topic should again gain attention. It is important to point out that the lower short-circuit withstand times are not an intrinsic lack of SiC MOSFETs. Even modern IGBT types exist that are not specified with regard to short circuits (e.g. those designed for soft-switching applications). SiC MOSFETs are also available with a cell structure that enables a short-circuit performance comparable to that of typical IGBTs. The key question is whether the resulting (negative) impact on the on-state performance is worth going down this path. Alternatively, external measures could be pursued to find a system-based solution for the problem. But what is the main reason for the observed difference between IGBTs and SiC MOSFETs? Two things are evident – the SiC devices have a much smaller area for the same power handling capability and the ratio between peak current and nominal current is significantly higher compared to IGBT's.

After analyzing the destruction mechanisms in SiC MOSFETs during short-circuit incidents, thermal limitations (melting of metal layers) have been detected for the most part as the root cause. During a short circuit, a high voltage is combined with a high current defined by the load and thus, an extreme power density peak has to be managed within the chip. Also, additional effects are reported for SiC MOSFETs during or after short-circuit operation, e.g. gate shorts after short-circuit pulse [33].

One important finding is that temperatures within the chip rise to much higher values during the managing of a short circuit, and affect areas closer to the surface when compared to IGBTs. Higher temperatures are a consequence of the current densities, which are substantially higher than for IGBTs due to less saturation (see Figure 15.26).

The MOSFET design focusses on optimizing low $R_{DS(on)}$, having typically short channels and reduced JFET pinch-off impact. The unipolar nature of the device itself leads to a later reduction of the current (see dark gray line in Figure 15.26), but the temperature increase can still continue. Due to the short time frame of µs, the heat cannot be transported away from the drift zone to the bottom-side heat sink, and is consequently concentrated in the active layer, which, in the case of SiC, is very thin and close to the surface. Thus, all surface elements like sensitive-isolating oxides and metals are affected by the temperature rise as well. Figure 15.27 compares the chip-internal vertical temperature distribution to an IGBT. The silicon device temperature peak has a lower maximum value and is located deeper in the body of the device.

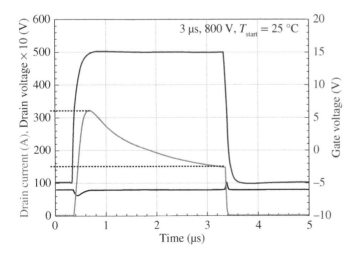

Figure 15.26 Typical short circuit waveforms for a SiC MOSFET.

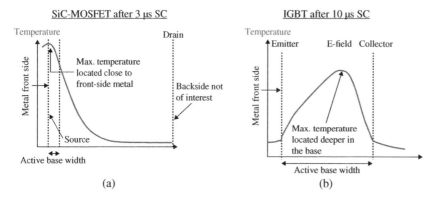

Figure 15.27 Sketch of the temperature distribution after short-circuit event of an IGBT (a) and a SiC MOSFET (b).

Consequently, the failure modes for SiC MOSFETs and IGBTs will be different. Furthermore, measures to influence the short circuit withstand time as long as it is facilitated at chip level will be not identical between both technologies.

In SiC MOSFETs, it is essential to limit the maximum current during short circuit. A straightforward design measure is to enhance the JFET effect between the p-body regions; from an operational point of view, one can reduce the applied $V_{GS(on)}$. In [34], further ideas are discussed. However, any action taken to improve the short-circuit withstand time in a given cell concept will have a negative impact on $R_{DS(on)}$ as a consequence. Alternatively, enabled by a smart assessment of the final system, one could pursue system innovations [35] instead of potential device-related measures to deal with short circuit, while keeping the attractive low-loss behavior.

Another extraordinary operating mode often connected to MOSFETs in certain applications is avalanche handling, e.g. when an unclamped inductive load is switched. In this mode, at least for a short period of time, the full nominal load current is applied to the device at the maximum drain-source voltage (the highest internal electric field in the semiconductor at the pn-junction), so it can be counted as one of the most demanding operating conditions for a power MOSFET.

If an unclamped inductive load is turned off, the energy stored in the inductance needs to be dissipated somewhere and therefore, pushes the current by driving the voltage across the (blocked) switch up to the avalanche bias (which is typically much higher than the maximum V_{DS} in the datasheet). In avalanche mode, the generated electron hole pairs now enable a current flow, and the energy stored in the inductance will be dissipated. In mechanical switches, this mode is recognizable by its typical arcing.

The current through the device starts at drain-current level before turn-off; thus, the peak power is very high. It can happen regularly, or intentionally, due to the circuit topology, e.g. fly-back in power supplies, in occasional fault modes (Low-Voltage Ride-Through in solar applications for example) or in unpredictable system fault conditions. The requirement is more or less always the same. The energy stored in the inductance needs to be dissipated by the active device without it being destroying by thermal overload, and it must not degrade other important device parameters. A sufficiently high avalanche energy rating of a power transistor enables some simplifications at the system side. As an example, snubber circuits can be eliminated, or the margin can be reduced between the applied maximum drain-source bias in the system and the rated V_{DSMAX}. MPS diodes are proven to deliver a solid avalanche capability, as described in [36] even under repetitive stress. It has been verified that beyond thermal limitations, no additional degradations,

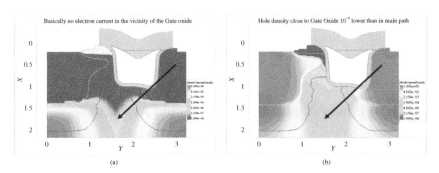

Figure 15.28 Electron (a) and hole (b) density during avalanche in 1200 V CoolSiC Trench MOSFETs.

e.g. stacking fault growth, are initiated in avalanche mode. Similar tests have been performed with the afore-mentioned trench MOSFET devices.

Compared to the diodes, additional failures might occur, caused mainly by generated electrons and holes, which, when injected into sensitive gate oxide, degrade the insulator up to destruction. In the development of the CoolSiC MOSFETs, one focus was to design the cell in such a way that electrons and holes flow through a part of the device that is at a secure distance from the gate oxide.

In Figure 15.28 the corresponding charge densities during avalanche are quantified. Obviously, the current density close to the channel is orders of magnitude lower than other parts of the device. Another parasitic effect during avalanche in power MOSFETs is the turn-on of a bipolar transistor (latch-up) formed between source, the p-body and the drift zone. It could be triggered by the forward voltage coming with the current flow in the base emitter part of this npn element. Theoretical and experimental studies, however, reveal that this latch-up is in fact unlikely with SiC. On the one hand, the low p-type conductivity might favor latch up, however, the gain of the bipolar transistor is very poor, and the base-emitter voltage needed to foster current flow is very high at >3 V.

Initial results prove that SiC MOSFETs are able to handle single-pulse avalanche energies at attractive levels [37]. This is remarkable since area and volume of the chip are small in comparison to silicon devices. Figure 15.29 displays typical waveforms of the discussed 1200 V 45 mΩ device.

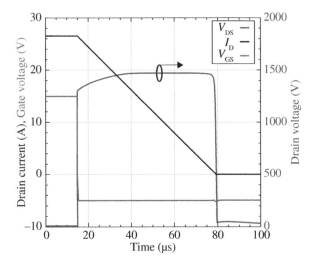

Figure 15.29 Avalanche behavior of a 1200 V SiC MOSFET, example 45 mΩ CoolSiC.

The limits in energy to be dissipated are once again defined by the thermal behavior of the device. Calculated temperatures at the destruction point approach the values being identified for short-circuit failures.

15.5 SiC MOSFETs in Power Applications – Selected Aspects and Prospects

Taking into account the technical benefits of SiC MOSFETs compared to IGBTs in high-voltage applications, one would indisputably select WBG components for all new designs. A few disadvantages still exist, however, including the following:

- The thermal performance, due to the very small die size, which results in much higher R_{th} values compared to silicon-based solutions handling the same power. In fact, even though total power losses are significantly smaller than in silicon chips, the loss power density is often much higher. Thus, smart thermal stacks are required.
- The power cycling performance of SiC chips, which is smaller for the same die attach technology, compared to silicon, reaching only about one-third of the silicon capability in power modules. The main reason is the higher Youngs modulus of SiC which causes higher mechanical stress to back-side joints.
- The already discussed limitations in short-circuit withstand times.

Some further, but less critical, aspects might be added to the list, e.g. the need of a stabilized power supply for SiC MOSFET drivers in order to fix the targeted $V_{GS(on)}$, which is mandatory due to the weaker transconductance of SiC as mentioned in the earlier chapters.

At any rate, the biggest obstacle for a broader rollout is still the cost of SiC compared to silicon, and some concerns about the maturity of the new technology. The latter issue has been successfully addressed in the last few years by an open discussion concerning the challenges and potential solutions to enable a field reliability similar to that of silicon technologies. In order to deal with the cost challenges, the key aspect is to identify system advantages which accompany the implementation of SiC that can justify the higher cost.

Those are mostly found by leveraging power density aspects (e.g. smaller inductive components enabled by higher switching frequencies, or smaller heat sinks due to lower losses to be dissipated). Prominent examples are solar power conversion systems [38], motor drive inverters for servo drives or fast charging systems in the e-mobility sector.

Efficiency advantages include direct benefits like energy-bill savings in UPS systems, train propulsion, and battery savings in electrical vehicles. The indirect impact comes in the form of smaller heat sinks or disruptive options to reduce cooling efforts by eliminating the need for forced cooling for example. Initially, cases, for which fast switching was not of interest, seemed to be outside the scope of application for SiC. Meanwhile it has become clear that even for low switching frequencies and comparable, commonly used switching slopes below $10\,kV/\mu s$ such as those in motor drives, a loss reduction can be seen, mostly due to savings in on-state mode under partial load, or to the lack of tail currents and Q_{rr} related turn-on losses. Furthermore, taking the example of motor drives, innovative ideas like the use of sinus output filters between the inverter and the motor might be re-assessed. The technical advantages are clear – no dv/dt related stress at the motor, no expensive shielded cables, and no discussions about the short-circuit capability of the used switches. In the silicon world, those approaches have not been successful, since the filters were prohibitively expensive and bulky assuming the limited switching frequencies with IGBTs. However, new opportunities will be possible with SiC MOSFETs.

A special application for SiC MOSFETs are auxiliary power supplies for power systems which are operated on DC links. In those solutions, the power supply is often powered directly out of the high-voltage DC link and thus, a high-voltage capability and fast switching are common requirements. Silicon-based solutions for such circuits are either expensive and have high losses (like high-voltage Si MOSFETs) or very complex if implemented in a multistage setup. Since the required ohmic ratings are also quite high (500 mΩ up to 1 Ω), the required SiC chip sizes are small. This is one of the very few examples where SiC can outperform silicon power devices in terms of cost at the component level.

SiC MOSFETs enable an easy fly-back based solution. Due to the low losses, the component can now also be implemented in an SMD package and thus, the assembly can be automated, and bulky heat sinks are no longer needed. Finally, with some adopted gate drive conditions, the system can be operated directly out of the controller, and a driver IC is no longer needed.

A further application where WBG switches are seen as an enabler are modern AC–DC stages in power supplies based on the totem pole topology [39]. The full-bridge totem pole topology is bridgeless, hence, the biggest loss contributor of traditional circuits – diode losses – are no longer relevant. The simple topology offers a bi-directional power flow and can deliver outstanding efficiency levels. However, the topology can be implemented only if the semiconductor switches exhibit certain features. Generally, these have to be switched with 45 ... 100 kHz, while one transistor acts as a boost switch and the second one as a synchronous rectifier. The transistors need to have extremely low Q_{rr} values and a low output charge Q_{OSS}. Finally, a smooth dependence of the output capacitance on the drain voltage is preferred. None of the three requirements can be delivered by silicon super-junction MOSFETs and thus, 650 V SiC MOSFETs or GaN HEMTs are the devices of choice to enable this new path toward low losses and high-power density PFC stages.

In conclusion, there are many motivations for implementing SiC MOSFETs in power circuits, as illustrated in the examples above. Since new operating conditions like higher frequencies and steeper voltage transients are mandatory – in combination with revolutionary system architecture such as power PCBs instead of bus bars, or passive cooling instead of complex cooling setups – to leverage the full potential of the new technology, it must be accepted that the changeover to SiC MOSFETs will require additional one-off efforts. Far fewer hurdles can be expected if the emerging applications do not rely on traditional system architecture, one of the most prominent examples being aircraft electric propulsion.

References

1 F. Bjoerk, J. Hancock, M. Treu, R. Rupp, and T. Reimann, "2nd Generation 600 V SiC schottky diodes use merged pn/Schottky structure for surge overload protection," in *Proceedings from Applied Power Electronics Conference and Exposition 2006*, 19–23 March 2006, CD Publication ISBN 0-7803-9548-4.

2 M. Domeij, A. Lindgren, C. Zaring, A. O. Konstantinov, J. O. Svedberg, K. Gumaelius, I. Keri, J. Grenell, M. Oestling, and M. Reimark, "1200 V 6 A SiC BJTs with very low V_{CESAT} and fast switching," *Proceedings of PCIM*, Nuremberg, 16–18 March 2010.

3 R. Kelley, G. Stewart, A. Ritenour, V. Bondarenko, and D. C. Sheridan, "1700 V Enhancement-mode SiC V_{JFET} for high voltage auxiliary flyback SMPS," in *Proceedings of PCIM*, Nuremberg, 2010.

4 M. Treu, R. Rupp, P. Blaschitz, K. Rueschenschmidt, T. Sekinger, P. Friedrichs, R. Elpelt, and D. Peters, "Strategic considerations for unipolar SiC switch options: JFET vs. MOSFET," in *Proceedings of IAS*, New Orleans, 23–27 Sept. 2007, pp. 324–330.

5 A. Agarwal and S. H. Ryu, "Status of SiC power devices and manufacturing issues," *Proceedings of CS MANTECH Conference*, Vancouver, 2006, pp. 215–218.

6 T. Nakamura, Y. Nakano, M. Aketa, R. Nakamura, S. Mitani, H. Sakairi, and Y. Yokotsuji, "High performance trench SiC devices with ultra-low ron," in *Proceedings of IEDM*, Washington, 5–7 Dec. 2011, pp. 599–601.

7 R. Nakamura, Y. Nakano, M. Aketa, N. Kawamoto, and K. Ino, 1200 V 4H-SiC trench devices," in *Proceedings of PCIM*, Nuremberg, 20–22 May 2014, pp. 441–447.

8 D. Peters, T. Basler, B. Zippelius, T. Aichinger, W. Bergner, R. Esteve, D. Kueck, and R. Siemieniec, "The new CoolSiC™ trench MOSFET technology for low gate oxide stress and high performance," in *Proceedings of the PCIM*, 16–18 May 2017, pp. 1–8, ISBN 978-3-8007-4424-4.

9 D. Heer, D. Domes, and D. Peters, "Switching performance of a 1200 V SiC-trench-MOSFET in a low-power module," in *Proceedings of PCIM*, Nuremberg, 10–12 May 2016, pp. 1–7, ISBN 978-3-8007-4186-1.

10 W. J. Schaffer, H. S. Kong, G. H. Negley, and J. W. Palmour, "Hall effect and CV measurements on epitaxial 6H-and 4H-SiC," in *Institute of Physics Conference Series*, Bristol, vol. 137, pp. 155–160, 1994.

11 G. Rescher, G. Pobegen, and T. Aichinger, "On the subthreshold drain current sweep hysteresis of 4H-SiC nMOSFETs," in *Proceedigs of IEDM*, San Francisco, 3-7 Dec. 2016, pp. 276–279.

12 G. Pobegen, J. Weisse, M. Hauck, H.B. Weber, and M. Krieger, "On the origin of threshold voltage instability under operating conditions of 4H-SiC n-channel MOSFETs," *Materials Science Forum*, vol. 858, pp. 473–476, 2016.

13 H. Yano, H. Nakao, T. Hatayama, Y. Uraoka, and T. Fuyuki, "Increased channel mobility in 4H-SiC UMOSFETs using on-axis substrates," *Materials Science Forum*, vols. 556, 557, pp. 807–811, 2007.

14 T. Kimoto, Y. Kanzaki, M. Noborio, H. Kawano and H. Matsunami.: Interface properties of metal-oxide-semiconductor structures on 4H-SiC{0001} and ("11" ‾("2") "0") formed by N_2O oxidation, *Journal of Applied and Physical Sciences*, vol. 44, no. 3, pp. 1213–1218, 2005

15 P. Hofstetter and M. Bakran, "Comparison of the surge current ruggedness between the body diode of SiC MOSFETs and Si diodes for IGBT", in *Proceedings of CIPS*, 20–22 March 2018, pp. 1–7.

16 T. Kimoto, A. Iijima, H. Tsuchida, T. Miyazawa, T. Tawara, A. Otsuki, T. Kato, and Y. Yonezawa, "Understanding and reduction of degradation phenomena in SiC power devices," in *IRPS Proceedings*, 2–6 April 2017, pp. 2A-1.1–1.7.

17 J. Nakashima, A. Fukumoto, Y. Obiraki, T. Oi, Y. Mitsui, H. Nakatake, Y. Toyoda, A. Nishizawa, K. Kawahara, S. Hino, and H. Watanabe, "6.5-kV Full-SiC Power Module (HV100) with SBD-embedded SiC-MOSFETs," *PCIM Europe 2018; International Exhibition and Conference for Power Electronics, Intelligent Motion, Renewable Energy and Energy Management*, Nuremberg, Germany, 5–7 June 2018, pp. 1–7.

18 H. Lin and P. Gueguen, "GaN and SiC Devices for Power Electronics Applications," Yole Development Market & Technology Report, 07/2015, www.yole.fr

19 M. Beier-Moebius and J. Lutz, "Breakdown of gate oxide of 1.2 kV SiC-MOSFETs under high temperature and high gate voltage," in *Proc. PCIM*, Nuremberg, 10–12 May 2016, pp. 1–8.

20 J. Lutz, T. Aichinger, and R. Rupp, Reliability evaluation. In K. Suganuma (ed.), *Wide Bandgap Power Semiconductor Packaging: Materials, Components, and Reliability*. Elsevier, to be published.

21 L. C. Yu, K. P. Cheung, G. Dunne, K. Matocha, J. S. Suehle, and K. Sheng, "Gate oxide long-term reliability of 4H-SiC MOS devices," *Materials Science Forum*, vols. 645–648, pp. 805–808, 2010.

22 J. Sameshima, O. Ishiyama, A. Shimozato, K. Tamura, H. Oshima, T. Yamashita, T. Tanaka, N. Sugiyama, H. Sako, J. Senzaki, H. Matsuhata, and M. Kitabatake, "Relation between defects on 4H-SiC epitaxial surface and gate oxide reliability," *Materials Science Forum*, vols. 740–742, pp. 745–748, 2013.

23 J. Senzaki, A. Shimozato, M. Okamoto, K. Kojima, K. Fukuda, H. Okumura, and K. Arai, "Gate-area dependence of SiC thermal oxides reliability," *Materials Science Forum*, vols. 600–603, pp. 787–790, 2007.

24 R. Singh and A. R. Hefner, "Reliability of SiC MOS devices," *Solid-State Electronics*, vol. 48, pp. 1717–1720, 2004.

25 J. W. McPherson and H. C. Mogul, "Underlying physics of the thermochemical E model in describing low-field time-dependent dielectric breakdown in SiO_2 thin films," *Journal of Applied Physics*, vol. 84, no. 3, pp. 1513–23, 1998.

26 P. Friedrichs, "Ruggedness of SiC devices under extreme conditions," in *2020 IEEE International Reliability Physics Symposium (IRPS)*, 28 April–30 May 2020, pp. 1–6.

27 T. Hatakeyama, H. Matsuhata, T. Suzuki, T. Shinohe, and H. Okumura, "Microscopic examination of SiO_2/4H-SiC interfaces," *Materials Science Forum*, vols. 679–680, pp. 330–333, 2011.

28 W. Weibull, "A statistical distribution function of wide applicability," *ASME Journal of Applied Mechanics*, pp. 293–297, 1951.

29 J. C. Lee, I. -C. Chen, and H. Chenming, "Modeling and characterization of gate oxide reliability," *IEEE Transactions on Electron Devices*, vol. 35, no. 12, pp. 2268–2278, 1988.

30 M. Beier-Moebius and J. Lutz, "Breakdown of gate oxide of SiC-MOSFETs and Si-IGBTs under high temperature and high gate voltage," in *Proceedings of PCIM*, Nuremberg, 10–12 May 2017, pp. 365–372.

31 R. Siemieniec. T. Aichinger, W. Jantscher, D. Kammerlander, R. Mente, and U. Wenzel, "650 V SiC trench MOSFET for high-efficiency power supplies," in *Proc. EPE*, Genua, Italien, 3–5 Sept. 2019, pp. P-1.

32 J. Berens and T. Aichinger, "A straightforward electrical method to determine screening capability of GOX extrinsics in arbitrary, commercially available SiC MOSFETs", to be published.

33 C. Chen, D. Labrousse, S. Lefebvre, M. Petit, C. Buttay, and H. Morel, "Study of short-circuit robustness of SiC MOSFETs, analysis of the failure modes and comparison with BJTs," *Microelectronics Reliability*, vol. 55, no. 5, 2015.

34 H. Hatta, T. Tominaga, S. Hino, N. Miura, S. Tomohisa, and S. Yamakawa, "Suppression of short circuit current with embedded source resistance in SiC MOSFETs", *Materials Science Forum*, vol. 924, pp. 727–730, 2018.

35 M. M. Bakran, S. Hain, "Integrating the new 2D – short circuit detection method into a power module with a power supply fed by the gate voltage," in *2016 IEEE 2nd Annual Southern Power Electronics Conference (SPEC)*, 5–8 Dec. 2016, pp. 1–6.

36 S. Palanisamy, Md. K. Ahmmed, J. Kowalsky, J. Lutz, and T. Basler, "Investigation of the avalanche ruggedness of SiC MPS diodes under repetitive unclamped-inductive-switching stress," *Microelectronics Reliability*, vols. 100, 101, 2019.

37 J. Wei, S. Liu, S. Li, J. Fang, T. Li and W. Sun, "Comprehensive investigations on degradations of dynamic characteristics for SiC power MOSFETs under repetitive avalanche shocks," *IEEE Transactions on Power Electronics*, vol. 34, no. 3, pp. 2748–2757, 2019.

38 https://www.pv-magazine.de/2018/11/14/pv-magazine-top-innovation-kacos-neuer-siliziumkarbid-wechselrichter

39 R. Mente, R. Siemieniec, W. Jantscher, D. Kammerlander, U. Wenzel, and G. Mattiussi, "A novel 650 V SiC Technology for high efficiency Totem Pole PFC topologies," in *PCIM Europe digital days 2020; International Exhibition and Conference for Power Electronics, Intelligent Motion, Renewable Energy and Energy Management*, Germany, 7–8 July 2020, pp. 1–8.

16

Multiphase VRM and Power Stage Evolution

Danny Clavette

Infineon Technologies, Americas Corporation, El Segundo, CA, USA

16.1 Evolution of the First Multiphase Controllers

Early 1990s Central Processing Unit (CPU) consumed <10 A and could be powered by linear regulators, often using a discrete Metal Oxide Field Effect Transistor (MOSFET) as the linear pass element driven by an operational amplifier. Most motherboards during this era were powered mainly from a distributed +5 V, while +12 V was used for miscellaneous bias needs. The 80386 processors core voltages were around 3.45 V making linear regulation from 5 V feasible with pass element power loss of 10–15 W when utilizing heat sinks. See below example of a typical linear regulator module utilized in early IBM® systems. Three parallel pass devices can be seen in the top right (Figure 16.1).

Increased reliability, efficiency, and reduced cooling requirements drove the transition from linear regulators to nonsynchronous buck regulators, with the first single phase buck controllers available in the late 1990s. Additionally, the buck regulator enabled the transition to 12 V input power delivery to relieve the heavily burdened 5 V supply.

In 1998/99, Intel's® Pentium® II Xeon™ 400 MHz processor was consuming 30 W with 2 V V_{core} at 15 A. Intel defined these Xeon systems to utilize voltage regulator modules (VRMs), which provided board design flexibility and simplicity. VRMs were used extensively in early computer systems and were produced by several module vendors such as Delta, Liteon, and Artesyn. These VRMs utilized printed circuit board (PCB) edge connectors and were plugged into sockets as shown in Figure 16.2.

Intel VRM-8.2 specification for Pentium II defined the VRM electrical requirements, dimensions, and socket pinout. VRM-8.3 [1] quickly followed with the addition of an input voltage sense pin enabling control loop feed forward techniques and output voltage remote sense pins for processor die voltage Kelvin sensing. Both these improvements were needed since power delivery network (PDN) voltage drops in the PCB were already beginning to limit processor computing speed during benchmark performance testing and could result in CPU "blue screen" lockup if processor timing violations occurred due to poor supply regulation accuracy.

In 2000/01, Intel's Pentium III current consumption increased to 23 A; a 35% increase in current and early indicator of future trends. To manage processor performance optimization, Intel introduced VRM-8.5 [2] which included an adjustable 5-bit digital to analog converter (DAC), referred to as Voltage Identification (VID), to adjust the processor regulation voltage in steps of 25 mV/bit. This VID enabled CPU performance optimization for various processor Stock Keeping Units (SKUs) and enabled processor binning to improve yields. For instance, some processors may have timing sensitivities or higher leakage at higher voltages and having the flexibility of optimizing V_{core} for a

Advances in Semiconductor Technologies: Selected Topics Beyond Conventional CMOS, First Edition. Edited by An Chen.
© 2023 The Institute of Electrical and Electronics Engineers, Inc. Published 2023 by John Wiley & Sons, Inc.

Figure 16.1 A linear regulator module used in 80386 generations.

(a) (b)

Figure 16.2 (a) A generation VRM-9.1 3-phase voltage regulator module. (b) 6-phase VRM-10 module plugged into a demonstration board.

specific processor helped reduce waste. The Pentium III typical V_{core} was 1.75 V and thus dissipated 40 W at full power.

To help manage CPU thermal energy, a new programmable output-resistance regulation technique often referred to as a load-line resistance or R_{LL}, was introduced to reduce CPU power loss at higher processing loads. Refer to the load line diagram in Figure 16.3a showing regulated voltage decreasing as the load current increases. This load line technique also benefits transient response by pre-positioning the regulated voltage prior to a load step release such that the resulting V_{core} voltage peak-to-peak transient is reduced for a given capacitor network and load step as shown in Figure 16.3b,c. Note the overall voltage excursion during a load transient can be halve, enabling output capacitor reduction for a given transient peak-to-peak. Load line is still used in systems today.

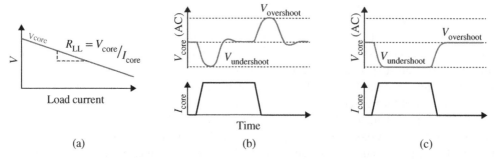

(a) (b) (c)

Figure 16.3 Explanation of load line and its transient benefits. (a) Load line resistance. (b) Load transient response without load line. (c) Load transient response with load line.

The implementation of load-line requires accurate current sensing to achieve the tight voltage regulation goals of the CPU. Regulation errors can lead to processor timing issues resulting in potential lockup. Early systems relied on current sense resistors, which dissipated additional power, added cost, and increased solution size. Some early systems attempted to use PCB trace resistance in order to save cost however process variations often led to very poor regulation accuracy.

Semtech® released the first monolithic "Bi-Phase" multiphase controller SC2442 [3] in 1998 to address Intel's increasing processor bias currents. The SC2442 integrated the control and the synchronous buck drivers into one package in order to drive discrete SO8 (Small Outline 8 pin) negative type MOSFET (NMOS) devices. SC2442 was the first two-phase interleaved synchronous buck regulator and relied on matched MOSFET drive event timing for current balance. This balance technique exhibited poor current balance when a transient was applied, often leading to power supply damage. An input current-mode scheme was introduced to balance the peak inductor magnetizing currents; however, this also proved challenging to implement with several application notes [4] to help tune for proper operation.

The SC2442 set the multiphase industry in motion and exposed the need for good phase-to-phase current balance. Input current sensing did reduce the number of sense elements but was very noisy and would not allow phase timing overlap. Per-phase current sensing enabled phase overlap; however, current sense resistors in each phase increased cost, power loss, and size. Multiphase controller developers naturally started using the inductor DCR (direct current resistance) as a current sense element. In 1998, Linfinity issued an application note [5] describing a simple inductor current sensing scheme which dramatically improved DCR sense quality. An RC-filter placed across the inductor stores the DC voltage developed by current flowing through the inductor winding resistance, depicted as RL in their schematic. See Figure 16.4 below. Linfinity also described how to properly select the RC-filter time constant parameters to match the inductor L/R time constant, $R * C_s = L/RL$, in order to achieve better transient sensing accuracy needed for load line implementation.

Further DCR accuracy enhancements included the introduction of a negative temperature coefficient (NTC) resistor in the RC network to compensate for increasing inductor winding resistance due to conductor temperature rise ($\sim$4000 ppm) at higher currents.

In 1999, Cherry Semiconductor introduced the CS5308 [6] two-phase controller with integrated inductor DCR sensing of each phase. Current balance was implemented by adding averaged current information to an artificial ramp at the input of each phase's pulse width modulator (PWM) comparator.

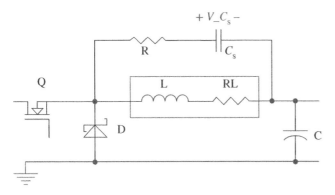

Figure 16.4 Inductor DCR sensing.

The benefits of multiphase control were very clear as described in a 1999 Application Note from Linear Technology, titled "High Efficiency, High Density, PolyPhase Converters for High Current Applications" [7]. The resulting multiphase control architecture proved very robust with good market success, paved the way to higher phase count systems.

16.2 Transition from VRMs to "Down" Solutions

Figure 16.5 provides a snapshot of typical server generational timeline with key voltage regulator (VR) trends, peak power values, and notable VR improvements related to process node advancements from Intel and advanced micro devices (AMD). Graphic and Tensor artificial intelligence (AI) processors are increasingly being adopted into future cloud and enterprise server systems.

Intel issued their VRM 9.0 specification [8] in 2002 to power their 100 W Pentium 4 Processor. The significant current increase to 65 A@1.5 V_{core} and 42 A@1.7 V which triggered several VR paralleling efforts in attempt to scale the output current capabilities of systems. In response to this critical lack of high current capability, Unitrode (now Texas Instruments) introduced a family of Load Share Controllers (UC1907, 2907, 3901) (Figure 16.6) [9]:

> The UCx907 family of load share controller integrated circuits (ICs) provides all the necessary features to allow multiple-independent-power modules to be paralleled such that each module supplies only its proportionate share to total-load current. … By monitoring the current from each module, the current share bus circuitry determines which paralleled module would normally have the highest output current and, with the designation of this unit as the master, adjusts all the other modules to increase their output current to within 2.5% of that of the master.

These load share devices enabled paralleling of existing 25 A VRs to support Intel's 65 A system requirements. With one load share controller required per VR, systems were quite complicated, costly, and space constrained. These challenges enabled several new integrated and simplified concepts, such as "Direct Duty Cycle Current Sharing" (US6642631B1) [10], concepts which continue to enable active current in systems today.

Higher phase count multiphase controllers with onboard drivers offered cost-effective alternatives with families of fixed phase count controllers optimized for processor current ranges. For example, Cherry Semiconductor offered two, three, and four phase solutions, with each device optimized for die size and cost due to the large onboard MOSFET drivers needed for each phase.

CS5332 Two-Phase using paralleled HS and LS MOSFETs [11]
CS5301 Three-Phase Buck Controller
CS5307 Four-Phase Buck Controller [12]

Early integrated high- and low-side bipolar drivers had limited slew rates due to large parasitic tub capacitances slowing down internal signal transitions. The high-side driver pull down NPNs were typically tied to ground instead of terminating to SW as they do today, where the high-side MOSFET gate charge is delivered to the load instead of ground. Additionally, the early SO8 and D2Pak N-channel MOSFETs used in these applications had fairly high gate-to-source and gate-to-drain charges. MOSFETs were typically doubled to reduce effective RDSon and offering improve thermal performance. VRM systems utilized large heat sinks mounted on front and back

Evolution and strategy of Power Management for Servers
Based on Intel and AMD Generational Enhancements

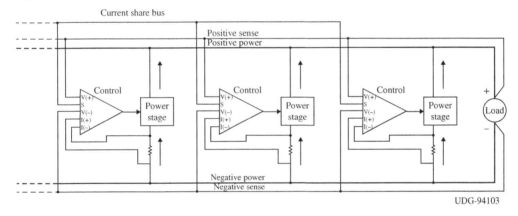

Power demand rises	Price Performance	Enhanced System Performance	Smart System with Telemetry	Customized Power System	Digital Connected Power System	Integrated AI
100 W	200 W	250 W	300 W	400 W	550 W	1000 W?
› First ever digital control solution for processors › Voltage Regulation Modules (VRM)	› Increased MOSFET efficiency and power density enable VR on motherboard	› Serial Voltage Identification SVID enables advanced power management functionality › Full system solution with optimized components	› Adoption of nonlinear control algorithms › Integrated powerstages with i and t sense › Adaptive phase control	› Digital control fully adopted › Embedded micro-controller enables use of firmware for customization for flexibility	› Software defined digital controller › Enables powerflow optimization on system level › Sophisticated powerstage fault monitoring	› AI enabled digital controller › Auto-tuning over the life of the system › System integration
2002	2006	2010	2012	2014	2021	202x

Figure 16.5 Evolution and strategy of power management in server CPU generations.

Current share bus

Positive sense
Positive power

Control
Control
Control

Power stage
Power stage
Power stage

+ Load −

V(+)
S
V(−)
I(+)
I(−)

Negative power
Negative sense

UDG-94103

Figure 16.6 Unitrode load share controller system diagram.

for thermal management as shown in Figure 16.7. The resulting optimal switching frequencies were around 100–200 kHz.

International Rectifier (IR) (now Infineon Technologies) introduced early samples of the DirectFET® package [13, 14] in late 2003. The DirectFET proved to be a very disruptive technology with reduced gate charge N-channel MOSFETs, 1 °C/W thermal resistance in a low profile 0.7 mm metal can package as seen in Figure 16.7. This technology allowed the reduction of the number of MOSFETs (no need to parallel) and enabled higher per-phase currents to reduce phase count.

With these technology improvements, the resulting VRM design complexity reduced substantially. Space and thermal management challenges reduced and the VR reliability improved significantly. These simplifications led to VR "Down" on motherboard solutions which provided cost reductions and down designs were driven by Enterprise Server companies such as Hewlet Packard (HP). This enabled lower profile server chassis, "Blade" type servers and resulted in better system utilization.

In 2004, Intel issued their next VR-10 [15] specification as both "Voltage Regulator Module (VRM) and Enterprise Voltage Regulator-Down (EVRD) 10.0," setting the stage for price optimized performance and reduced use of VRMs.

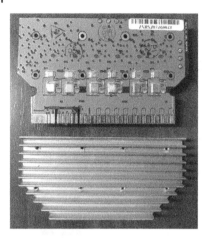

Figure 16.7 Six phase VRM with the heat sink removed, using DirectFET IRF6622 and IRF6628.

16.3 Intel Xeon Generations Challenges Moore's Law

VR-10 powered the first Xeon processors which consumed 100 A with 1.25 mΩ load line with SKUs up to 250 W. This was another significant jump in performance and power dissipation which required advances in thermal management. The 5 bit VID DAC least significant bit (LSB) resolution was halved to 12.5 mV, and Xeon was the first processor to support dynamic voltage identification (DVID) where the processor could actively change its supply voltage while operating to actively manage processor performance and thermal dissipation.

The number of phases needed to support various processor SKUs is also dependent on the phase's maximum current capability, where higher current phases can enable the reduction of phase count, solution cost, and size. Even today, lower wattage processor SKUs common in cost optimized desktop and laptop systems, typically use three and four phase controllers with integrated drivers.

As Xeon SKUs begin to exceed 100 W and phase counts exceeded four phases, VR systems transitioned to external discrete drivers to

- Keep the controller die size reasonable (drivers require large die area)
- Maintain reasonable controller package sizes (needed more than four pins per driver)
- Simplify routing challenges to high phase count MOSFETs
- Prevent high controller temperatures which makes meeting critical control system accuracy specification easier
- Utilizing external drivers in server power also helped reduce the controller cost overhead in high phase count systems.

Some control architectures relocated the PWM and current sense circuitry to the discrete driver to help reduce PCB routing challenges and improve noise sensitivity of inductor DCR current sensing. Partitioning the PWM control into the driver also enabled a scalable high phase count system by allowing phases to easily be added to a design without impacting the controller pinout or package size. ON/Cherry Semiconductor investigated these "Poly-Phase" architectures to allow scalability and International Rectifier introduced the X-Phase™ [16] architecture which combined PWM management with local current sensing in the MOSFET driver.

Local current sensing improved accuracy while enabling lower DCR inductors for higher system efficiency. Figure 16.8 below shows an X-Phase application where there are no PWMs between the controller and phases. Instead, the PHASE_TIMING bus provided the timing mechanism to start

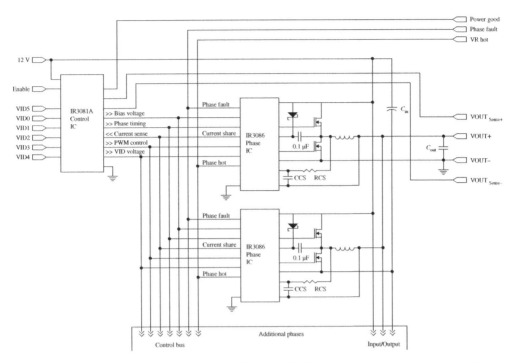

Figure 16.8 X-phase® bus architecture supporting up to 32 phases. C_{CS}, capacitor current sense; R_{CS}, resistor current sense.

interleaved PWM pulses for up to 32 phases. The VID_VOLTAGE bus communicates the amplified and compensated regulation error to the PWM comparators located in each driver. When combined with the International Rectifier DirectFET, four to five phases offered the best combination of density and efficiency and was a very successful solution in these early Xeon systems.

In 2007, Intel released the initial spec for their next generation VR11 Xeon processor which again supported both VRM and EVRD. Peak processor current reached 150 A for V_{core}, utilized a 0.8 mΩ load line to assist with load transients, and utilized an 8-bit parallel VID with a 6.25 mV LSB supporting DVID.

The VR11.1 [17] update added the system "IOUT" to support the new LGA1366 processor's self-monitoring capability, added Power on Configuration (POC) and Market Segment Identification (MSID) functions which were simple truth-table options configured onto the 8-bit VID lines during power up. VR11.1 also introduced the Power State Indicator Input (PSI#) which places unused phases into low power modes. Phases could be "shed" or turned off when load currents reduced in order to boost overall system efficiency. For example, 1 phase supporting 10 A is much more efficient than five phases each supporting 2 A.

By VR11.1, multiphase server power solutions were finding commonality with most systems utilizing six phases for V_{core}, with X-Phase and DirectFET being the most common design.

16.4 Increased System Digitization Enables Digital Control

Until VR12, analog control was adequate for nearly all CPU multiphase VR needs. Analog solutions were extensively tuned for optimizing transient responses. High current "Swing" inductors, whose inductance purposely reduced at higher currents through the use of stepped gaps, found a few successes at improving transient performance but never gained broad adoption. The basic two phase "Coupled" inductor was introduced by Virginia Tech University in the late 1990s to help improve transients by reducing the inductance under current imbalance events which occur during load transients. Volterra (now Maxim) focus their solutions around coupled inductors and continue to expand their coupled inductor patent portfolio today. Both swing and coupled typically introduce undesirable side effects and limitations.

Primarion was a successful digital control pioneer, releasing its first digital multiphase controller in 2001. Their digital PID filters (Proportional, Integral, and Derivative) intended to simplify control loop optimization while offering improved nonlinear transient control, which is nearly impossible to implement in analog control systems. Nonlinear control utilizes system triggers, such as regulation error levels or rates, to instantaneously alter the PID coefficients in order to improve transient responses. Analogous nonlinear control in analog type-3 type voltage mode systems would need to alter the control loop operating point by instantaneously altering charge in compensation capacitors to achieve just the right amount of voltage needed for the new operating point – a nontrivial task with finite response time.

Continued system digitization encouraged "going digital" decisions. Introduction of I2C digital communication and the adoption of PMBus as a common Power Management Bus language in Enterprise Server power systems, began to challenge analog systems. "Digital Wrapper" solutions marry analog control loops with analog to digital converters (ADCs) for telemetry and DACs for control required by PMBus. Additional customer-specific fault communication, system diagnostic, and optimization requirements further increase digital content.

VR12 introduced the first Intel Serial Voltage Identification (SVID) in 2007/2008 and was quickly followed by AMD® with their SVI2 specification in 2010. These CPU specific digital communication standards provide respective CPUs a method to directly communicate with VR control systems. These serial VID specifications further challenge analog wrapper designs, with additional digital requirements, accelerated the need to go digital.

Infineon Technologies acquired Primarion in April 2008, while International Rectifier acquired Chil Semiconductor (another digital controller startup) in February 2011 with their CHL8316 and CHL8318 becoming the first mainstream multiphase digital controller in VR12. Chil's initial success was mainly driven by overclocking designs in desktop systems where registers could be manipulated to achieve higher performance targets.

Digital control is now mainstream in server and desktop CPU power where Texas Instruments and Renesas also offer multiphase digital control while other multiphase VR vendors continue to develop their own digital wrapper solutions. Digital control is slowly being adopted in laptop as digital controller quiescent current consumption associated with ADC sampling continues to improve. The available digital features and autonomous power saving mode capabilities encourage digital control adoption into mobile systems.

16.5 DrMOS 1.0: Driver + MOSFETs

DrMOS is a package concept which integrates a driver and two MOSFETs in a Buck half-bridge configuration as shown in Figure 16.9.

DrMOS originated with EVRD-10 and was driven by a goal to reduce "down" solution area with the understanding that microprocessor power requirements were going to continue increasing every process generation. During this period, it was common to utilize four SO8 MOSFETs per

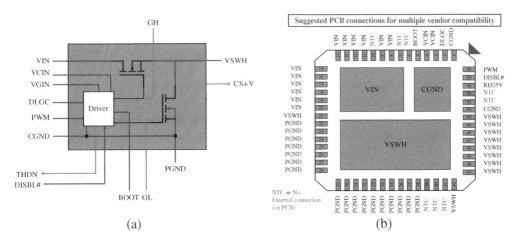

Figure 16.9 DrMOS-1.0 integration schematic and suggested pinout.

phase, consuming large PCB area. Integration would reduce the number of packages from 5 to 1 (Driver IC + four SO8).

Inductors were very large in size with inductor values around 680 nH to 1 μH, often constructed using heavy gauge wound toroidal structures. Integration could also enable increased switching frequencies beyond the typical 150 kHz of that day and enable inductor dimension reduction.

DrMOS standardization of power delivery was also a desired outcome, with the potential for improved costs through the reduction of packages and reduced price due to commoditization.

Other objectives included increased efficiency at higher switching frequencies due to tighter PCB layouts and more efficient MOSFET drive, better transient response with lower inductance (enabled by higher frequencies), and easier PCB layout routing.

Intel began DrMOS discussions in 2004 and issued a specification [18] in November with the following introduction;

> This document summarizes the different requirements needed to enable a standard integrated Driver-MOSFET (DrMOS) specification for implementation in a typical personal computer (PC) platform. The aim of this specification is to enable features that are necessary to develop an integrated device such that inter-operability between various devices and controllers is feasible.

The first 8 × 8 mm DrMOS-1.0 package targeted a 25 A thermal design rating and utilized bond wires for high current interconnect of the MOSFETs, which were sized to support ~10% duty cycles of most applications. Although DrMOS certainly addressed density, early devices were still quite costly and performance was not much better than the discrete solution which offered a much larger thermal footprint. Furthermore, the number of vendors initially participating in DrMOS was limited which further limited adoption.

DrMOS required the combination of two specialties: driver ICs and MOSFETs, and there were few companies that had both capabilities in-house. There were many SO8 MOSFET choices and power engineers focused on finding the best high-side (HS) and low-side (LS) MOSFET combinations to give them a performance advantage over competition.

Several MOSFET vendors were also improving package and MOSFET technologies in parallel to DrMOS efforts. Some SO8 vendors replaced bond wires with clips to improve performance.

The power quad flat no-lead package (PQFN), also known as Micro Leadframe Package (MLP), is a good replacement vehicle of paralleled SO8 devices. The PQFN supports larger MOSFET die and often uses clips to replace the high current bond wires resulting in better thermal and RDSon performance.

International Rectifier developed the proprietary DirectFET, providing the best system efficiency and thermal performance, especially when using heat sinks. 30–40 A per-phase systems were not uncommon with DirectFET and thus became the lead solution for both VRM and EVRD down solutions. Infineon did license this technology, marketed as the "CanPak."

16.6 DrMOS 4.0 and International Rectifier's Power Stage Alternative

Vendors investigating alternate package concepts to compete against the DirectFET continued to improve the DrMOS performance. In 2008, Intel began discussing DrMOS-4.0 with a new 6×6 mm common footprint. Several vendors utilized copper clips to improve performance however the defined footprint could only be realized with multiple clips; first clip from the HS source to the switch node and second clip from the LS source to ground. See Figure 16.10 below of Infineon's TDA21220 DrMOS.

DrMOS 4.0 efficiency performance is significantly better than DrMOS 1.0 due to improvements in MOSFET technologies combined with stronger drive and the introduction of the dedicated PHASE pin to remove source inductance from the high side drive loop. Source inductance is represented as L parasitic (L_p) in Figure 16.11. $V(L_p)$, defined as $V(\text{PHASE}) - V(\text{SW})$, increases when Q_1 initially turns on and a change in Q_1 source current occurs. Voltage must be expressed across the inductor for current to change and $V(L_p)$ returns to 0 V once Q_1 is fully on and the source current remains fairly constant.

Note C_{BOOT} can be connected to either PHASE or SW, significantly altering MOSFET drive performance.

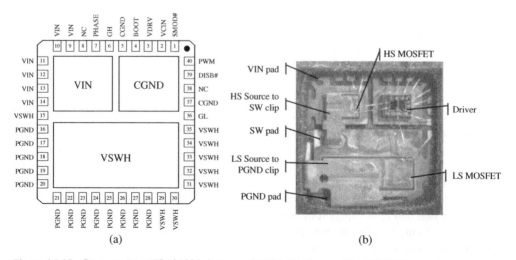

Figure 16.10 De-capsulated TDA21220 6×6 mm DrMOS-4.0. Source: TDA21220 / Infineon Technologies AG.

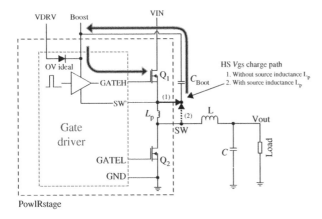

Figure 16.11 Source inductance L_p included (1) or excluded (2) from HS MOSFET V_{gs} charge path.

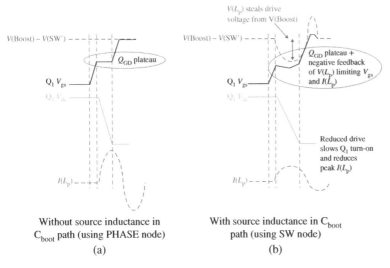

Figure 16.12 Source inductance impact on HS MOSFET turn-on.

- Connecting C_{BOOT} to PHASE removes $V(L_p)$ from the V_{gs} charge loop resulting in full available drive bias to charge the gate of Q_1. Refer to the left diagrams in Figure 16.12. A plateau can be seen in V_{gs} due to "Miller Capacitance" gate bias injected by the drain-gate capacitance when the drain voltage falls. In this scenario, Q_1 turn-on time is determined by drive strength for a given MOSFET and large voltage ringing can occur on SW.
- Connecting C_{BOOT} to SW includes the source inductance $V(L_p)$ in the V_{gs} charge path loop. An increase in $V(L_p)$, during a rising switch node event, effectively reduces the C_{BOOT} drive voltage (with respect to Q_1 source) and weakens drive strength. The larger the change in current during turn-on, the weaker the drive. Note this is a negative feedback loop and the MOSFET operates in sustained linear mode resulting in slower turn-on times at higher load currents. Ringing and efficiency typically reduce as load current increases in these systems.

The efficiency improvements from the PHASE node enabled increased adoption, mostly in cost sensitive systems. There are few remaining discrete solutions powering processors.

The DrMOS multi-clip solution still has several limitations

- VIN to PGND pads are ideally close together for effective C_{IN} capacitor decoupling and reduced ringing. In the DrMOS package, the loop is extended by the multiple clips and the large SW pad located between VIN and PGND.
- The SW pad is very noisy and typically relegated to the top PCB layer only. Yet in DrMOS, the top layer SW path to the inductor is blocked by PGND and must be routed internally to the PCB.
- The PGND pad is quite small and does not offer the most effective thermal path to internal PCB PGND layers for thermal distribution within the motherboard. As a result, DrMOS runs hotter than discrete systems.
- Multiple clip solder connections, plus long package lateral interconnect distances increased package RDSon.
- The driver bond wires are quite long resulting in nonoptimal drive
- Inductor DCR current sense needed to be routed to the controller

International Rectifier (IR) realized these many deficiencies and introduced an alternative 6×6 mm package concept: the IR3550 Power Stage [19] family with a scalable footprint to support the 5×6 mm IR3551 and 4×6 mm IR3553 as shown in Figure 16.13.

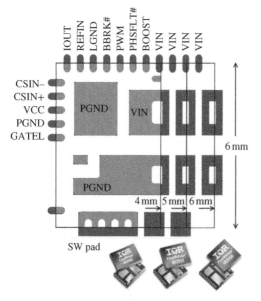

Figure 16.13 IR3550 scalable family alternative to DrMOS-4.0.

This new IR3550 package concept corrected all the DrMOS non-idealities by flipping the LS MOSFET and thus swapping the SW and PGND pad locations of DrMOS as shown in Figure 16.14 below. Note the LS Gate solder pad on the left side of the LS MOSFET. The MOSFET gate pad is soldered down onto the lead frame which connects to the GATEL pad for wire bonding. The gray area between these connections identifies bottom half-etching of the lead frame to form a bridge. Once the product is molded, this area under the bridge is filled.

The resulting IR3550 improvements included

- Enabling effective VIN decoupling with PGND and VIN pads close together and minimizing the current flow loop

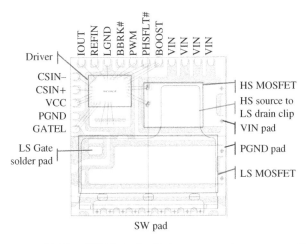

Figure 16.14 IR3550 single clip interconnect improvement. Source: Based on IR3550 DrMOS-4.0.

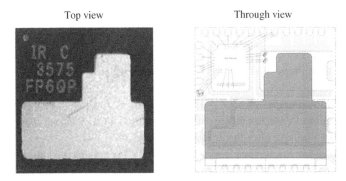

Figure 16.15 IR3575 single clip interconnect improvement. Source: Based on IR3575 DrMOS-4.0.

- Correctly minimize the SW pad size and keeping SW at top PCB motherboard routing to mini-mize noise
- Providing a large PGND thermal pad to help distribute the power stage heat into the multi-layer PCB ground plane
- Using a single clip interconnect between the HS and LS MOSFETs to reduce package distribution resistance and parasitic inductance
- Shorter gate driver bond wires
- Introduced a local current sense amplifier with onboard temperature compensation to eliminate the inductor sensing NTC. The local amplifier minimized inductor DCR sense routing noise thus enabling lower DCR inductors for improved efficiency

The omission of the PHASE node did not significantly impact efficiency due to all the other improvements and the ability to increase VDRV from 5 to 6.8 V. Customers also saw value in reduced ringing at higher currents.

An improved top side cooled IR3575 version was also produced to round off the DrMOS alterna-tive product portfolio, where the single clip was exposed (Figure 16.15) to provide a 1 °C/W thermal impedance from die to the package top resulting in thermal performance similar to the DirectFET.

16.7 International Rectifier's "Smart" Power Stage

Intel VR12.5 system definition requested increased current sense accuracy to improve processor performance. Inaccuracies in current telemetry can result in underperformance in the case of over-reporting, or can cause CPU overheating and thermal shutdown if under-reporting. The following race car analogy can be used to better understand the importance of accurate telemetry:

- Current sense telemetry acts like a tachometer in a race car and allows a processor to run right up to the red line for maximum performance. Like a driver, the processor can use current information as feedback to optimize the processing speed.
- Consider a scenario where your tachometer has a +10% error with a 10 K redline; you would lose the benchmark performance race running at 9000 RPM.
- For a −10% error, you lose the race because you have been racing at 11000 RPM and overheat your engine… Same is true for processor benchmarking where you want to squeeze out the maximum performance without failing.

Inductor DCR sensing for current sense telemetry has many issues.

- Standard DCR values can vary 10%. 5% tolerance is the best case available at additional binning cost.
- DCR values less than 300 μΩ offer poor system accuracy due to the very small signal levels produced in a very noisy switching environment. The result is increased power loss for the sake of current sensing.
- NTC and other thermal compensation strategies provide coarse correction but can deviate with a change of airflow, thermal environment or temperature sense element location. For instance, changes in airflow direction or magnitude can alter the amount of correction.
- The $L/R_L = R*C$ time constant matching can deviate under high current inductance variation and temperature.
- Multi-sourced inductors with similar time constants are difficult to match. These time constant mismatches can negatively impact transient response performance due to errors in load-line response. This can lead to AC spec violations with excessive output voltage overshoot or undershoot.

IR introduced the industry's first "Smart" Power Stage (SPS) concept which set the standard for most recent SPS products. The IR3555 [20] product family integrated current and temperature telemetry into the driver, relieving the inductor DCR requirements while improving current reporting accuracies. These new feature pins are highlighted in Figure 16.16 below.

- The IR3555 originated the 5 mV/A (and future 5 μA/A), 10 MHz bandwidth current telemetry standard now followed by all SPS devices.
- 8 mV/C° temperature telemetry reporting was also created and standardized by IR along with the wired ORed architecture where the hottest power stage drives the TMON bus. Powerstage temperature reporting allows notification of an improperly cooled powerstage – perhaps a heat sink is not well connected… Without real-time telemetry, a powerstage over-temperature shutdown would cause the whole system to fail.

Similar to the prior generation, the pin compatible IR3578 included an enhanced top-side cooling option which exposed the MOSFET interconnect clip. The SPS became widely adopted by customers needing to achieve highest CPU performance with highest system efficiency.

Figure 16.16 IR3555 first smart power stage with current and temperature telemetry. Source: IR3555 / Infineon Technologies AG.

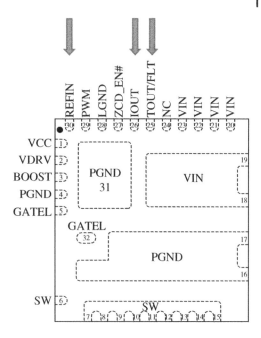

16.8   DrMOS 5 × 5 mm and 4 × 4 mm De-standardization

During the same VR12.5 generation, new 5 × 5 mm DrMOS products were initiated by several industry vendors. Lack of a common target specification resulted in footprint deviations in this DrMOS generation.

As seen in Figure 16.17, Infineon TDA21231 [21], Alpha & Omega Semiconductor™ AOZ5131 [22], and Vishay Siliconix™ SiC645 [23] utilize International Rectifier's concept of low-side MOS-FET source-down packaging to provide optimized PGND cooling through a larger pad while providing a direct SW top level PCB trace to the inductor. All three packages are shown in top through view. Note the Vishay footprint is mirror image of the more closely matched Infineon and Alpha & Omega footprints.

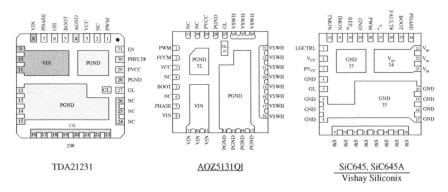

Figure 16.17 Variations in 5 × 5 mm DrMOS top view footprints. Source: Based on Refs. [21, 22]; Ref. [23].

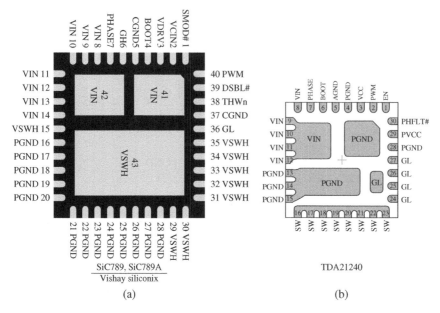

Figure 16.18 Vishay Siliconix SiC789 5 × 5 mm and Infineon TDA21240 4 × 4 mm DrMOS top view footprints. Source: Vishay Siliconix SiC789 / VISHAY INTERTECHNOLOGY.

In addition, Vishay offers an additional SiC789 product in a 5 × 5 mm footprint which is a scaled shrink of DrMOS-4.0, while Infineon offers TDA21240 4 × 4 mm uDrMOS which utilizes the low-side source-down concept as seen in Figure 16.18.

Without clear standardization requirements, DrMOS 5 × 5 mm demonstrated how quickly vendors can deviate from common footprints.

16.9   5 × 6 mm Smart Power Stage: Industry Driven Standardization

The need for power stage standardization had become very obvious during VR11.1, VR12, and VR12.5 generations with ever increasing volume, driven by higher processor currents, and the push toward higher current per phase challenged system reliability. Power stage designs are continually optimized each generation, driven by customer demand for maximum performance, resulting in vendor issues such as continuity of supply or product issues delaying or impacting customer production ramp up.

Hewlet Packard Enterprise (HPE) attempted to establish Power Stage common footprint VR12.5 by partnering two vendors on several different footprints; however, process and assembly variations along with proprietary concerns prevented vendors to adapt fast enough for this development cycle.

VR13 brought another opportunity for standardization and HPE is credited for successfully convincing three lead suppliers to support the "HPMOS" definition, and the rest followed. Dell was also driving "DellMOS" standardization, converging onto the 5 × 6 footprint with minor pinout and feature differences. Both definitions heavily leveraged International Rectifier IR3555 smart power stage functionality in a new 5 × 6 mm footprint which was already in development.

TDA21472

TDA21472 OptiMOS™ Powerstage

DC–DC converter voltage regulator

Features

- Co-packaged driver, Schottky diode, and high-side and low-side MOSFETs
- 5-mV/A on-chip MOSFET current sensing with temperature compensated reporting
- Input voltage (VIN) range of 4.25–16 V
- VCC and VDRV supply of 4.25–5.5 V
- Output voltage range from 0.25 up to 5.5 V
- Output current capability of 70 A
- Operation up to 1.5 MHz
- VCC/VDRV under voltage lockout (UVLO)
- 8-mV/°C temperature analog output
- Thermal shutdown and fault flag
- Cycle-by-cycle over current protection with programmable threshold and fault flag
- MOSFET phase fault detection and flag
- Auto-replenishment of bootstrap capacitor
- Deep-sleep mode for power saving
- Compatible with 3.3-V tri-state PWM input
- Body-braking load transient support
- Small 5 mm × 6 mm × 1 mm PQFN package
- Lead free RoHS compliant package

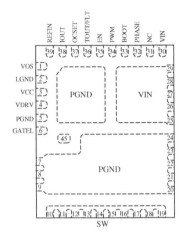

Figure 16.19 Industry standard 5 × 6 mm Infineon TDA21472. Source: Infineon TDA21472 / Infineon Technologies AG.

Infineon Technologies acquired International Rectifier in 2014, merging OptiMOS™ high performance MOSFETs into IR's 5 × 6 mm SPS, resulting in the very successful TDA2147X [24] "Voyager" family of smart power stages as shown in Figure 16.19.

Standardization enables participating vendors with the common footprint on nearly all processor VR designs, providing the opportunity to win on performance, cost, and merit. As planned, this 5 × 6 mm package is supported by many vendors such as ON Semiconductor, Texas Instruments, Renasas, Vishay Siliconix, Monolithic Power Systems, Enperion, and more.

16.10 Latest SPS Activities

The latest round of smart power stage definitions, as of February 2021, are listed below.

- **4 × 6 mm 90 A peak SPS:** The 4 mm width enables tighter phase-pitch layouts to address high density and high phase count systems when paired with narrow inductors. This definition leverages early Infineon TDA21590 prototypes. Dell, HPE, and Intel worked together toward commonality with the aid of a few lead vendors. A fault identification scheme has been agreed upon and will allow systems to identify which fault in which phase without any major changes from the 5 × 6 mm pinout functions. Intel has issued a specification to ensure compatibility.

  For this 4 × 6 mm, Infineon is re-introducing the chip embedded package technology, offering a very thin, low inductance package with very good thermal impedance. A version with onboard

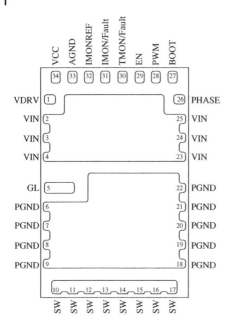

Figure 16.20 Industry standard 4 × 6 mm 90 A SPS, Infineon TDA21590. Source: Infineon TDA21590/ Infineon Technologies AG.

C_{IN}, C_{BOOT}, C_{VCC}, and C_{DRV} decoupling capacitors to provide low noise and higher density solutions (Figure 16.20).

- **5 × 6 mm 70 A peak Automotive SPS:** This is a derivative of the server power 5 × 6 version and includes auto specific changes.
    - 0.5 mm pin pitch instead of the industrial version 0.45 mm pitch. This is required by customers to minimize tin whisker and dendrite formation in humid environments.
    - The package reverts to a two-clip DrMOS type layout to increase reliability of LS MOSFET LGATE interconnect. Soldering of the GATEL to the lead frame can lead to long-term degradation under extreme temperature cycling which is not seen in server systems.
    - AEC-Q100 Rev H grade 1 compliant
    - 30 V automotive grade MOSFETs (Figure 16.21)

16.11 Trending Back to VRMs

Several processor roadmaps continue to increase current demand projections, with some ASICs (Application Specific Integrated Circuits) requiring over 800 A and up to 1200 A. These current levels cannot be supported by standard "down" solutions due to their remoteness to the processor and the resulting PDN losses in the PCB which is $= I^2 * R_{PCB}$. Even with a 100 μΩ PCB resistance, PCB power loss would be 100 W at 1000 A.

Furthermore, the 12 V ecosystem losses also limit system efficiencies, and thus a 48 V distribution trend has evolved which can reduce input PCB losses by 16× due to a 4× drop in input current with the 4× increase in input voltage. These 48 V systems typically utilize transformers in step down conversion and resonant topologies to maximize efficiency.

These application complexities and the need for voltage regulation much closer to the processor have enabled a new generation of VRMs which can be placed on the motherboard backside and

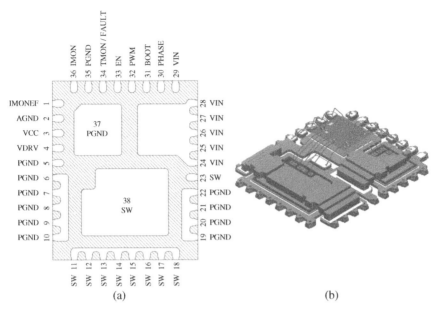

Figure 16.21 Automotive standard 5 × 6 mm 60 A SPS, Infineon TLF12501. Source: Infineon TLF12501/ Infineon Technologies AG.

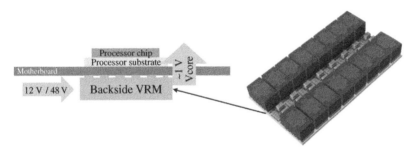

Figure 16.22 Concept of a high current VRM placed on the motherboard backside to reduced 1 V distribution losses.

directly under the processor as depicted in Figure 16.22. VRMs are also solving high density challenges by stacking inductors and power stages to decrease the solution area and increasing power volume density.

16.12 Summary

Generational increases in CPU power consumption have been a consistent trend, first driven by Moore's law with the number of transistors doubling about every two years and followed by compute architecture improvements such as multi-core parallel computing with increasing integrated memories.

The multiphase buck regulator has been the dominant topology by offering easily scalable and highly efficient solutions. Many companies and their differentiating contributions have enabled a very robust 12 V ecosystem which has both enabled and supported significant advancements in computational and networked processing capability.

Integration has also been a consistent trend, with discrete power systems transitioning to the truly visionary DrMOS-1.0, and eventually successful DrMOS-4.0 which provides a sufficient-performing solution uses in cost sensitive systems today. The SPS (smart power stage), offering higher efficiency and additional current/temperature telemetry, further enabled processor performance advancements. And increased processor current trends are driving higher power density requirements, enabling power modules which integrate capacitors, inductors, and powerstages.

Driven by high-volume market demands and supply continuity risks, common footprints are now commonplace in various power stage, multiphase controller, and Point of Load (PoL) devices.

Some feel standardization limits creativity and forces infrastructure on vendors, such as package technologies or processes. This may be accurate for incremental advances; however, history has shown significant improvements in performance and function can drive new rounds of standardization.

References

1 Intel®. (1999). VRM 8.3 DC-DC Converter Design Guidelines. Order Number 243870-002.

2 Intel®. (2001). VRM 8.5 DC-DC Converter Design Guidelines. Order Number 249659-001

3 SEMTECH Corp. (2000). SC2422A Data Sheet, Biphase Current Mode Controller. Available: http://pdf.datasheetcatalog.com/datasheets/120/273779_DS.pdf [verified July 2022]

4 Mehrzad Koohian. (2001). Input Current Mode Controller Delivers Dynamic Performance. Available: https://www.electronicdesign.com/content/article/21186077/input-current-mode-controller-delivers-dynamic-performance [verified July 2022]

5 Linfinity Application Note 7. (1998). A Simple Current-Sense Technique Eliminating a Sense Resistor. Available: https://www.microsemi.com/document-portal/doc_download/14646-an-7-a-simple-current-sense-technique-eliminating-a-sense-resistor [verified July 2022]

6 ONSEMI(TM). (2006). CS5308 Data Sheet, Two-phase PWM Controller with Integrated Gate Drivers for VRM 8.5. Available: https://www.onsemi.com/pub/Collateral/CS5308-D.PDF [verified July 2022]

7 Linear Technology. (1999). Application Note 77: High Efficiency, High Density, PolyPhase Converters for High Current Applications. Available: https://xdevs.com/doc/Linear/AppNotes/AN77 %20-%20High%20Efficiency%2C%20High%20Density%2C%20PolyPhase%20Converters%20for %20High%20Current%20Applications.pdf [verified July 2022]

8 Intel®. (2002). VRM 9.0 DC-DC Converter Design Guidelines. Order Number 249205-004. Available: https://www.intel.com/content/dam/www/public/us/en/documents/design-guides/voltage-regulator-module-9.0-dc-dc-converter-guidelines.pdf [verified July 2022]

9 Unitrode. (1999). UC1907, UC2907, UC3907 Data Sheet, Load Share Controller. Available: https://www.ti.com/lit/ds/symlink/uc3907.pdf?ts=1613159937979&ref_url=https%253A%252F %252Fwww.google.com%252F [verified July 2022]

10 Danny Clavette. (2000). Circuit and method of direct duty cycle current sharing (US6642631B1). Available: https://patents.google.com/patent/US6642631 [verified July 2022]

11 ONSEMI(TM). (2006). CS5332 Data Sheet, Two-Phase Buck Controller with Integrated Gate Drivers for VRM 9.0. Available: https://www.onsemi.com/pdf/datasheet/cs5332-d.pdf [verified July 2022]

12 ONSEMI$^{(TM)}$. (2006). CS5307 Data Sheet, Four–Phase VRM 9.0 Buck Controller. Available: https://www.onsemi.com/pub/Collateral/CS5307-D.PDF [verified July 2022]

13 International Rectifier. (2004). DirectFET® MOSFETs for Switching Applications. Available: http://www.irf.com/product-info/fact_sheet/farnell/10771.pdf [verified July 2022]

14 John Larking. (2003). Direct Fet device for high frequency application (US20040104489A1). Available: https://patents.google.com/patent/US20040104489 [verified July 2022]

15 Intel®. (2005). Voltage Regulator Module (VRM) and Enterprise Voltage Regulator-Down (EVRD) 10.0. Available: https://www.intel.com/content/dam/www/public/us/en/documents/design-guides/voltage-regulator-module-enterprise-voltage-regulator-down-10-0-guidelines.pdf [verified July 2022]

16 International Rectifier. (2009). IR3084 Data Sheet, X-Phase™ VR 10/11 Control IC. Available: http://www.irf.com/product-info/datasheets/data/ir3084mpbf.pdf [verified July 2022]

17 Intel®. (2009). Voltage Regulator Module (VRM) and Enterprise Voltage Regulator-Down (EVRD) 11.1. Available: https://www.intel.it/content/dam/doc/design-guide/voltage-regulator-module-enterprise-voltage-regulator-down-11-1-guidelines.pdf [verified July 2022]

18 Intel®. (2004). DrMOS Specification 1.0. Available: http://www.ebvnews.ru/doc11/drmos.pdf [verified July 2022]

19 International Rectifier. (2014). IR3550 Data Sheet, 60A Integrated Power Stage. Available: https://www.infineon.com/dgdl/ir3550.pdf?fileId=5546d462533600a4015355cd7c831761 [verified July 2022]

20 International Rectifier. (2014). IR3555 Data Sheet, 60A Integrated Power Stage. Available: https://www.infineon.com/cms/en/product/power/dc-dc-converters/integrated-power-stages/ir3555 [verified July 2022]

21 Infineon Technologies. (2015). Product Brief DrMOS 5 × 5 Power MOSFET and Driver in one Small 5 × 5 Package. Available: https://www.infineon.com/dgdl/Infineon-ProductBrief_DrMOS5x5_Power%20Stage-PB-v01_00-EN.pdf?fileId=5546d4624bcaebcf014c09da2be423f2 [verified July 2022]

22 Alpha & Omega Semiconductor. (2017). AOZ5131QI Data Sheet, High-Current, High-Performance DrMOS Power Module. Available: http://www.aosmd.com/res/data_sheets/AOZ5131QI.pdf [verified July 2022]

23 Vishay Siliconix. (2020). SiC645, SiC645A Data Sheet, 60 A VRPower Smart Power Stage (SPS) Module with Integrated High Accuracy Current and Temperature Monitors. Available: https://www.vishay.com/docs/65424/sic645.pdf [verified July 2022]

24 Infineon Technologies. (2020). TDA21472 Data Sheet, Optimos$^{(TM)}$ Powerstage. Available: https://www.infineon.com/dgdl/Infineon-TDA21472-DataSheet-v01_00-EN.pdf?fileId=5546d4626cb27db2016d175ca2e1448e [verified July 2022]

Abbreviations

ADC	analog to digital converter
AI	artificial intelligence
AMD	advanced micro devices
ASIC	application specific integrated circuit
C_{IN}	capacitance at input
CPU	Central Processing Unit

DAC	digital to analog converter
DCR	direct current resistance
DrMOS	driver-MOSFET
DVID	dynamic voltage identification
HP	Hewlet Packard
HPE	Hewlet Packard Enterprise
HS	high-side MOSFET
IC	integrated circuit
IR	international rectifier
LS	low-side MOSFET
LSB	least significant bit
MOSFET	metal oxide field effect transistor
MSID	market segment identification
NMOS	negative type MOSFET
NTC	negative temperature coefficient
PC	personal computer
PCB	printed circuit board
PGND	power ground
POC	power on configuration
PoL	point of load
PSI(#)	power state indicator (active low)
QFN	quad flat no-lead
RDSon	resistance from drain to source when on
R_{LL}	load-line resistance
SKU	Stock Keeping Unit
SO8	Small Outline 8 pin package
SPS	smart power stage
SW	switched node
VID	voltage identification, or DAC target setpoint
VIN	voltage input
VR	voltage regulator
VRM	voltage regulator module

Index

Advances in Semiconductor Technologies: Selected Topics Beyond Conventional CMOS, First Edition. Edited by An Chen.
© 2023 The Institute of Electrical and Electronics Engineers, Inc. Published 2023 by John Wiley & Sons, Inc.

Printed and bound by CPI Group (UK) Ltd, Croydon, CR0 4YY

16/04/2025

14658590-0003